W9-ALW-047

ELEMENTARY ALGEBRA

ELEMENTARY ALGEBRA
Structure and Use

Raymond A. Barnett
DEPARTMENT OF MATHEMATICS
MERRITT COLLEGE
OAKLAND, CALIFORNIA

Second Edition

McGraw-Hill Book Company
New York St. Louis San Francisco Auckland
Düsseldorf Johannesburg Kuala Lumpur
London Mexico Montreal New Delhi
Panama Paris São Paulo Singapore
Sydney Tokyo Toronto

ELEMENTARY ALGEBRA: STRUCTURE AND USE

4 5 6 7 8 9 0 VHVH 7 9 8 7 6

This book was set in Laurel by York Graphic Services, Inc.
The editors were A. Anthony Arthur, Shelly Levine Langman, and Andrea Stryker-Rodda;
the designer was Ben Kann;
the production supervisor was Joe Campanella.
New drawings were done by Eric G. Hieber Associates Inc.
Von Hoffmann Press, Inc., was printer and binder.

Library of Congress Cataloging in Publication Data

Barnett, Raymond A
Elementary algebra.

1. Algebra. I. Title.
QA152.2.B37 1975 512.9′042 74-13229
ISBN 0-07-003781-7

Contents

Preface

This new edition of *Elementary Algebra: Structure and Use,* reflecting recommendations from many users, embodies much of the same spirit of the first edition, but in a substantially simplified form. This is still an introductory text in algebra, written for students with no background in algebra and for those who need a review of algebra from a contemporary point of view. The principle changes from the first edition are:

1 Explanations have been simplified and the amount of verbal material has been reduced substantially. Chapters on function and formal systems have been deleted as well as a few other sections of a more advanced nature.
2 A larger number of easier exercises are included. In fact, the exercises are divided into *A*, *B*, and *C* groupings with the *A* problems easy and routine; the *B* problems a little more challenging, but still emphasizing mechanics; and the *C* problems a mixture of theoretical and difficult mechanics.
3 Set ideas and notation are kept on an informal basis and are used only where clarity results and not otherwise. Set-builder notation appears only in the *C* exercises and can be avoided altogether, if desired.

4 More examples are included and each example is followed by a matching problem with answer to encourage an active rather than passive reading of the text.
5 The many and varied realistic applications from the first edition have been retained, but a greater number of easier word problems supplement the list.
6 An informal postulational approach is used with emphasis on the spirit of the approach rather than each logical detail. Theoretical developments are left mostly to the *C* exercises.
7 The text continues to make use of spiraling techniques. By the end of the first three chapters most of the basic algebraic processes have been introduced in a relatively simple framework. These ideas are returned to again and extended as the text progresses. Consequently, the material in the last five chapters can be covered at a more rapid pace if the first three chapters are covered carefully.
8 A Keller-type plan for individualized instruction is to be found in the Instructor's Manual. Included are pre- and posttests for each of 3 units, 6 progress tests (3 forms of each) for each of the 3 units, and a final examination (2 forms). All of these tests have easy-to-grade answer keys. An $8\frac{1}{2}$- by 11-in. format is used for ease of reproduction. Also included are brief discussions of Keller-type plans and a possible grading procedure.

I would like to thank the many users of the first edition for their many kind comments and helpful suggestions that were incorporated into the second edition. I am particularly indebted to Prof. Harold Engelsohn of Kingsborough Community College, City University of New York, for his thorough readings of the first edition and the manuscript for the second edition. His detailed comments and suggestions were most useful and appreciated.

A special thanks is due to Mrs. Iku Workman and the author's daughters Janet and Margaret for their expert typing of the final manuscript, and to Margaret for her thorough checking of examples and exercises.

Raymond A. Barnett

USE OF A, B, and C EXERCISES

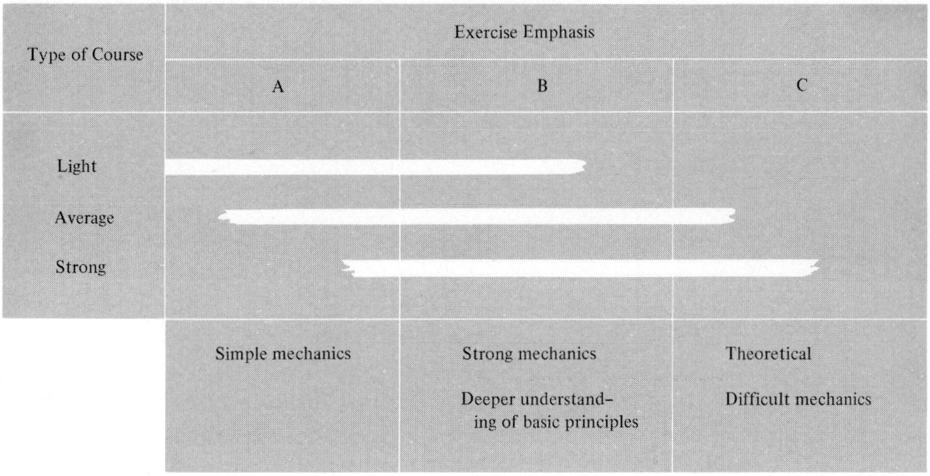

Type of Course	Exercise Emphasis		
	A	B	C
Light			
Average			
Strong			
	Simple mechanics	Strong mechanics Deeper understand- ing of basic principles	Theoretical Difficult mechanics

INDIVIDUALIZED ASSIGNMENTS (AVERAGE COURSE)

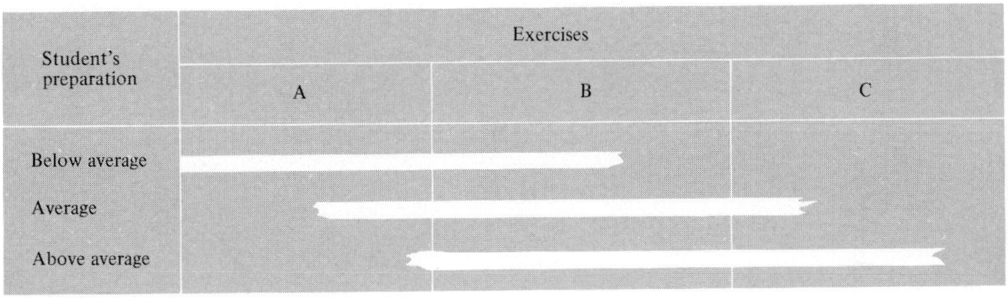

Student's preparation	Exercises		
	A	B	C
Below average			
Average			
Above average			

To the Student

The following suggestions are made to help you get the most out of the course and the most out of your efforts.

Using the text is a five-step process.

For each section:

1 Read a mathematical development.
2 Read an illustrative example.
3 Work the matched problem.

} Repeat 1-2-3 cycle until section is finished.

4 Review the main ideas in the section.
5 Work the assigned exercises at end of the section.

All of the above should be done with plenty of inexpensive paper, pencils, and a waste basket. No mathematics text should be read without pencil and paper in hand. This is not a spectator's sport!

If you have difficulty with the course, then, in addition to the regular assignments:

1 Spend more time on the examples and matched problems.
2 Work more *A* exercises, even if they are not assigned.
3 If the *A* exercises continue to be difficult for you, then you probably should take an arithmetic review course before attempting this one.

If you find the course too easy:

1 Work more problems from the *C* exercises, even if they are not assigned.
2 If the *C* exercises are consistently easy for you, then you should probably be in an intermediate algebra class.

Raymond A. Barnett

ELEMENTARY ALGEBRA

CHAPTER 1
Natural Numbers

1.1 Sets and Subsets

To begin the study of algebra, we will start with a very simple but important mathematical idea, the concept of set. Our use of this word will not differ appreciably from the way it is used in everyday language. Words such as "set," "collection," "bunch," and "flock" all convey the same idea. Thus, we think of a *set* as any collection of objects with the important property that given any object, it is either a member of the set or it is not. If an object a is in a set A, we say that a is a *member* or *element* of A.

Sets are often represented by listing their elements within braces { }. For example,

$$\{2, 4, 6\}$$

represents the set with elements 2, 4, and 6.

Two sets A and B are said to be *equal*, and we write

$$A = B$$

if the two sets have exactly the same elements; the order does not matter.

From time to time we will be interested in sets within sets called subsets. We say that a set A is a *subset* of a set B if every element in A is in B. For example, in a class of boys and girls, all of the girls would form a subset of the set of all students in the class.

1.2 The Set of Natural Numbers

Since algebra has to do with manipulating symbols that represent numbers, it is essential that we go back and take a careful look at some of the properties of numbers that you might have overlooked or have taken for granted.

The set of *counting numbers*

$\{1, 2, 3, \ldots\}$

will be our starting place. (The three dots tell us that the numbers go on without end following the pattern indicated by the first three numbers. This is a convenient way to represent some infinite sets.) The set of counting numbers is also referred to as the set of *natural numbers*, and these two names will be used interchangeably.

EXAMPLE 1 Some natural numbers: 1, 5, 17, 54, 525

Other numbers: $\frac{2}{3}$, $5\frac{7}{8}$, 7.63, $\sqrt{2}$

PROBLEM 1 Select the natural numbers out of the following list: 4, $\frac{3}{4}$, 19, 305, $4\frac{2}{3}$, 7.32, $\sqrt{3}$

ANSWER 4, 19, 305

ASSUMPTION We assume that you know what natural numbers are, how to add and multiply them, and how to subtract and divide them when permitted.

IMPORTANT SUBSETS OF THE NATURAL NUMBERS

The natural numbers can be divided into two subsets called even numbers and odd numbers. A natural number is an *even* number if it is exactly divisible by 2; if it is not exactly divisible by 2, then it is an *odd* number.

EXAMPLE 2 The set of even numbers: $\{2, 4, 6, \ldots\}$

The set of odd numbers: $\{1, 3, 5, \ldots\}$

PROBLEM 2 Divide the following list into even and odd numbers: 8, 13, 7, 32, 57, 625, 532.

ANSWER Even: 8, 32, 532 odd: 13, 7, 57, 625

When we add or subtract two or more numbers, the numbers are called *terms;* when we multiply two or more numbers, the numbers are called *factors.* In the sum

$$3 + 5 + 8$$

3, 5, and 8 are terms; in the product

$$3 \times 5 \times 8$$

3, 5, and 8 are called factors.

In mathematics, at the level of algebra and higher, parentheses ( ) or the dot "$\cdot$" are usually used in place of the times sign "$\times$," since the latter is easily confused with the letter "x," a letter that finds frequent use in algebra.
Thus

$$(3)(5)(8)$$

$$3 \cdot 5 \cdot 8$$

also represent the product of 3, 5, and 8.

The natural numbers, excluding 1, can also be divided into two subsets called composite numbers and prime numbers. A natural number is a *composite number* if it can be represented as a product of two or more natural numbers other than itself and 1; all other natural numbers, excluding 1, are called *prime numbers*.

EXAMPLE 3 (*A*) The first eight composite numbers are: 4, 6, 8, 9, 10, 12, 14, 15 since

$4 = 2 \cdot 2$ $\quad$ $10 = 2 \cdot 5$

$6 = 2 \cdot 3$ $\quad$ $12 = 3 \cdot 4 = 2 \cdot 6$

$8 = 2 \cdot 4$ $\quad$ $14 = 2 \cdot 7$

$9 = 3 \cdot 3$ $\quad$ $15 = 3 \cdot 5$

(*B*) The first eight prime numbers are: 2, 3, 5, 7, 11, 13, 17, 19.

PROBLEM 3 Divide the following list into composite and prime numbers: 6, 9, 11, 21, 23, 25, 27, 29.

ANSWER Composite numbers: 6, 9, 21, 25, 27

Prime numbers: 11, 23, 29

EXAMPLE 4 Write each composite number as a product of prime factors: 8, 36, 60.

SOLUTION

$8 = 2 \cdot 4 = 2 \cdot 2 \cdot 2$

$36 = 6 \cdot 6 = 2 \cdot 3 \cdot 2 \cdot 3$

$60 = 10 \cdot 6 = 2 \cdot 5 \cdot 2 \cdot 3$

PROBLEM 4 Repeat Example 4 using 12, 26, and 72.

12 = 2·6 = 2·2·3
26 = 2·13
72 = 2·36 = 2·2·18 = 2·2·2·3·3

ANSWER $2 \cdot 2 \cdot 3$ $\quad 2 \cdot 13$ $\quad 2 \cdot 2 \cdot 2 \cdot 3 \cdot 3$

The order of the prime factors does not matter; in fact, an important property of the naturals numbers is the property that each composite number has, except for order, one and only one set of prime factors.

Exercise 1

A *Select the natural numbers out of each list.*

1. 6, 13, 3.5, $\frac{2}{3}$

2. 4, $\frac{1}{8}$, 22, 6.5

3. $3\frac{1}{2}$, 67, 402, 22.35

4. 203.17, 63, $\frac{33}{5}$, 999

Divide each list into even and odd numbers.

5. 9, 14, 28, 33

6. 8, 24, 1, 41

7. 23, 105, 77, 426

8. 68, 530, 421, 72

Divide each list into composite and prime numbers.

9. 2, 6, 9, 11

10. 3, 4, 7, 15

11. 12, 17, 23, 27

12. 16, 19, 25, 39

B *Let M be the set of natural numbers between 20 and 30 inclusive and N the set of natural numbers between 40 and 50 inclusive. List the following:*

13. Even numbers in M - 20, 22, 24, 26, 28, 30

14. Even numbers in N - 40, 42, 44, 46, 48, 50

15. Odd numbers in M - 21, 23, 25, 27, 29

16. Odd numbers in N - 41, 43, 45, 47, 49

17. Composite numbers in M - 20, 21, 22, 24, 25, 26, 27, 28, 30

18. Composite numbers in N

19. Prime numbers in M

20. Prime numbers in N

Write each of the following composite numbers as a product of prime factors.

21. 10 **22.** 21 **23.** 30

24. 90 **25.** 210 **26.** 252

C **27.** Is every even number a prime number? Is every odd number a prime number? Is every prime number an odd number? Is every prime number except 2 an odd number?

28. Is every even number a composite number? Is every odd number a composite number? Is every even number except 2 a composite number?

29. Suppose that you know that 3 is not a factor of a given natural number, is it possible that 6 is a factor of that number?

30. Suppose that you know that 6 is not a factor of a given number, is it possible that 3 is a factor of that number?

Intuitively, a set is said to have a finite number of elements if the elements can be counted; otherwise, the set is said to be infinite. Tell which of the following sets are finite or infinite.

31. The set of natural numbers between 1 and 1 million.

32. The set of even numbers between 1 and 1 million.

33. The set of composite numbers between 1 and 1 million.

34. The set of prime numbers between 1 and 1 million.

35. The set of all natural numbers.

36. The set of all even numbers.

37. The set of all composite numbers.

38. The set of all natural numbers exactly divisible by 3.

39. The set of all the grains of sand on all of the beaches in the world.

1.3 Formulas and Other Algebraic Expressions

Above it was suggested that algebra has to do with symbols and the manipulation of symbols that represent numbers. This idea should not be new to you. Who has not worked with formulas where letters are used as place holders for numerals? All students by this time should be familiar with most if not all of the following formulas:

AREA OF A RECTANGLE (Fig. 1):

$A = ab$ area = base × height

PERIMETER OF A RECTANGLE (Fig. 1):

$P = 2a + 2b$ perimeter = 2 × base + 2 × height

Figure 1

VOLUME OF A RECTANGULAR SOLID (Fig. 2):

$V = abc$ volume = length × width × height

Figure 2

CIRCUMFERENCE OF A CIRCLE (Fig. 3):

$C = \pi D$ circumference = pi × diameter

AREA OF A CIRCLE (Fig. 3):

$A = \pi r^2$ area = pi × radius × radius

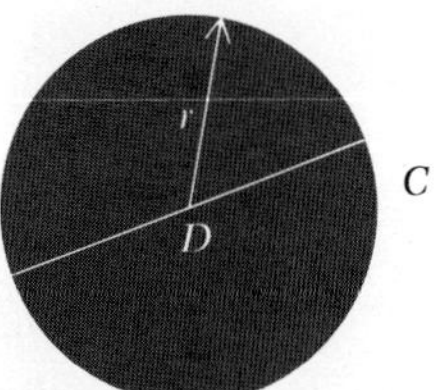

Figure 3

SIMPLE INTEREST:

$I = prt$ interest = principal × rate × time

DISTANCE FORMULA:

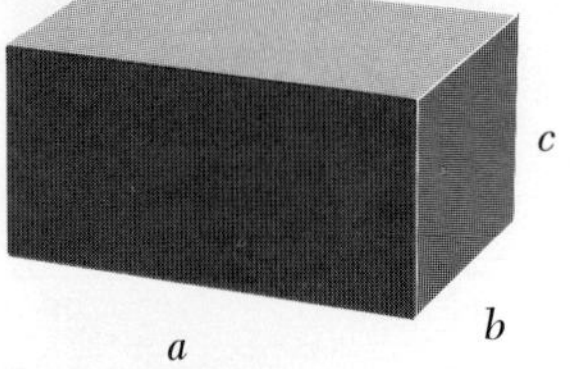

$d = rt$ distance = rate × time

EXAMPLE 5

(*A*) The area of a 5- by 7-in. rectangle is

$A = 5 \cdot 7 = 35$ sq in.

(*B*) The perimeter of a 5- by 7-in. rectangle is

$P = 2 \cdot 5 + 2 \cdot 7 = 10 + 14 = 24$ in.

(*C*) The volume of a 3- by 8- by 2-in. rectangular solid is

$V = 3 \cdot 8 \cdot 2 = 48$ cu in.

(*D*) If a car goes 50 mph for 8 hr, it will travel

$d = 50 \cdot 8 = 400$ miles

PROBLEM 5

Find:

(*A*) The area of a 2- by 9-ft rectangle

(*B*) The perimeter of a 2- by 9-ft rectangle

(*C*) The volume of a 5- by 3- by 2-ft rectangular solid

(*D*) The distance that a car travels in 6 hr going 65 mph

ANSWER (*A*) 18 sq ft (*B*) 22 ft (*C*) 30 cu ft (*D*) 390 miles

In the formula for the perimeter of a rectangle

$$P = 2a + 2b$$

the three letters P, a, and b can be replaced with many different numerals, depending on the size of the rectangle; hence these letters are called variables. The symbol "2" names only one number and is consequently called a constant. In general, a *constant* is defined to be any symbol that names one particular thing; a *variable* is a symbol that holds a place for constants. In most cases in this book, constants will name numbers.

EXAMPLE 6

Constants: IV, 3, $5 - 2$, π, $5 \cdot 7$

Variables: C and D in $C = \pi D$

A and r in $A = \pi r^2$

PROBLEM 6

List the constants and variables in each formula:

(*A*) $P = 4s$ perimeter of a square

(*B*) $A = s^2$ area of a square

ANSWER Constants: (*A*) 4; (*B*) 2

Variables: (*A*) P and s; (*B*) A and s

The introduction of variables into mathematics as a systematic notational device occurred about A.D. 1600 or a little before. A Frenchman, Francois Vieta (1540–1603), is singled out as the one mainly responsible for this innovation. It is now clear that mathematics would not have developed nearly as far as it has without the important notion of the variable; in fact, many mark this point as the beginning of modern mathematics.

ALGEBRAIC EXPRESSIONS—THEIR FORMULATION AND EVALUATION

An *algebraic expression* is a symbolic form involving constants; variables; mathematical operations such as addition, subtraction, multiplication, and division (other operations will be added later); and grouping symbols such as parentheses ( ), brackets [ ], and braces { }. For example,

$$3 + 5 \qquad 2 + 5 \cdot 6 \qquad 7 - 3(4 - 2)$$

$$2x + 3y \qquad 5(x + 2y) \qquad 2\{x + 3[x - 2(x + 1)]\}$$

are all algebraic expressions. It is important to note that unless otherwise indicated by symbols of grouping, multiplication and division precede addition and subtraction.

EXAMPLE 7

(*A*) $2 + 5 \cdot 6 = 2 + 30 = 32$

(*B*) $3 \cdot 5 - 2 \cdot 4 = 15 - 8 = 7$

(*C*) $7 - 3(4 - 2) = 7 - 3 \cdot 2 = 7 - 6 = 1$

(*D*) $5[8 - 2(5 - 3)] = 5[8 - 2 \cdot 2] = 5[8 - 4] = 5 \cdot 4 = 20$

Note that we performed the operations within the grouping symbols, when present, first starting with the parentheses ( ), then the brackets [ ], and if braces { } were present, we would end with them.

PROBLEM 7

Evaluate each expression:

(*A*) $9 - 4 \cdot 2$ $= 9-8=1$ (*B*) $5 \cdot 7 + 6 \cdot 5$ $= 35+30 = 65$

(*C*) $8 - 2(4 - 1)$ $8-2 \cdot 3 = 8-6=2$ (*D*) $2[9 - 3(4 - 2)]$ $= 2[9-3 \cdot 2] = 2[9-6]$ $2 \cdot 3 = 6$

ANSWER (*A*) 1 (*B*) 65 (*C*) 2 (*D*) 6

EXAMPLE 8

Evaluate each algebraic expression for $x = 9$ and $y = 2$:

(*A*) $x - 2y$ $= 9-2 \cdot 2 = 9-4=5$ (*B*) $3x + 5y$ $= 3 \cdot 9+5 \cdot 2 = 27+10 = 37$

(*C*) $x - 2(y + 1)$ $= 9-2(2+1) = 9-2 \cdot 3 = 9-6=3$ (*D*) $5[22 - x(x - 7)]$ $= 5[22-9(9-7)] = 5[22-9 \cdot 2] = 5[22-18] = 5 \cdot 4 = 20$

SOLUTION

(*A*) $x - 2y$
$9 - 2(2) = 9 - 4 = 5$

(*B*) $3x + 5y$
$3(9) + 5(2) = 27 + 10 = 37$

(*C*) $x - 2(y + 1)$
$9 - 2(2 + 1) = 9 - 2(3) = 9 - 6 = 3$

(*D*) $5[22 - x(x - 7)]$
$5[22 - 9(9 - 7)] = 5[22 - 9(2)] = 5[22 - 18] = 5(4) = 20$

PROBLEM 8

Evaluate each algebraic expression for $x = 12$ and $y = 3$:

(*A*) $x - 3y$ $= 12-3 \cdot 3 = 12-9=3$ (*B*) $2x - 2y$ $= 2 \cdot 12-2 \cdot 3 = 24-6 = 18$

(*C*) $x - 4(y - 1)$ $= 12-4(3-1) = 12-4 \cdot 2 = 12-8=4$ (*D*) $3[x - 2(x - 9)]$ $= 3[12-2(12-9)] = 3[12-2 \cdot 3] = 3[12-6] = 3 \cdot 6 = 18$

ANSWER (*A*) 3 (*B*) 18 (*C*) 4 (*D*) 18

EXAMPLE 9

If x represents a natural number, write an algebraic expression that represents each of the expressed numbers.

(*A*) A number 3 times as large as x

SOLUTION $3x$

(*B*) A number 3 more than x

SOLUTION $x + 3$ or $3 + x$

(*C*) A number 5 less than twice x

SOLUTION $2x - 5$

(*D*) A number twice a number 3 larger than x

SOLUTION $2(x + 3)$

PROBLEM 9 If y represents a natural number, write an algebraic expression that represents each of the expressed numbers:

(*A*) A number 7 times as large as y

(*B*) A number 7 more than y

(*C*) A number 9 less than 4 times y

(*D*) A number 5 times the number 4 less than y

ANSWER (*A*) $7y$ (*B*) $y + 7$ (*C*) $4y - 9$ (*D*) $5(y - 4)$

Exercise 2

A

1. Find the area of a 3- by 7-in. rectangle.
2. Find the area of a 2- by 6-ft rectangle.
3. Find the perimeter of a 3- by 7-in. rectangle.
4. Find the perimeter of a 2- by 6-ft rectangle.
5. Find the volume of a 12- by 4- by 2-in. rectangular solid.
6. Find the volume of a 10- by 5- by 3-ft rectangular solid.
7. How far can you travel in 9 hr at 43 mph?
8. How far can you travel in 12 hr at 57 mph?

Evaluate each expression.

9. $7 + 3 \cdot 2$

10. $5 + 6 \cdot 3$

11. $8 - 2 \cdot 3$

12. $20 - 5 \cdot 3$

13. $7 \cdot 6 - 5 \cdot 5$

14. $8 \cdot 9 - 6 \cdot 11$

15. $(2 + 9) - (3 + 6)$

16. $(8 - 3) + (7 - 2)$

17. $10 - 3(7 - 4)$

18. $20 - 5(12 - 9)$

Evaluate each algebraic expression for $x = 8$ and $y = 3$.

19. $x + 2$

20. $y + 5$

21. $x - y$

22. $22 - x$

23. $6y - x$

24. $x - 2y$

25. $3x - 2y$

26. $9y - xy$

27. $x - 2(y - 1)$

28. $x - y(x - 7)$

B *If in a rectangular solid (Fig. 2), $a = 5$ in., $b = 3$ in., and $c = 2$ in., find:*

29. The area of the base.

30. The area of an end.

31. The perimeter of the base.

32. The perimeter of an end.

Evaluate each.

33. $2[(7 + 2) - (5 - 3)]$

34. $6[(8 - 3) + (4 - 2)]$

35. $5(8 - 3) - 3 \cdot 6$

36. $7 \cdot 9 - 6(8 - 3)$

37. $4[15 - 10(9 - 8)]$

38. $6[22 - 3(13 - 7)]$

Evaluate each for $w = 2$, $x = 5$, $y = 1$, and $z = 3$.

39. $w(y + z)$

40. $wy + wz$

41. $wy + z$

42. $y + wz$

43. $(z - y) + (z - w)$

44. $4(y + w) - 2z$

45. $2[x + 3(z - y)]$

46. $6[(x + z) - 3(z - w)]$

47. $9(x - y) - 3[x - 2(x - z)]$

48. $2[(y + z) + 3(z - 1) + wz]$

Identify the constants and variables in the following algebraic expressions.

49. $2x + 3y$

50. $5x - y$

51. $3(u + v) + 2u$

52. $2(x + 1) + 3(w + 5z)$

If x represents a natural number, write an algebraic expression that represents each of the following numbers:

53. A number 5 more than x

54. A number 5 times as large as x

55. A number 5 less than x

56. A number x less than 33

57. A number 3 more than twice x

58. A number 4 less than 5 times x

59. A number 3 times a number 3 smaller than x

60. A number 7 times a number 5 larger than x

C *If in a rectangular solid (Fig. 2), $a = 12$ ft, $b = 7$ ft, and $c = 4$ ft, find:*

61. The total surface area.

62. The total length of all edges.

Evaluate each expression.

63. $2\{26 - 3[12 - 2(8 - 5)]\}$

64. $5\{32 - 5[(10 - 2) - 2 \cdot 3]\}$

Evaluate for $u = 2$, $v = 3$, $w = 4$, and $x = 5$.

65. $2\{w + 2[7 - (u + v)]\}$

66. $3\{(u + v) + 3[x - 2(w - u)] + uv\}$

If t represents an even number, write an algebraic expression that represents each of the following:

67. A number 3 times the first even number larger than t

68. A number 5 times the first even number smaller than t

69. The sum of three consecutive even numbers starting with t

70. The sum of three consecutive odd numbers following t

71. An earthquake produces several types of waves that travel through the earth. One is called a shear wave, which travels at about 2 miles per sec and causes the earth to move at right angles to the direction of motion of the wave. (*A*) Write a formula that indicates the distance d that the wave travels in t sec. (*B*) Identify the constants and variables. (*C*) How far will the wave travel in 10 sec?

72. The distance s that an object falls in t sec (Fig. 4) is 16 times the square of the time. (*A*) Write a formula that indicates the distance s that the object falls in t sec. (*B*) Identify the constants and variables. (*C*) How far will the object have fallen at the end of 8 sec?

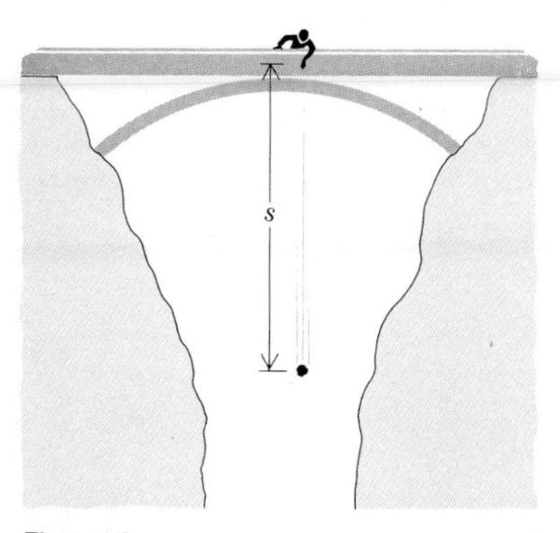

Figure 4

1.4 Equality

In the preceding sections the equality sign "=" was used in a number of places. You are probably most familiar with its use in formulas such as

$$d = rt$$

$$A = ab$$

$$I = prt$$

This sign is very important in mathematics, and you will be using it frequently. Its mathematical meaning, however, is not as obvious as it first might seem. For this reason we are devoting one section of this chapter solely to this sign so that you will use it correctly from the beginning.

We will always use *equality* in the sense of logical identity. That is, an *equality sign* will be used to join two expressions if the two expressions are names or descriptions of exactly the same thing. In this sense "=" means "is identical with." Since

$$a = b$$

means a and b are names for the same object, it is natural that we define

$$a \neq b$$

to mean a and b do not name the same thing; that is, *a is not equal to b.*

EXAMPLE 10 (*A*) $\text{VI} = 6$

(*B*) $4 - 3 \neq 2$

(*C*) $3 + 4 \cdot 2 = 11$

PROBLEM 10 True or false?

(*A*) $\text{III} = 3$ (*B*) $\text{V} \neq 6$

(*C*) $8 - 6 = 2$ (*D*) $8 - 3 \cdot 2 = 10$

ANSWER (*A*) T (*B*) T (*C*) T (*D*) F

If two algebraic expressions involving at least one variable are joined with an equal sign, the resulting form is called an *algebraic equation*. The following are algebraic equations in one or more variables:

$$x + 3 = 2(x + 1)$$

$$a + b = b + a$$

$$2x + 3y = 5$$

Since a variable is a place holder for constants, an equation is neither true nor false as it stands; it does not become so until the variable has been replaced by a constant. Formulating algebraic equations is an important first step in solving certain types of problems using algebraic methods. We address ourself to this problem now.

EXAMPLE 11 Translate each statement into an algebraic equation using only one variable:

(*A*) 15 is 9 more than a certain number

SOLUTION Let x represent the certain number, then

$$15 = 9 + x$$

(*B*) 3 times a certain number is 7 more than twice the number

SOLUTION Let n represent the certain number, then

$$3n = 7 + 2n$$

PROBLEM 11 Repeat Example 11 for:

(*A*) 8 is 5 more than a certain number

(*B*) 6 is 4 less than a certain number

(*C*) 4 times a certain number is 3 more than twice the number

(*D*) If 12 is added to a certain number, the sum is twice the number 3 larger than the original number.

ANSWER (*A*) $8 = 5 + x$ (*B*) $6 = x - 4$ (*C*) $4x = 2x + 3$
(*D*) $x + 12 = 2(x + 3)$

From the logical meaning of the equality sign a number of rules or laws can easily be established for its use. We state these laws now and will rely on them many times throughout the book, since they control a great deal of activity related to manipulating and solving equations.

LAWS OF EQUALITY

If a, b, and c are names of objects, then:

1. $a = a$ reflexive law
2. If $a = b$, then $b = a$. symmetric law
3. If $a = b$ and $b = c$, then $a = c$. transitive law
4. If $a = b$, then either may replace the other in any expression without changing the truth or falsity of the statement. substitution principle

The importance of these four laws will not be fully appreciated until we start solving equations and simplifying algebraic expressions.

Exercise 3

A *Indicate which of the following are true (T) or false (F).*

1. $10 - 6 = 4$

2. $12 + 7 = 19$

3. $6 \cdot 9 = 56$

4. $8 \cdot 7 = 54$

5. $12 - 9 \neq 2$

6. $15 - 8 \neq 6$

7. $\text{II} = 2$

8. $\text{IX} \neq 11$

9. $9 - 2 \cdot 3 \neq 21$

10. $4 + 2 \cdot 5 \neq 30$

Translate each statement into an algebraic equation using only the variable x.

11. 5 is 3 more than a certain number

12. 10 is 7 more than a certain number

13. 8 is 3 less than a certain number

14. 14 is 6 less than a certain number

15. 18 is 3 times a certain number

16. 25 is 5 times a certain number

B **17.** 49 is 7 more than twice a certain number

18. 52 is 8 less than 5 times a certain number

19. 4 times a given number is 3 more than 3 times that number

20. 8 times a number is 20 more than 4 times the number

21. The sum of four consecutive natural numbers is 54.

22. The sum of two consecutive even numbers is 54.

23. Pythagoras found that the octave chord could be produced (Fig. 5) by placing the movable bridge so that a taut string is divided into two parts with the longer piece twice the length of the shorter piece. If the total string is 27 in. long and we let x represent the length of the shorter piece, write an equation relating the lengths of the two pieces and the total length of the string.

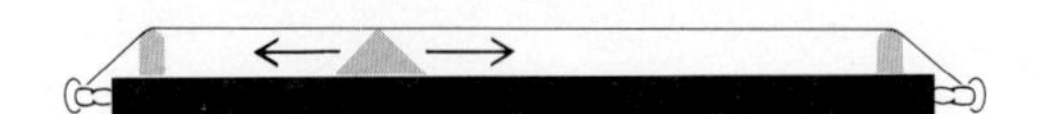

Figure 5 Monochord.

24. A board 16 ft long is cut into two pieces so that the longer piece is 2 ft less than twice the length of the shorter piece. Write an equation relating the lengths of the two pieces with the total length.

25. In a rectangle of area 50 sq ft the length is 10 ft more than the width. (NOTE: Length and width are other names for base and height.) Write an equation that relates the area with the length and the width.

26. In a rectangle of area 75 sq yd its length is 5 more than 3 times the width. Write an equation relating the area with the length and the width.

C **27.** What is wrong with the following argument? The number 5 is a prime number and the number 7 is a prime number. Hence, we can write 5 = prime number and 7 = prime number. By the symmetric law for equality we can write prime number = 7 and conclude, using the transitive law for equality (since 5 = prime number and prime number = 7), that $5 = 7$.

28. What is wrong with the following argument? John is a human and Mary is a human. Hence, we can write John = human and Mary = human. By the symmetric law for equality we can write human = Mary and conclude, using the transitive law for equality (since John = human and human = Mary), that John = Mary.

29. Supply the reasons for the proof that for natural numbers x and y, if $x = y$, then $x + 3 = y + 3$.

PROOF

STATEMENT	REASON
1 $x + 3 = x + 3$	*1*
2 $x = y$	*2*
3 $x + 3 = y + 3$	*3*

30. Supply the reasons for the proof that for natural numbers x and y, if $x = y$, then $3x = 3y$.

PROOF

STATEMENT	REASON
1 $3x = 3y$	*1*
2 $x = y$	*2*
3 $3x = 3y$	*3*

31. Show, using the laws of equality, that for any natural numbers a, b, and c, if $a = b$, then $a + c = b + c$.

32. Show, using the laws of equality, that for any natural numbers a, b, and c, if $a = b$, then $ac = bc$.

1.5 Axioms for Addition and Multiplication

Algebra in many ways can be thought of as a game, a game that requires the changing of symbols that represent numbers from one arrangement to another. Like any game, to play, one must know the rules of the game. We are going to uncover a short basic list of properties that all numbers within your experience share. These properties, plus a few others we will add later, govern all of the manipulation of symbols that represent numbers in algebra.

REMARK: Do not let the names of the three properties discussed in this section frighten you. The properties are really very simple in spite of their names. After a little use, the names will be less forbidding.

Given a set of numbers, one important question we can ask about the set is, "Can we perform certain operations on the elements in the set and stay in the set?" If we can, then the set is said to be *closed* with respect to the given operation.

Is the set of natural numbers

$$N = \{1, 2, 3, \ldots\}$$

closed with respect to any of the basic arithmetic operations: addition, multiplication, subtraction, or division? Let us first investigate addition and multiplication. If we take any two elements of N, say 4 and 12, and form their sum and product

$$4 + 12 = 16$$

$$4 \cdot 12 - 48$$

we see that 16 and 48 are still in N. In fact, we cannot think of a single case where the sum or product of any two natural numbers is not a natural number. However, since we are not in a position to prove that this is so, we state the result as an *axiom,* an unproved assumption. (Statements requiring proofs are called *theorems.*)

CLOSURE AXIOMS FOR ADDITION AND MULTIPLICATION

For all natural numbers a and b,

$a + b$ is a natural number

ab is a natural number

Are the natural numbers closed under subtraction or division? We see that

$$5 - 3 = 2$$

is a natural number, but

$$4 - 7$$

is not defined in the natural numbers. Similarly,

$$6 \div 2 = 3$$

is a natural number, but

$$5 \div 2$$

is not defined in the natural numbers. We, therefore, conclude that the natural numbers are not closed under subtraction or division.

EXAMPLE 12 $M = \{1, 2, 3, 4, 5\}$ is not closed under addition or multiplication since

$$5 + 3 = 8$$

$$3 \cdot 5 = 15$$

are neither in M.

PROBLEM 12 Which of the following sets are closed under:

(*A*) addition (*B*) multiplication

$D = \{2, 4, 6, 8\}$

$E = \{2, 4, 6, 8, \ldots\}$

$F = \{1, 3, 5, 7\}$

$G = \{1, 3, 5, 7, \ldots\}$

ANSWER (*A*) E (*B*) E, G

Is the result of applying an operation on any two numbers of a set the same irrespective of the order in which the numbers are chosen? If so, the set is said to be *commutative* relative to the given operation.

Is the set of natural numbers

$$N = \{1, 2, 3, \ldots\}$$

commutative with respect to addition, multiplication, subtraction, or division? If we take any two elements of N, say 5 and 7, then

$$5 + 7 = 7 + 5$$

$$5 \cdot 7 = 7 \cdot 5$$

and we see that the order in which the two numbers are added or multiplied does not make any difference. In fact, we will not be able to find two natural numbers in which the order in which we add or multiply them will make a difference. But since we are not in a position to prove this, we state the result as an axiom.

COMMUTATIVE AXIOMS FOR ADDITION AND MULTIPLICATION

For all natural numbers a and b,

$$a + b = b + a$$

$$ab = ba$$

Is subtraction or division commutative in the set of natural numbers? The answer is no, since, for example,

$$7 - 5 \neq 5 - 7$$

$$8 \div 4 \neq 4 \div 8$$

The commutative axioms for addition and multiplication provides us with one of our first tools for changing algebraic expressions from one form to an equivalent form.

EXAMPLE 13 If x and y are natural numbers, then

(A) $x + 7 = 7 + x$

(B) $5y = y5$

(C) $xy = yx$

(D) $3 + 2x = 2x + 3$

PROBLEM 13 If a and b are natural numbers, use the commutative axioms for addition or multiplication to write each of the following in an equivalent form:

(A) $a + 3$ (B) $5b$

(C) ab (D) $5 + 7b$

ANSWER (A) $3 + a$ (B) $b5$ (C) ba (D) $7b + 5$ or $5 + b7$ or $b7 + 5$

When computing

$$5 + 8 + 7$$

$$5 \cdot 8 \cdot 7$$

why do we not need parentheses to show us which two numbers are to be added or multiplied first? You may answer that because of our past experience we assume that the grouping does not make any difference; we "know" that

$$(5 + 8) + 7 = 5 + (8 + 7)$$

$$(5 \cdot 8) \cdot 7 = 5 \cdot (8 \cdot 7)$$

We seem to have uncovered another important property of the natural numbers. As before, we are not in a position to prove this result for all natural numbers so we state it as an axiom:

ASSOCIATIVE AXIOMS FOR ADDITION AND SUBTRACTION

For all natural numbers a, b, and c,

$$(a + b) + c = a + (b + c)$$

$$(ab)c = a(bc)$$

Not all arithmetic operations are associative relative to the natural numbers.

$$(8 - 4) - 2 \neq 8 - (4 - 2)$$

$$(8 \div 4) \div 2 \neq 8 \div (4 \div 2)$$

showing that subtraction and division are not associative.

We now have another tool for transforming algebraic expressions into other equivalent forms.

EXAMPLE 14 If x, y, and z are natural numbers, then

(A) $(x + 3) + 5 = x + (3 + 5)$

(B) $2(3x) = (2 \cdot 3)x$

(C) $x + (x + 2) = (x + x) + 2$

(D) $(2x)x = 2(xx)$

(E) $(x + y) + z = x + (y + z)$

(F) $(xy)z = x(yz)$

PROBLEM 14 If a, b, and c are natural numbers, replace each question mark with an appropriate expression:

(A) $(a + 5) + 7 = a + (?)$

(B) $5(9b) = (?)b$

(C) $(3 + c) + c = 3 + (?)$

(D) $(5a)a = 5(?)$

(E) $(a + b) + c = a + (?)$

(F) $(ab)c = a(?)$

ANSWER (A) $5 + 7$ (B) $5 \cdot 9$ (C) $c + c$ (D) aa (E) $b + c$ (F) bc

CONCLUSION

It should now be clear that relative to addition the above properties permit us to rearrange and regroup symbols that represent natural numbers as we please; that is, we may change the order of addition at will and insert or remove symbols of grouping as we please. And the same thing is true about multiplication, but not about subtraction and division. Notice the use of these properties in the following example.

EXAMPLE 15 (A)

$$\begin{aligned}(x + 3) + (y + 5) &= x + 3 + y + 5 \\ &= x + y + 3 + 5 \\ &= x + y + 8\end{aligned}$$

(B)

$$\begin{aligned}(2x)(3y) &= 2x3y \\ &= 2 \cdot 3xy \\ &= 6xy\end{aligned}$$

PROBLEM 15 Simplify:

(A) $(a + 5) + (b + 6)$ (B) $(7a)(3b)$

ANSWER (A) $a + b + 11$ (B) $21ab$

Exercise 4

All variables represent natural numbers.

A *State the justifying axiom for each statement.*

1. $a + y$ is a natural number

2. ry is a natural number

3. $5 + z = z + 5$

4. $bc = cb$

5. $(5x)y = 5(xy)$

6. $(a + 5) + 7 = a + (5 + 7)$

7. $11 + x = x + 11$

8. $(14c)d = 14(cd)$

9. yz is a natural number

10. $u + v$ is a natural number

11. $az = za$

12. $(xy)z = x(yz)$

13. $c + x$ is a natural number

14. $(x + 3) + 9 = x + (3 + 9)$

15. $2(3x) = (2 \cdot 3)x$

16. $(35 + a) + b = 35 + (a + b)$

Remove parentheses and simplify.

17. $(x + 3) + 2$

18. $2 + (7 + y)$

19. $2(6x)$

20. $5(9y)$

21. $(7 + x) + 3$

22. $(5 + z) + 12$

23. $(3x)5$

24. $(6y)7$

B *Assuming closure axioms without statement, state the justifying axiom for each equation.*

25. $3 + (x + 2) = 3 + (2 + x)$

26. $5(x8) = 5(8x)$

27. $(x + 2) + (y + 3) = (x + 2) + (3 + y)$

28. $3x + x5 = 3x + 5x$

29. $5 + (x + 3) = (x + 3) + 5$

30. $(5x)y = y(5x)$

Remove parentheses and simplify.

31. $(x + 7) + (y + 8)$

32. $(7 + a) + (9 + b)$

33. $(3x)(4y)$

34. $(7a)(4b)$

35. $(a + 3) + (b + 5) + (c + 2)$

36. $(x + 2) + (y + 4) + (z + 8)$

37. $(2x)(8y)(3z)$

38. $(3a)(5b)(2c)$

39. Given the sets:

A = the set of natural numbers
B = the set of odd numbers
$C = \{1, 2, 3, 4, 5, 6\}$
$D = \{1, 3, 5, 7, 9, 11\}$

(A) Which sets are closed with respect to addition?
(B) Which sets are closed with respect to multiplication?

40. Given the sets:

A = the set of natural numbers larger than 10
B = the set of even numbers
C = the set of prime numbers

(A) Which sets are closed with respect to addition?
(B) Which sets are closed with respect to multiplication?

C **41.** If a statement is not true for all natural numbers a and b, find replacements for a and b that show that the statement is false.

(A) $a + b = b + a$
(B) $ab = ba$
(C) $a - b = b - a$
(D) $a \div b = b \div a$

42. Repeat the preceding problem for

(A) $(a + b) + c = a + (b + c)$
(B) $(ab)c = a(bc)$
(C) $(a - b) - c = a - (b - c)$
(D) $(a \div b) \div c = a \div (b \div c)$

Supply the reason for each step.

43.

	STATEMENT	REASON
1	$(x + 8) + 7 = x + (8 + 7)$	*1*
2	$= x + 15$	2

44.

STATEMENT	REASON
1 $5(4x) = (5 \cdot 4)x$	*1*
2 $\quad = 20x$	2

45. $(z + x) + (c + a) = z + [x + (c + a)]$ is justified by which axioms?

46. $(zx)(ca) = z[x(ca)]$ is justified by which axioms?

1.6 Exponents

In a preceding section we encountered the form

$$r^2 = rr$$

There is obviously no reason to stop here: you no doubt can guess how r^3 and r^4 should be defined. If you guessed

$$r^3 = rrr$$

$$r^4 = rrrr$$

then you have anticipated the following general definition of b^n where n is any natural number and b is any number:

$$b^n = \underbrace{bbb \cdots b}_{n \text{ factors of } b}$$

b is called the *base* and n the *power* or *exponent.* In addition, we define

$$b^1 = b$$

and usually use b in place of b^1.

EXAMPLE 16 (A) $x^2 = xx$ $\qquad t^1 = t$

$3^4 = 3 \cdot 3 \cdot 3 \cdot 3$ $\qquad 5x^3y^5 = 5xxxyyyyy$

(B) $xxx = x^3$ $\qquad 2xxy = 2x^2y$

$2 \cdot 2 \cdot 2 \cdot 2 = 2^4$ $\qquad 3xxxyy = 3x^3y^2$

PROBLEM 16 (A) Write in nonpower form: y^3, 2^4, $3x^3y^4$

(B) Write in exponent form: uu, $5 \cdot 5 \cdot 5 \cdot 5$, $7xxxxyyy$

ANSWER (A) yyy, $2 \cdot 2 \cdot 2 \cdot 2$, $3xxxyyyy$ $\qquad$ (B) u^2, 5^4, $7x^4y^3$

If we observe that

$$10^1 = 10$$

$10^2 = 100$

$10^3 = 1,000$

$10^4 = 10,000$

. . . .

and note that the number of zeros on the right is the same as the power of ten on the left, then we can represent some large numbers very conveniently in a *power-of-ten* form.

EXAMPLE 17

$$90 = 9 \times 10$$

$$900 = 9 \times 10^2$$

$$63,000 = 63 \times 10^3$$

$$93,000,000 = 93 \times 10^6$$

PROBLEM 17 (*A*) Write in power-of-ten form: 80, 800, 750,000.

(*B*) Write in nonpower-of-ten form: 7×10, 6×10^3, 83×10^6.

ANSWER (*A*) 8×10, 8×10^2, 75×10^4 (*B*) 70, 6,000, 83,000,000

The following three statements are more or less typical of what one is likely to encounter in scientific writing. Knowing the meaning of the power symbol certainly adds to the meaning and interest of these statements.

1 Due to radiation of energy, the sun loses approximately 42×10^5 tons of solar mass per sec.

2 In 1929 a biologist named Vernadsky suggested that all the free oxygen of the earth, about 15×10^{20} grams, is produced by living organisms alone.

3 Edgar Altenburg wrote in his book on genetics, "A human begins life as a single cell, the fertilized egg, which by successive cell divisions forms the 10^{13} cells contained in a grown man. The size of the fertilized egg is approximately the size of the cross section of a human hair. Yet a single fertilized egg, despite its minuteness, contains all of the potentialities of a Shakespeare or a Darwin. . . ."

In addition to increasing your understanding of scientific writings, the power symbol is very useful in algebra and will be used often. Because of its frequency of use we will establish several rules for manipulating exponent forms to produce new exponent forms. The first law of exponents will be introduced in this section; others will follow later.

We know that a natural number exponent indicates how many times a base is to be taken as a factor. Thus

$$x^3y^5 = xxxyyyyy$$

Something interesting happens if we multiply two exponent forms with the same base:

$$\begin{aligned} x^3x^5 &= (xxx)(xxxxx) \\ &= xxxxxxxx \\ &= x^8 \end{aligned}$$

which we could get by simply adding the exponents in x^3x^5.

In general, for any natural numbers m and n and any number b

$$b^mb^n = b^{m+n}$$

This is the *first law of exponents,* one of five very important exponent laws you will get to know as well as your multiplication tables before the end of this book.

NOTE: In the following example, and throughout the text, dotted boxes are used to indicate steps that are usually done mentally.

EXAMPLE 18

(A) $x^3x^4 \quad \boxed{= x^{3+4}} \quad = x^7$

(B) $a^{10}a^{23} \quad \boxed{= a^{10+23}} \quad = a^{33}$

(C) $(2y^2)(3y^5) \quad \boxed{= 2 \cdot 3y^{2+5}} \quad = 6y^7$

(D) $(3x^2y)(4x^4y^5) \quad \boxed{= 3 \cdot 4x^{2+4}y^{1+5}} \quad = 12x^6y^6$

(E) $(4 \times 10^7)(6 \times 10^6) \quad \boxed{= (4 \cdot 6) \times 10^{7+6}} \quad = 24 \times 10^{13}$

PROBLEM 18

Simplify as in Example 18:

(A) y^5y^3 (B) $x^{17}x^{20}$ (C) $(3a^6)(5a^3)$

(D) $(2x^2y^4)(3xy^2)$ (E) $(8 \times 10^5)(7 \times 10^{12})$

ANSWER (A) y^8 (B) x^{37} (C) $15a^9$ (D) $6x^3y^6$ (E) 56×10^{17}

Exercise 5

A *Write in nonexponent form.*

1. x^3

2. y^4

3. a^5

4. u^7

5. $2x^3y^2$

6. $5a^2b^3$

7. $3w^2xy^3$

8. $7ab^3c^2$

Write in exponent form.

9. xxx

10. $yyyy$

11. $3aaaaaa$

12. $5uuuuuuu$

13. $2xxxyy$

14. $7uuvvvvv$

15. $3xyyzzz$

16. $9aabccc$

Use the first law of exponents to simplify.

17. x^3x^2

18. y^2y^3

19. $u^{10}u^4$

20. m^8m^7

21. aa^5

22. b^7b

23. $w^{12}w^7$

24. $n^{23}n^{10}$

25. $7^3 \cdot 7^5$

26. $9^5 \cdot 9^6$

B *Write in power-of-ten form.*

27. 40

28. 400

29. 4,000

30. 40,000

31. 47,000,000

32. 27,000,000,000

Write in nonpower-of-ten form.

33. 5×10

34. 5×10^3

35. 5×10^5

36. 5×10^7

37. 23×10^4

38. 39×10^6

Use the first law of exponents to simplify.

39. $(2x)(3x^4)$

40. $(5y^4)(2y)$

41. $(7u^9)(5u^7)$

42. $(8x^{11})(3x^9)$

43. $(3 \times 10^5)(2 \times 10^4)$

44. $(5 \times 10^2)(9 \times 10^4)$

45. $(22 \times 10^4)(3 \times 10^6)$

46. $(7 \times 10^8)(13 \times 10^3)$

47. x^2xx^4

48. mm^3m^4

49. $(2x^3)(3x)(4x^5)$

50. $(3u^4)(2u^5)(u^7)$

51. $(2 \times 10^5)(3 \times 10^2)(4 \times 10)$

52. $(5 \times 10^3)(3 \times 10^5)(7 \times 10^2)$

53. $(a^2b)(ab^2)$

54. $(cd^2)(c^2d^2)$

55. $(4x)(3xy^2)$

56. $(5b)(2a^2b^3)$

57. $(2xy^3)(3x^3y)$

58. $(3xy^2z^3)(5xyz^2)$

Write each quantity in each statement as a single number in nonexponent form.

59. The sun is 93×10^6 miles from earth.

60. A grown man contains 10^{13} cells.

61. The estimated age of the earth is 5×10^9 years.

62. The distance that light travels in 1 year (called 1 light-year) is 588×10^{10} miles (approximately).

C **63.** Our galaxy is estimated to have a diameter of about 10^5 light-years and a thickness of about 2×10^4 light-years. If a light-year, the distance light travels in 1 year, is approximately 588×10^{10} miles, what is the diameter and thickness of our galaxy in miles? Express the answer in power-of-ten notation.

64. The 200-in. telescope at Palomar, California, can distinguish objects up to an estimated 6×10^6 light-years away. Express this distance in miles (see the preceding problem) using the power-of-ten notation.

65. The mass of the earth is approximately 66×10^{20} tons. If one ton is 2×10^3 lb, what is the mass of the earth in pounds? Express the answer using power-of-ten notation.

66. If the area of the earth is approximately 197×10^6 sq miles, find the area of the earth in sq ft if 1 sq mile contains approximately 279×10^5 sq ft. Express the answer using power-of-ten notation.

1.7 An Axiom Involving Multiplication and Addition

We now introduce another important property of the natural numbers, a property that involves both multiplication and addition, called the distributive axiom. To discover this property let us compute $3(5 + 2)$ and $3 \cdot 5 + 3 \cdot 2$:

$$3(5 + 2) = 3 \cdot 7 = 21$$

$$3 \cdot 5 + 3 \cdot 2 = 15 + 6 = 21$$

Thus

$$3(5 + 2) = 3 \cdot 5 + 3 \cdot 2$$

Note that the right side of the last equality is obtained from the left by multiplying each term within the parentheses by 3. Is the fact that these are equal just a coincidence? Let us try another set of numbers:

$$7(2 + 6) = 7 \cdot 8 = 56$$

$$7 \cdot 2 + 7 \cdot 6 = 14 + 42 = 56$$

Thus

$$7(2 + 6) = 7 \cdot 2 + 7 \cdot 6$$

Again we see that if we multiply each term within the parentheses by 7 first and add, we get the same result as adding the terms first, and then multiplying by 7. If we continue testing this apparent relationship for various other sets of natural numbers, we will not be able to find any for which it does not hold. But once again we are not in a position to prove it; therefore, we state the result as an axiom.

DISTRIBUTIVE AXIOM

For all natural numbers a, b, and c

$$a(b + c) = ab + ac$$

Verbally, the axiom states that multiplication distributes over addition.

A simple geometric example shows that the relationship is at least reasonable. Let a, b, and c be any three natural numbers. Form a rectangle with the dimensions as indicated in Fig. 6. The areas are related as follows:

area of large rectangle	=	area of left rectangle	+	area of right rectangle
$a(b + c)$	=	ab	+	ac

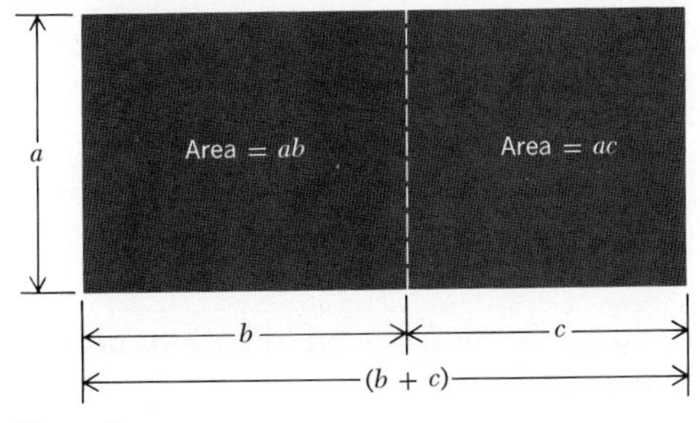

Figure 6

EXAMPLE 19

(A) $3(x + y) = 3x + 3y$

(B) $4(w + 2) = 4w + 8$

(C) $x(x + 1) = x^2 + x$

(D) $2x^2(3x + 2y) = 6x^3 + 4x^2y$

PROBLEM 19

Multiply using the distributive axiom:

(A) $2(a + b)$ (B) $5(x + 3)$

(C) $u(u^2 + 1)$ (D) $3n^2(2m^2 + 3n)$

ANSWER (A) $2a + 2b$ (B) $5x + 15$ (C) $u^3 + u$ (D) $6m^2n^2 + 9n^3$

It is important to realize that the distributive axiom holds in "both directions." That is, if $a(b + c) = ab + ac$, then using the symmetric law for equality,

$$ab + ac = a(b + c)$$

EXAMPLE 20

(A) $2x + 2y = 2(x + y)$

(B) $3w + 6 = 3(w + 2)$

(C) $x^2 + x = x(x + 1)$

(D) $4x^3 + 6x^2 = 2x^2(2x + 3)$

PROBLEM 20

Take out factors common to all terms:

(A) $5x + 5y$ (B) $2w + 8$

(C) $u^2 + u$ (D) $6y^2 + 4y^3$

ANSWER (A) $5(x + y)$ (B) $2(w + 4)$ (C) $u(u + 1)$ (D) $2y^2(3 + 2y)$

Using the commutative axiom in combination with the distributive axiom, we can obtain what is often referred to as the *right-hand distributive law:*

$$(b + c)a = ba + ca$$

which also can be written as

$$ba + ca = (b + c)a$$

Thus, we conclude that a factor common to both terms in a sum can be taken out either on the left or on the right.

EXAMPLE 21

(A) $3x + 5x = (3 + 5)x = 8x$

(B) $7y + 2y = (7 + 2)y = 9y$

(C) $3x^2y^2 + 4x^2y^2 = (3 + 4)x^2y^2 = 7x^2y^2$

Notice the marked simplification obtained. This is a significant result that will be expanded upon in the next section.

PROBLEM 21

Use the right-hand distributive law to write as a single term:

(A) $2x + 3x$ (B) $8u + 3u$ (C) $5uv^2 + 7uv^2$

ANSWER (A) $5x$ (B) $11u$ (C) $12uv^2$

By repeated use of the distributive axiom we can show that multiplication distributes over any finite sum. Thus

$$a(b + c + d) = ab + ac + ad$$

$$a(b + c + d + e) = ab + ac + ad + ae$$

and so on.

EXAMPLE 22

(A) $3(x + y + z) = 3x + 3y + 3z$

(B) $ma + mb + mc = m(a + b + c)$

(C) $2x(x + 3y + 2) = 2x^2 + 6xy + 4x$

(D) $4x^2 + 2xy + xz = x(4x + 2y + z)$

PROBLEM 22

(A) Multiply: $5(a + b + c)$

(B) Take out factors common to all terms: $ax + ay + az$

(C) Multiply: $3x(2x + 3y + 5)$

(D) Take out factors common to all terms: $4x^3 + 2xy + 6x^2$

ANSWER (A) $5a + 5b + 5c$ (B) $a(x + y + z)$ (C) $6x^2 + 9xy + 15x$
(D) $2x(2x^2 + y + 3x)$

The axioms that we have considered regulate a considerable amount of the activity in algebra. These are, so to speak, some of the rules of the game. However, like chess, knowing the rules of the game does not make one a good chess player. A great deal of practice in using the rules in a large variety of situations is necessary.

Exercise 6

A *Compute:*

1. $2(1 + 5)$ and $2 \cdot 1 + 2 \cdot 5$

2. $3(4 + 2)$ and $3 \cdot 4 + 3 \cdot 2$

3. $5(2 + 7)$ and $5 \cdot 2 + 5 \cdot 7$

4. $7(3 + 2)$ and $7 \cdot 3 + 7 \cdot 2$

Multiply, using the distributive axiom.

5. $4(x + y)$

6. $5(a + b)$

7. $7(m + n)$

8. $9(u + v)$

9. $6(x + 2)$

10. $3(y + 7)$

11. $5(2 + m)$

12. $8(3 + n)$

Take out factors common to all terms.

13. $3x + 3y$

14. $2a + 2b$

15. $5m + 5n$

16. $7u + 7v$

17. $ax + ay$

18. $mu + mv$

19. $2x + 4$

20. $3y + 9$

Multiply, using the distributive property.

21. $2(x + y + z)$

22. $5(a + b + c)$

23. $3(x + y + z)$

24. $4(a + b + 3)$

Take out factors common to all terms.

25. $7x + 7y + 7z$

26. $9a + 9b + 9c$

27. $2m + 2n + 6$

28. $3x + 3y + 12$

B *Multiply, using the distributive property.*

29. $x(1 + x)$

30. $y(y + 7)$

31. $y(1 + y^2)$

32. $x(x^2 + 3)$

33. $3x(2x + 5)$

34. $5y(2y + 7)$

35. $2m^2(m^2 + 3m)$

36. $3a^2(a^3 + 2a^2)$

37. $3x(2x^2 + 3x + 1)$

38. $2y(y^2 + 2y + 3)$

39. $5(2x^3 + 3x^2 + x + 2)$

40. $4(y^4 + 2y^3 + y^2 + 3y + 1)$

41. $3x^2(2x^3 + 3x^2 + x + 2)$

42. $7m^3(m^3 + 2m^2 + m + 4)$

Use the right-hand distributive law to write as a single term.

43. $3x + 7x$

44. $4y + 5y$

45. $2u + 9u$

46. $8m + 5m$

47. $2xy + 3xy$

48. $5mn + 7mn$

49. $2x^2y + 8x^2y$

50. $3uv^2 + 5uv^2$

51. $7x + 2x + 5x$

52. $8y + 3y + 4y$

Take out factors common to all terms.

53. $x^2 + 2x$

54. $y^2 + 3y$

55. $u^2 + u$

56. $m^2 + m$

57. $(2x^3 + 4x)$

58. $3u^5 + 6u^3$

59. $x^2 + xy + xz$

60. $y^3 + y^2 + y$

61. $3m^3 + 6m^2 + 9m$

62. $12x^3 + 9x^2 + 3x$

63. $u^2v + uv^2$

64. $2x^3y^2 + 4x^2y^3$

C *Multiply, using the distributive property.*

65. $4m^2n^3(2m^3n + mn^2)$

66. $5uv^2(2u^3v + 3uv^2)$

67. $3x^2y(2xy^3 + 4x + y^2)$

68. $2cd^3(c^2d + 2cd + 4c^3d^2)$

69. $(u + v)(c + d)$

70. $(m + n)(x + y)$

71. $(x + 3)(x + 2)$

72. $(m + 5)(m + 3)$

Take out factors common to all terms.

73. $a^2bc + ab^2c + abc^2$

74. $m^3n + mn^2 + m^2n^2$

75. $16x^3yz^2 + 4x^2y^2z + 12xy^2z^3$

76. $27u^5v^2w^2 + 9u^2v^3w^4 + 12u^3v^2w^5$

1.8 Combining Like Terms

A constant present as a factor in a term is called the *numerical coefficient* (or simply the *coefficient*) of the term. If no constant factor appears in the term, then the coefficient is understood to be 1.

EXAMPLE 23 In the expression

$$2x^3 + x^2y + 3xy^2 + y^3$$

the coefficient of the first term is 2, the second 1, the third 3, and the fourth 1.

PROBLEM 23 Given the algebraic expression

$$5x^4 + 2x^3y + x^2y^2 + 4xy^3 + y^4$$

what is the coefficient of each term?

ANSWER 5, 2, 1, 4, and 1

If two or more terms are exactly alike except for numerical coefficients, they are called *like terms.*

EXAMPLE 24 (A) In $4x + 2y + 3x$, $4x$ and $3x$ are like terms.

(B) In $9x^2y + 3xy + 2x^2y + x^2y$, the first, third, and fourth terms are like terms.

PROBLEM 24 List the like terms in

(A) $5m + 6n + 2n$ (B) $2xy + 3xy^3 + xy + 2xy^3$

ANSWER (A) $6n$ and $2n$ (B) $2xy$ and xy, $3xy^3$ and $2xy^3$

If an algebraic expression contains two or more like terms, these terms can always be combined into a single term. The distributive axiom (as was seen in the last section) is the principle tool behind the process.

EXAMPLE 25 (A) $3x + 5x = (3 + 5)x = 8x$

(B) $5t + 4s + 7t + s = 5t + 7t + 4s + s = (5 + 7)t + (4 + 1)s = 12t + 5s$

PROBLEM 25 Combine like terms proceeding as in Example 25:

(A) $6y + 5y$ (B) $4x + 7y + x + 2y$

ANSWER (A) $6y + 5y = (6 + 5)y = 11y$
(B) $4x + 7y + x + 2y = 4x + x + 7y + 2y = (4 + 1)x + (7 + 2)y = 5x + 9y$

It should be clear that free use was made of the axioms discussed earlier. Most of the steps in the dashed boxes are done mentally. In fact, the process is quickly mechanized as follows:

Like terms are combined by adding their numerical coefficients.

EXAMPLE 26 Combine like terms mentally:

(A) $7x + 2y + 3x + y = 10x + 3y$

(B) $2uv + 3w + 5uv = 7uv + 3w$

(C) $(2x^2 + 3x + 1) + (7x^2 + x + 5) = 9x^2 + 4x + 6$

(D) $3u^2 + 2u + u^2 + 4u^2 = 8u^2 + 2u$

PROBLEM 26 Combine like terms mentally:

(A) $4x + 7y + 9x$ (B) $3x^2 + y^2 + 2x^2 + 3y^2$

(C) $(2m^2 + 3m + 5) + (m^2 + 4m + 2)$ (D) $x^2 + 3x^2 + 4x + 2x^2$

ANSWER (A) $13x + 7y$ (B) $5x^2 + 4y^2$ (C) $3m^2 + 7m + 7$ (D) $6x^2 + 4x$

Exercise 7

A *Indicate the numerical coefficient of each term.*

1. $4x$ **2.** $7ab$ **3.** $8x^2y$

4. $9uv^2$ **5.** x^3 **6.** y^5

7. u^2v^3 **8.** m^3n^5

Given the algebraic expression $2x^3 + 3x^2 + x + 5$, *indicate:*

9. The coefficient of the second term.

10. The coefficient of the first term.

11. The exponent of the variable in the second term.

12. The exponent of the variable in the first term.

13. The coefficient of the third term.

14. The exponent of the variable in the third term.

Select like terms in each group of terms.

15. $3x, 2y, 4x, 5y$

16. $3m, 2n, 5m, 7n$

17. $6x^2, x^3, 3x^2, x^2, 4x^3$

18. $2y^2, 3y^4, 5y^4, y^2, y^4$

19. $2u^2v, 3uv^2, u^2v, 5uv^2$

20. $5mn^2, m^2n, 2m^2n, 3mn^2$

Combine like terms.

21. $5x + 4x$

22. $2m + 3m$

23. $3u + u$

24. $x + 7x$

25. $7x^2 + 2x^2$

26. $4y^3 + 6y^3$

27. $2x + 3x + 5x$

28. $4u + 5u + u$

29. $2x + 3y + 5x + y$

30. $m + 2n + 3m + 4n$

31. $2x + 3y + 5 + x + 2y + 1$

32. $3a + b + 1 + a + 4b + 2$

B *Select like terms in each group.*

33. $m^2n, 4mn^2, 2mn, 3mn, 5m^2n, mn^2$

34. $3u^2v, 2uv, u^2v, 2uv^2, 4uv, uv^2$

Combine like terms.

35. $2t^2 + t^2 + 3t^2$

36. $6x^3 + 3x^3 + x^3$

37. $3x + 5y + x + 4z + 2y + 3z$

38. $2r + 7t + r + 4s + r + 3t + s$

39. $9x^3 + 4x^2 + 3x + 2x^3 + x$

40. $y^3 + 2y + 3y^2 + 4y^5 + 2y^2 + y + 5$

41. $x^2 + xy + y^2 + 3x^2 + 2xy + y^2$

42. $3x^2 + 2x + 1 + x^2 + 3x + 4$

43. $(2x + 1) + (2x + 3) + (2x + 5)$

44. $(4x + 1) + (3x + 2) + (2x + 5)$

45. $(t^2 + 5t + 3) + (3t^2 + t) + (2t + 7)$

46. $(4x^4 + 2x^2 + 3) + (x^4 + 3x^2 + 1)$

47. $(x^3 + 3x^2y + xy^2 + y^3) + (2x^3 + 3xy^2 + y^3)$

48. $(2u^3 + uv^2 + v^3) + (u^3 + v^3) + (u^3 + 3u^2v)$

Multiply, using the distributive axiom, and combine like terms.

49. $2(x + 5) + 3(2x + 7)$

50. $5(m + 7) + 2(3m + 6)$

51. $x(x + 1) + x(2x + 3)$

52. $2t(3t + 5) + 3t(4t + 1)$

53. $5(t^2 + 2t + 1) + 3(2t^2 + t + 4)$

54. $4(u^2 + 3u + 2) + 2(2u^2 + u + 1)$

55. $y(y^2 + 2y + 3) + (y^3 + y) + y^2(y + 1)$

56. $2y(y^2 + 2y + 5) + 7y(3y + 2) + y(y^2 + 1)$

57. $2x(3x + y) + 3y(x + 2y)$

58. $3m(2m + n) + 2n(3m + 2n)$

59. If x represents a natural number, write an algebraic expression for the sum of four consecutive natural numbers starting with x. Simplify the expression by combining like terms.

60. If t represents an even number, write an algebraic expression for the sum of three consecutive even numbers starting with t. Simplify.

C *Multiply, using the distributive axiom, and combine like terms.*

61. $2xy^2(3x + x^2y) + 3x^2y(y + xy^2)$

62. $3s^2t^3(2s^3t + s^2t^2) + 2s^3t^2(3s^2t^2 + st^3)$

63. $(2x + 3)(3x + 2)$

64. $(x + 2)(2x + 3)$

65. $(x + 2y)(2x + y)$

66. $(3x + y)(x + 3y)$

67. $(x + 3)(x^2 + 2x + 5)$

68. $(r + s + t)(r + s + t)$

69. If y represents an odd number, write an algebraic expression for the product of y and the next odd number. Write as the sum of two terms.

70. If y represents the first of four consecutive even numbers, write an algebraic expression that would represent the product of the first two added to the product of the last two. Simplify.

71. An even number plus the product of it and the next even number is 180. Introduce a variable, and write this sentence as an algebraic equation. Simplify the left and right sides of the equation where possible.

72. There exist at least two consecutive odd numbers such that 5 times the first plus twice the second is equal to twice the first plus 3 times the second. Introduce a variable and write this sentence as an algebraic equation. Simplify the left and right sides of the equation where possible.

1.9 Inequality Symbols

Just as we found it convenient to replace "is equal to" with "=," we will also find it useful to replace "is less than" and "is greater than" with appropriate symbols,* namely

$<$ and $>$

and "is less than or equal to" and "is greater than or equal to" with

$\leq$ and $\geq$

To avoid confusing $<$ with $>$, remember that the small end (the point) is directed toward the smaller of two numbers.

*Formally, we define "$<$" as follows: We write $a < b$ if and only if there is a natural number n such that $a + n = b$. We write $a > b$ if and only if $b < a$. This definition and its consequences will be discussed in greater detail in Chap. 2.

EXAMPLE 27

SYMBOLIC STATEMENTS	VERBAL STATEMENTS
$8 < 12$	8 is less than 12
$103 > 37$	103 is greater than 37
$5 \geq 2$	5 is greater than or equal to 2
$5 \geq 5$	5 is greater than or equal to 5
$4 \leq 7$	4 is less than or equal to 7
$4 \leq 4$	4 is less than or equal to 4
$6 \neq 1$	6 is not equal to 1
$x < 5$	x is less than 5
$x \geq 2$	x is greater than or equal to 2

PROBLEM 27 (*A*) Write each in symbolic form: 5 is less than 7, 8 is greater than 3, x is less than or equal to 5, x is greater than or equal to 9, and 7 is not equal to 5.

(*B*) Write each in verbal form: $5 > 2$, $8 \leq 10$, $x < 5$, $x \geq 6$.

ANSWER (*A*) $5 < 7, 8 > 3, x \leq 5, x \geq 9, 7 \neq 5$ (*B*) 5 is greater than 2, 8 is less than or equal to 10, x is less than 5, x is greater than or equal to 6.

If in an equation such as

$$2x + 3 = 5(x + 2)$$

we replace the equal sign with an inequality symbol, say $>$, we obtain the *inequality statement*

$$2x + 3 > 5(x + 2)$$

One of the objectives of algebra is to learn to solve equations and inequalities; that is, to learn to find all replacements of the variable from a certain set of numbers that will make the equation or inequality true. The set of all solutions is called the *solution set* of the equation or inequality.

EXAMPLE 28 Find the solution set for each equation or inequality if the replacements for x are restricted to $M = \{1, 2, 3, 4, 5\}$.

(*A*) $x < 3$

SOLUTION Test each number in M: $1 < 3$ true, $2 < 3$ true, $3 < 3$ false, $4 < 3$ false, $5 < 3$ false. Therefore,

$$\text{Solution set} = \{1, 2\}$$

(*B*) $x \geq 2$

SOLUTION $1 \geq 2$ false, $2 \geq 2$ true, $3 \geq 2$ true, $4 \geq 2$ true, $5 \geq 2$ true. Therefore,

$$\text{Solution set} = \{2, 3, 4, 5\}$$

(*C*) $3x = 12$

SOLUTION $3 \cdot 1 = 12$ false, $3 \cdot 2 = 12$ false, $3 \cdot 3 = 12$ false, $3 \cdot 4 = 12$ true, $3 \cdot 5 = 12$ false. Therefore,

Solution set $= \{4\}$

(*D*) $x + 2 = 8$

SOLUTION $1 + 2 =$ false, $2 + 2 = 8$ false, $3 + 2 = 8$ false, $4 + 2 = 8$ false, $5 + 2 = 8$ false. Therefore,

Solution set = the empty set (denoted by $\varnothing$)

PROBLEM 28 Find the solution set for each equation or inequality if the replacements for x are restricted to $\{1, 2, 3, 4, 5\}$:

(*A*) $x < 4$ (*B*) $x \geq 3$

(*C*) $2x = 12$ (*D*) $x + 2 = 4$

ANSWER (*A*) $\{1, 2, 3\}$ (*B*) $\{3, 4, 5\}$ (*C*) $\varnothing$ (*D*) $\{2\}$

One often encounters *double-inequality statements* of the form

$2 < x < 7$

This is a short and convenient way of writing

$2 < x$ and $x < 7$

which means that x is between 2 and 7, not including 2 or 7. The statement

$2 < x \leq 7$

means that x is between 2 and 7 and includes 7 but not 2. One does not see statements such as

$2 < x > 7$ or $2 > x < 7$

Why?

EXAMPLE 29 Find the solution set for each double inequality if the replacements for x are restricted to the set of natural numbers N.

(*A*) $2 < x < 6$

SOLUTION It is clear that this statement is true only for the natural numbers 3, 4, and 5. Thus, the

Solution set $= \{3, 4, 5\}$

(*B*) $2 \leq x \leq 6$

SOLUTION 2 and 6 can be added to the solution set for (*A*). Thus, the

Solution set $= \{2, 3, 4, 5, 6\}$

(*C*) $2 < x \leq 6$

SOLUTION In this case, 6 can be added to the solution set in (*A*). Thus, the

Solution set $= \{3, 4, 5, 6\}$

PROBLEM 29 Repeat Example 29 for:

(*A*) $3 < x < 7$ (*B*) $3 \leq x < 7$ (*C*) $3 \leq x \leq 7$

ANSWER (*A*) $\{4, 5, 6\}$ (*B*) $\{3, 4, 5, 6\}$ (*C*) $\{3, 4, 5, 6, 7\}$

Exercise 8

A *Write in symbolic form.*

1. 7 is greater than 3

2. 18 is greater than 4

3. 11 is less than 12

4. 5 is less than 9

5. x is greater than or equal to 5

6. x is greater than or equal to 8

7. x is less than or equal to 12

8. x is less than or equal to 3

True (T) or false (F)?

9. $7 < 10$ **10.** $3 < 6$

11. $10 > 7$ **12.** $6 > 3$

13. $8 > 9$ **14.** $6 > 8$

15. $8 \geq 9$ **16.** $6 \geq 8$

17. $6 \geq 5$ **18.** $9 \geq 7$

19. $5 \geq 5$ **20.** $7 \geq 7$

Find the solution set for each equation or inequality for x restricted to $\{2, 3, 4, 5, 6\}$.

21. $x > 5$ **22.** $x < 5$

23. $x < 6$ **24.** $x > 4$

25. $x \leq 6$ **26.** $x \geq 4$

27. $3x = 6$ **28.** $x + 4 = 8$

29. $3x \neq 6$

30. $x + 4 \neq 8$

31. $4 < x < 7$

32. $2 < x < 6$

33. $3 < x < 9$

34. $1 < x < 7$

35. $4 \leq x < 7$

36. $2 < x \leq 6$

37. $4 \leq x \leq 6$

38. $2 \leq x \leq 6$

B *Write in verbal form.*

39. $5 > 2$

40. $12 > 11$

41. $2 < 5$

42. $11 < 12$

43. $x \geq 3$

44. $x \leq 4$

True (T) or false (F)?

45. $75 \geq 75$

46. $75 \leq 75$

47. $5 \cdot 3 + 1 > 5 \cdot 1 + 3$

48. $7 + 9 \cdot 5 > 9 + 7 \cdot 5$

49. $6(2 + 3) \geq 6 \cdot 2 + 6 \cdot 3$

50. $6(2 + 3) \leq 6 \cdot 2 + 6 \cdot 3$

Find the solution set for each equation or inequality for x restricted to $\{2, 4, 6, 8, 10\}$.

51. $3x > 12$

52. $x + 4 > 10$

53. $3x \leq 12$

54. $x + 4 \leq 10$

55. $2x - 4 \geq 12$

56. $4x - 6 \leq 18$

Find all natural number solutions to each double inequality.

57. $2 < 2t < 8$

58. $4 < t + 3 < 7$

59. $2 < 2t \leq 8$

60. $4 \leq t + 3 < 7$

61. $2 \leq 2t \leq 8$

62. $4 \leq t + 3 \leq 7$

C **63.** A flat sheet of cardboard in the shape of an 18- by 12-in. rectangle is to be used to make an open-topped box by cutting an x- by x-in. square out of each corner and folding the remaining part appropriately. *(A)* Write an equation for the volume of the box in terms of x. *(B)* What natural number values can x assume? *(C)* Formulate the answer to *(B)* in terms of a double inequality statement.

64. A rectangular lot is to be fenced with 400 ft of wire. *(A)* If the lot is x ft wide, write a formula for the area of the lot in terms of x. *(B)* What natural number values can x assume? *(C)* Formulate the answer to *(B)* in terms of a double inequality statement.

1.10 Line Graphs

"A picture is worth a thousand words." How many times have you heard that statement? Did you ever think that it might apply to mathematics? We are now going to represent "pictorially" sets of numbers by means of graphs on number lines.

To start, we construct a number line. This is simply a line with certain points labeled with numbers. A ruler and household thermometer provide good examples already within your experience. In general, to construct a *number line* (see Fig. 7).

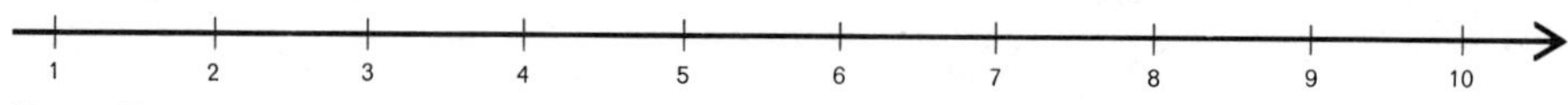

Figure 7

Select an arbitrary point on a line and mark it. Next divide the line to the right into equal line segments. The size and number of line segments to be included on a sheet of paper is a matter of choice and will depend on how many natural numbers you may wish to associate with points on the line. Finally, the set of natural numbers or one of its subsets is associated with the set of endpoints of the line segments. Often it is only necessary to show a few numbers associated with selected endpoints to establish a *scale* for the number line. Figure 8 should make the process clear.

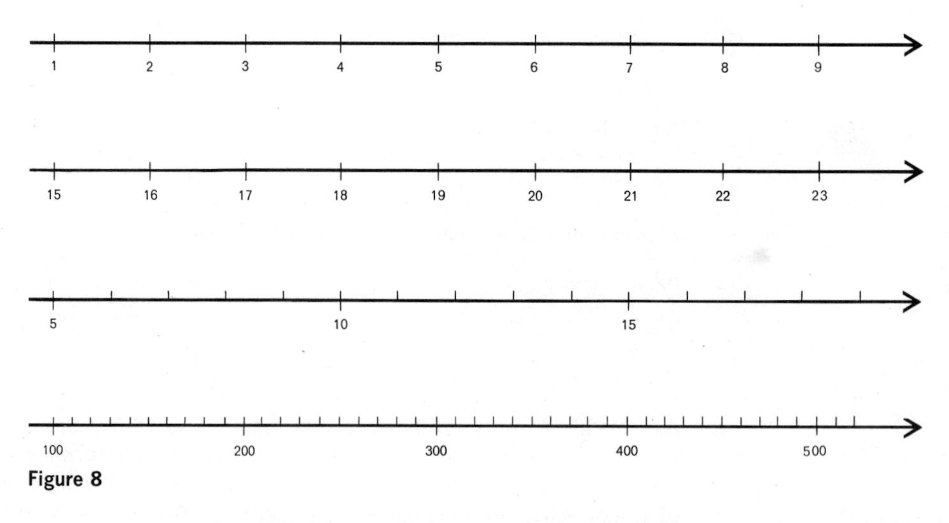

Figure 8

Once a number line is specified, any given natural number can be associated in one and only one way with a point on this line. The point associated with a particular number is called the *graph of the number* and is indicated by a darkened circle on the number line. On the other hand, the number associated with a point on a number line is called the *coordinate* of the point. The *graph of a set of numbers* is the set of points on the number line that are the graphs of the individual numbers.

EXAMPLE 30 (*A*) The graph of the set $\{3, 5, 8, 9\}$ is

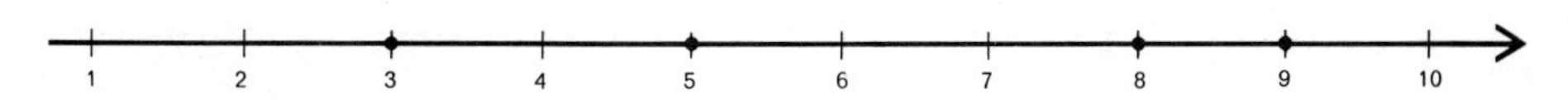

(*B*) The coordinates of points a, b, c, and d below are 22, 23, 25, and 28, respectively

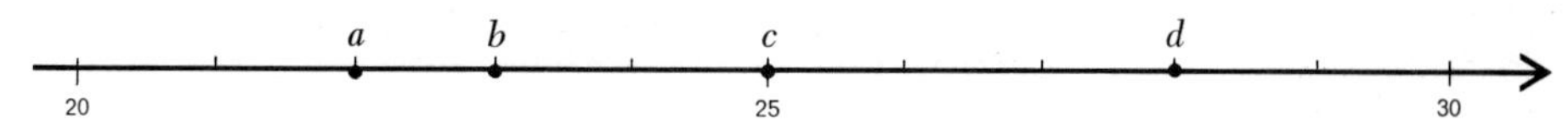

(*C*) For x, a natural number, the graph of $3 \le x < 8$ is

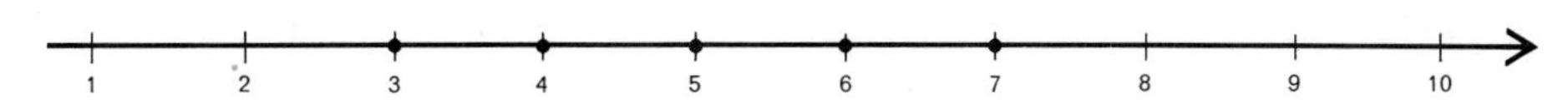

(*D*) For n, a natural number, the graph of $n \ge 6$ is

PROBLEM 30 (*A*) Graph $\{5, 7, 8, 10\}$ (*B*) What are the coordinates of a, b, c, and d?

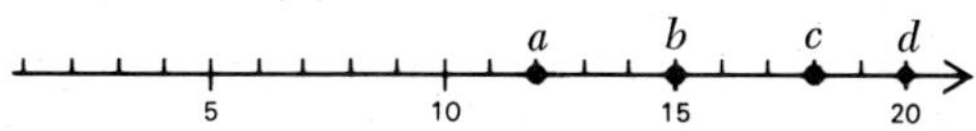

(*C*) For x a natural number, graph $4 < x \le 8$. (*D*) For n a natural number, graph $n > 7$.

ANSWERS (*A*) 5 10 (*B*) 12, 15, 18, 2

(*C*) 5 10 (*D*) 5 10

The inequality symbols $<$ and $>$ have a very clear geometric interpretation on a number line. If $a < b$, then the graph of a is to the left of b; if $a > b$, then the graph of a is to the right of b.

You no doubt noticed that we have ignored points to the left of the point with coordinate 1. In the next chapter we will extend the set of natural numbers to include 0 and a new set of numbers called the negative integers. These new numbers will be associated with points to the left of 1 in much the same way as the points to the right of 1 have been associated with natural numbers.

Exercise 9

A *Name the coordinate of a, b, and c respectively.*

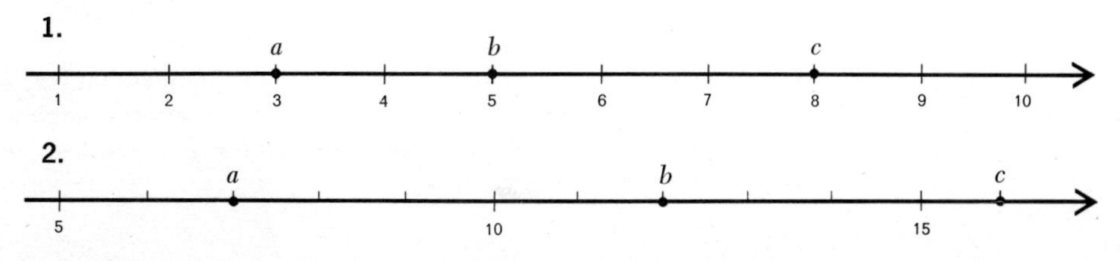

3.

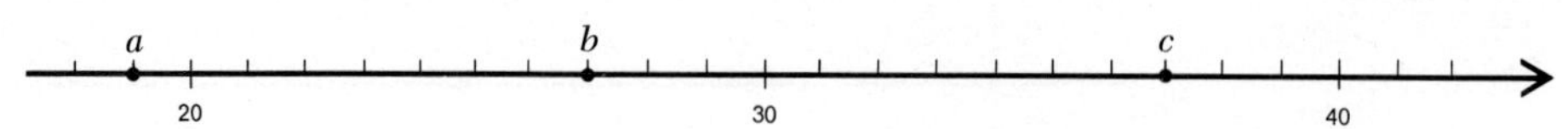

4.

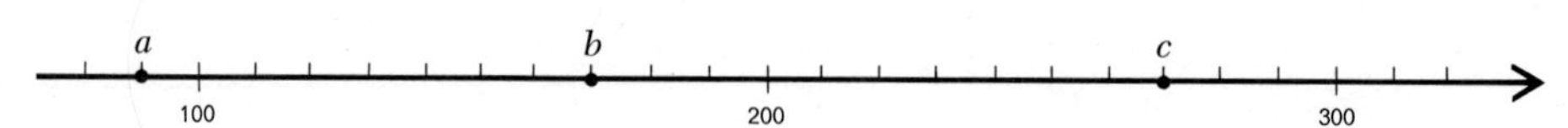

Graph each set.

5. $\{1, 2, 3, 4\}$

6. $\{1, 2, 3, 4, 5, 6\}$

7. $\{2, 4, 6, 8\}$

8. $\{3, 5, 7, 9\}$

9. $\{5, 6, 7, \ldots\}$

10. $\{8, 9, 10, \ldots\}$

For each variable restricted to natural numbers, graph:

11. $x < 6$

12. $x \leq 5$

13. $3 < x < 6$

14. $5 < n < 9$

15. $3 \leq y < 6$

16. $5 < t \leq 9$

17. $x > 3$

18. $x \geq 4$

B *Graph each set.*

19. $\{30, 35, 40, 45\}$

20. $\{65, 70, 75, 80\}$

21. $\{106, 108, 110, 112\}$

22. $\{210, 212, 214, 216\}$

23. $\{100, 101, 102, \ldots\}$

24. $\{1{,}001, 1{,}002, 1{,}003, \ldots\}$

For each variable restricted to natural numbers, graph:

25. $40 < x \leq 46$

26. $55 \leq x < 60$

27. $103 < y \leq 107$

28. $225 < m \leq 229$

29. $2 \leq x < 9$, x even

30. $5 \leq y \leq 11$, y odd

31. $2 \leq x < 9$, x composite

32. $5 \leq y \leq 11$, y prime

33. If the graph of set S is as indicated, then $S = \{(\text{list elements})\}$.

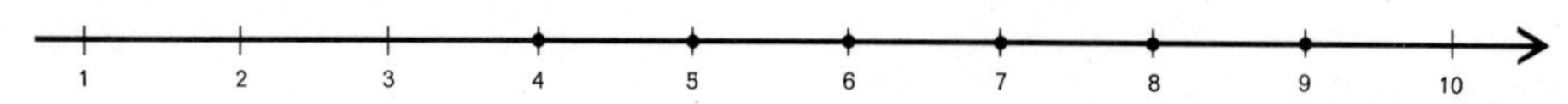

34. If the graph of set M is as indicated, then $M = \{(\text{list elements})\}$.

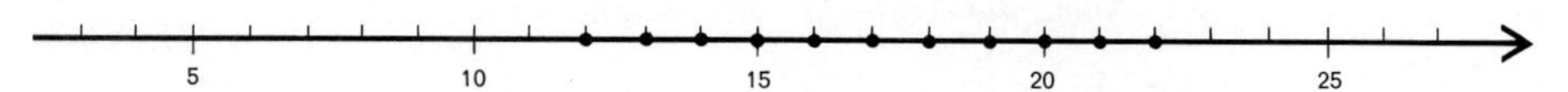

C The statement "A is the set of all natural numbers n such that $5 < n < 100$" is symbolized $A = \{n \in N | 5 < n < 100\}$ where N is the set of natural numbers and the symbol "$\in$" means "is an element of." This is not as forbidding as it first looks; each part of the symbol has a meaning as indicated below

Set A
is
the set of all
natural numbers n
such that
n is between 5 and 100

$$A = \{n \in N | 5 < n < 100\}$$

EXAMPLE (A) $\{n \in N | n < 8\} = \{1, 2, 3, 4, 5, 6, 7\}$

(B) $\{x \in N | 4 < x \leq 11\} = \{5, 6, 7, 8, 9, 10, 11\}$

(C) $\{x \in N | x + 5 = 8\} = \{3\}$

(D) $\{y \in N | y + 7 = 3\} = \varnothing$

Graph each set.

35. $\{n \in N | n < 5\}$

36. $\{x \in N | x \geq 7\}$

37. $\{t \in N | 3 \leq t < 7\}$

38. $\{x \in N | 2 < x \leq 8\}$

39. $\{y \in N | 2 + y = 6\}$

40. $\{x \in N | 3x = 11\}$

Exercise 10 Chapter Review

All variables are restricted to natural numbers.

A *Given the set of natural numbers $G = \{10, 11, 12, 13, 14, 15\}$, write as sets:*

1. The set of odd numbers in G.

2. The set of composite numbers in G.

3. True (T) or false (F): (A) $3 \cdot 7 - 4 = 9$, (B) V $\neq 6$, (C) $7 + 2 \cdot 3 = 13$?

4. Evaluate $w - 2z$ for $w = 9$ and $z = 3$.

5. If y is a certain number, write an algebraic expression for a number 8 times as large as y.

6. State the justifying axiom for (A) $3x = x3$ and (B) $(x + 3) + 2 = x + (3 + 2)$.

7. Remove parentheses and simplify:

(A) $7 + (x + 3)$

(B) $5(x2)$

8. (A) Write in nonexponent form: $7m^3n^2$
(B) Simplify: $y^{20}y^{12}$

9. Multiply, using the distributive axiom:

(A) $3(m + n)$

(B) $5(w + x + y)$

10. Take out factors common to all terms:

(*A*) $3m + 3n$ (*B*) $8u + 8v + 8w$

11. In the algebraic expression $5x^3 + 3x^2 + x + 7$ indicate the (*A*) coefficient of the second term, (*B*) coefficient of the third term, (*C*) exponent of the variable in the third term.

12. Combine like terms: (*A*) $3y + 6y$, (*B*) $2m + 5n + 3m$.

13. True (T) or false (F): (*A*) $11 > 4$, (*B*) $4 \leq 4$, (*C*) $6 < 5$?

14. Find the solution set for each inequality for x restricted to $\{1, 2, 3, 4, 5\}$: (*A*) $x > 3$, (*B*) $x + 2 \leq 5$, (*C*) $2 \leq x < 5$.

15. Graph $\{3, 5, 10, 12\}$ on a number line.

B **16.** Let M be the set of natural numbers from 24 to 31. Let P be the set of prime numbers in M. Write the elements of P within braces { }.

17. Write 120 as a product of prime factors.

18. Evaluate $2[9 - 3(3 - 1)]$.

19. Evaluate $3[14 - x(x + 1)]$ for $x = 3$.

20. Translate into an algebraic equation using only x as a variable: 3 times a certain number is 8 more than that number.

21. State justifying axioms for (*A*) $(2x)(y3) = (2x)(3y)$ and (*B*) $5(x + 3)$ is a natural number.

22. Remove parentheses and simplify:

(*A*) $(m + 5) + (n + 7) + (p + 2)$ (*B*) $(4x)(7y)$

23. Write in power-of-ten form and multiply: (120,000)(3,000,000)

24. Multiply, using the distributive axiom: $3y^3(2y^2 + y + 5)$

25. Take out factors common to all terms: $3m^5 + 6m^4 + 15m^2$

26. Combine like terms: $(5u^2 + 2u + 1) + (3u^2 + u + 5)$

27. Multiply, using the distributive axiom, and combine like terms: $3x(x + 5) + 2x(2x + 3) + x(x + 1)$

28. True (T) or false (F): (*A*) $2 \cdot 3 + 5 > 15$, (*B*) $12 - 5 \cdot 2 \leq 2$

29. Find all natural number solutions to (*A*) $3x < 15$, (*B*) $2 \leq x + 4 < 7$.

30. For x a natural number, graph $3 < x \leq 6$ on a number line.

C *Given the set* $M = \{27, 51, 61\}$, *write as a set:*

31. The odd numbers in M.

32. The prime numbers in M.

33. Evaluate $2\{9 - 2[(3 + 1) - (1 + 1)]\}$.

34. Evaluate $7\{x + 2[(x + y) - (x - y)]\}$ for $x = 3$ and $y = 1$.

35. If x represents the first of 3 consecutive odd numbers, write an algebraic equation that represents the fact that 4 times the first is equal to the sum of the second and third.

36. State justifying axioms for (*A*) $x + (y + z) = (y + z) + x$ and (*B*) $(3x)2 = 2(3x)$.

37. Which of the following sets are closed under (*A*) multiplication, (*B*) division?

$Q = \{2, 4, 6, 8\}$ $\qquad R = \{2, 4, 6, 8, \ldots\}$

38. State the reasons for each step (assume closure throughout without statement):

$$
\begin{array}{lrl}
1 & (2x + 7) + (3x + 5) & = (2x + 7) + (5 + 3x) \\
2 & & = 2x + [7 + (5 + 3x)] \\
3 & & = 2x + [(7 + 5) + 3x] \\
4 & & = 2x + (12 + 3x) \\
5 & & = 2x + (3x + 12) \\
6 & & = (2x + 3x) + 12 \\
7 & & = (2 + 3)x + 12 \\
8 & & = 5x + 12
\end{array}
$$

39. Multiply, using the distributive axiom: $5u^3v^2(2u^2v^2 + uv + 2)$.

40. Take out factors common to all terms: $20x^3y^2 + 5x^2y^3 + 15x^2y^2$.

41. Multiply, using the distributive axiom, and combine like terms: $2x^3(2x^2 + 1) + 3x^2(x^3 + 3x + 2)$.

42. Multiply, using the distributive axiom, and combine like terms: $(4x + 3)(2x + 1)$.

43. If the length of a rectangle is 5 in. more than twice the width x, (*A*) write an algebraic expression for its perimeter, and (*B*) simplify it.

44. Find all natural number solutions to $7 < 4x + 1 \leq 17$.

45. Replace the question marks in $C = \{x \in N \mid ? \leq x \, ? \, 7\} = \{3, 4, 5, 6\}$ with appropriate symbols.

CHAPTER 2 Integers

By limiting ourselves in the first chapter to the simplest number system within your experience, the natural numbers, we were able to develop many basic algebraic processes without the distracting influence of more complicated numbers such as decimals, fractions, and radicals. We will find that most of these processes carry on without change to the more involved number systems to be presented in this and the next chapter.

2.1 The Set of Integers

If the set of natural numbers were our only number system, we would be severely restricted. Even though it is closed under addition and multiplication, and has other desirable properties, it is not closed under subtraction and division. For example,

$$3 - 5 \text{ and } 3 \div 5$$

do not name natural numbers. As a step towards remedying these deficiencies and at the same time increasing our manipulative power, we now extend the natural numbers to the integers.

We start by giving the natural numbers another name. From now on they will also be called *positive integers*. To help us emphasize the difference between the positive integers (natural numbers) and the negative integers that are to be introduced shortly, we will often place a plus sign in front of a numeral used to name a natural number. Thus, we may use either

$+3$ or 3, $+25$ or 25, $+372$ or 372

and so on.

If we form a number line using the positive integers, and divide the line to the left into line segments equal to those used on the right, how should the endpoints of segments on the left be labeled?

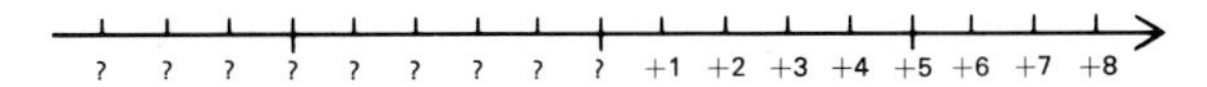

As you no doubt will guess, we label the first point to the left of $+1$ with *zero*

0

and the other points in succession with

$-1, -2, -3, \ldots$

These last numbers are called *negative integers*.

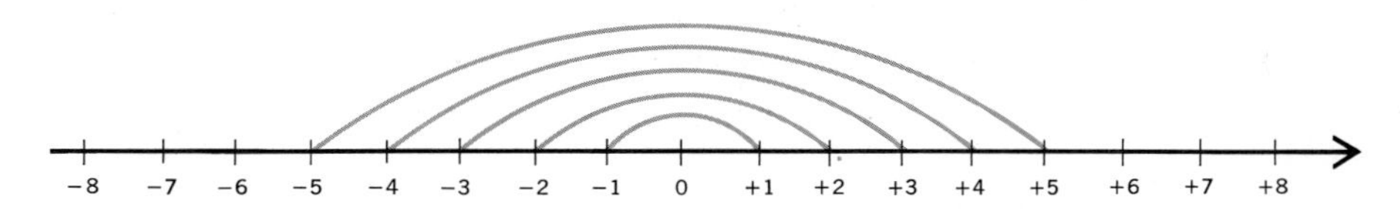

In general, to each positive integer there corresponds a unique (one and only one) number called a negative integer: -1 to $+1$, -2 to $+2$, -3 to $+3$, and so on. The minus sign is part of the number symbol.

By collecting the positive integers, 0, and the negative integers into one set we obtain the set of *integers*, *I*, the subject matter of this chapter. Thus

$$I = \{\ldots, -4, -3, -2, -1, 0, +1, +2, +3, +4, \ldots\}$$

We do not attempt to give a precise definition of each integer. We do, however, postulate the existance of this set of numbers, and we will learn to manipulate the symbols that name them according to certain rules.

EXAMPLE 1

The coordinates of points a, b, c, and d on are -8, -6, 0, and $+7$, respectively.

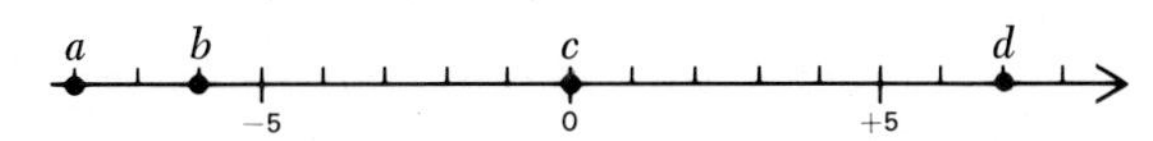

PROBLEM 1 What are the coordinates of a, b, c, and d on

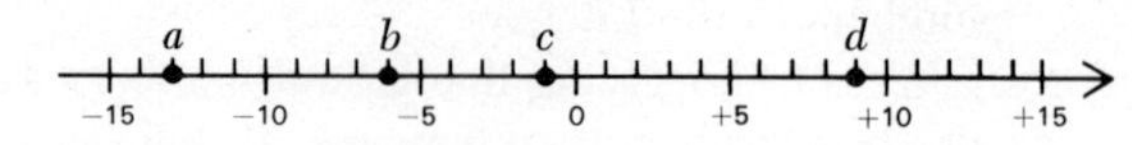

ANSWER $-13, -6, -1, +9$

Zero and the negative integers are relatively young historically speaking. Both were introduced as numbers in their own right between A.D. 600 and A.D. 700. Hindu mathematicians in India are given credit for their invention. The growing importance of commercial activities seemed to be the stimulus. Since business transactions involve decreases as well as increases, it was found that both transactions could be treated at once if the positive integers represented amounts received and the negative integers represented amounts paid out. Of course, since then negative numbers have been put to many other uses such as recording temperatures below 0, indicating altitudes below sea level, and representing deficits in financial statements, to name a few that the reader is probably already aware of. In addition, without negative numbers it is not possible to perform the operation

$$7 - 12$$

or to solve the simple equation

$$8 + x = 2$$

Before this course is over many more uses of negative numbers will be considered. In the next several sections we will learn how to add, subtract, multiply, and divide integers—an essential step to their many uses.

Exercise 11

A **1.** Write down the coordinates of a, b, c, and d.

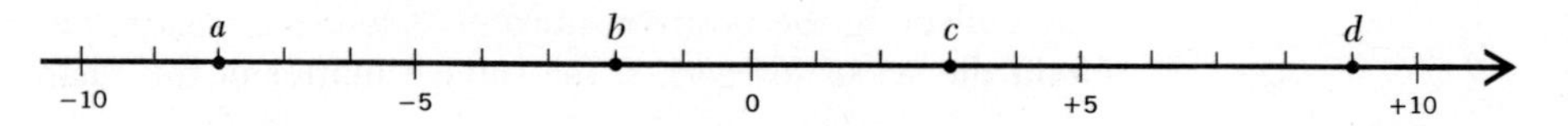

2. Write down the coordinates of a, b, c, d, and e.

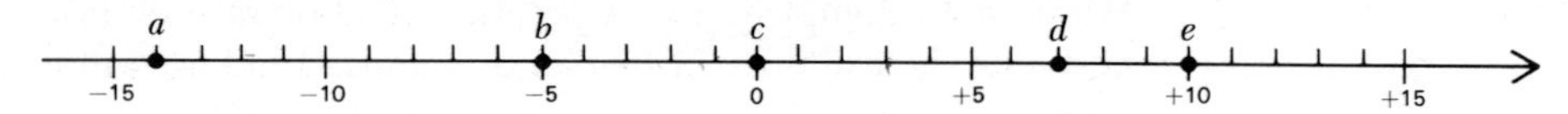

Graph each set of integers.

3. $\{-4, -2, 0, +2, +4\}$

4. $\{-7, -4, 0, +4, +8\}$

5. $\{-25, -20, -15, +5, +15\}$

6. $\{-30, -20, -5, +10, +15\}$

Using the figure for Prob. 2, write down the coordinate of each point.

7. 3 units to the left of d

8. 4 units to the right of e

9. 4 units to the right of a

10. 2 units to the left of b

11. 10 units to the left of d

12. 20 units to the right of a

B *Let* $P =$ *the set of positive integers*
$M =$ *the set of negative integers*
$I =$ *the set of integers*

indicate which are true (T) *or false* (F).

13. $+5$ is in P

14. -3 is in M

15. -4 is in P

16. $+3$ is in M

17. -7 is in I

18. $+4$ is in I

19. 0 is in P

20. 0 is in M

21. 0 is in I

22. -14 is in I

23. P is a subset of I.

24. I is a subset of M.

Referring to Fig. 1, express each of the following quantities by means of an appropriate integer.

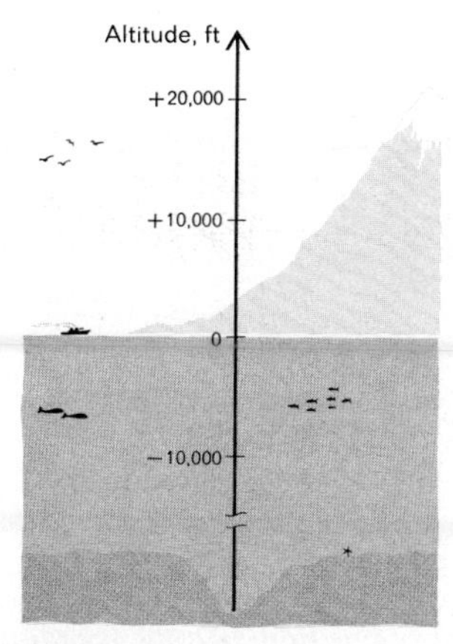

Figure 1

25. A mountain height of 20,270 ft (Mount McKinley, highest mountain in the United States).

26. A mountain peak 29,141 ft high (Mount Everest, highest point on earth).

27. A valley depth of 280 ft below sea level (Death Valley, the lowest point above water in the Western Hemisphere).

28. An ocean depth of 35,800 ft (Marianas Trench in the Western Pacific, greatest known depth in the world).

Referring to Fig. 2, express each of the following quantities by means of an appropriate integer.

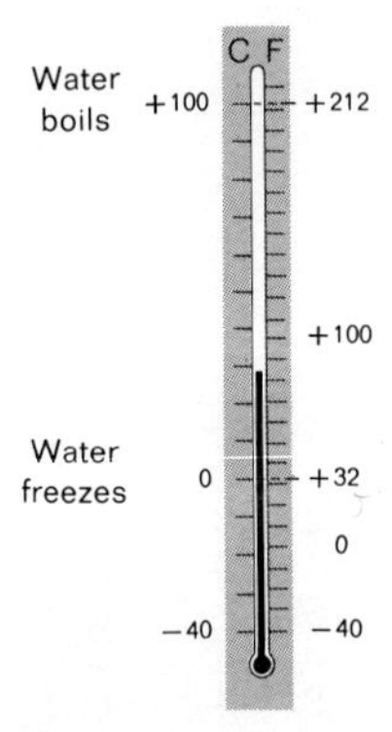

Figure 2

29. 5° below freezing on Celsius scale

30. 35° below freezing on Celsius scale

31. 5° below freezing on Fahrenheit scale

32. 100° below boiling on Fahrenheit scale

33. 35° below freezing on Fahrenheit scale

34. 220° below boiling on Fahrenheit scale

Express each of the following quantities by means of an appropriate integer.

35. A bank deposit of $25

36. A bank balance of $237

37. A bank withdrawal of $10

38. An overdrawn checking account of $17

39. A 9-yd loss in football

40. A 23-yd gain in football

C *In each problem start at 0 on a number line and give the coordinate at the final position.*

41. Move 7 units in the positive direction, 4 units in the negative direction, 5 more units in the negative direction, and finally, 3 units in the positive direction.

42. Move 4 units in the negative direction, 7 units in the positive direction, and 13 units in the negative direction.

Express the net gain or loss by means of an appropriate integer.

43. In banking: a \$23 deposit, a \$20 withdrawal, a \$14 deposit

44. In banking: a \$32 deposit, a \$15 withdrawal, an \$18 withdrawal

45. In football: 5-yd gain, a 3-yd loss, a 4-yd loss, an 8-yd gain, a 9-yd loss

46. In an elevator: up 2 floors, down 7 floors, up 3 floors, down 5 floors, down 2 floors

2.2 The Negative and Absolute Value of a Number

One of the important activities in algebra and in the uses of algebra in the real world is the evaluation of algebraic expressions for various replacements of variables by constants. An algebraic expression is like a recipe in that it contains symbolic instructions on how to proceed in its evaluation. For example, if we were to evaluate

$$5(x + 2y)$$

for $x = 10$ and $y = 3$, we now know that we would multiply 2 and 3, add the product to 10, and then multiply the sum by 5.

In this section we are going to define two more operations on numbers called "the negative of" and "the absolute value of," and you will get additional practice in following symbolic instructions. These two new operations are widely used in mathematics and its applications and will be used in the following sections to formulate definitions for addition, subtraction, multiplication, and division on integers.

We start by defining "the negative of a number" (which is not to be confused with "a negative number"): *The negative of a number* x is an operation on x, symbolized by

$$-x$$

that produces another number; namely, it changes the sign of x if x is not 0, and if x is 0 it leaves it alone.

EXAMPLE 2

(A) $-(+3) = -3$

(B) $-(-5) = +5$

(C) $-(0) = 0$

(D) $-[-(+3)] = -(-3) = +3$

PROBLEM 2

Find:

(A) $-(+7)$ (B) $-(-6)$

(C) $-(0)$ (D) $-[-(-4)]$

ANSWER (A) -7 (B) $+6$ (C) 0 (D) -4

Graphically, the negative of a number is its "mirror image" relative to 0 (see Fig. 3).

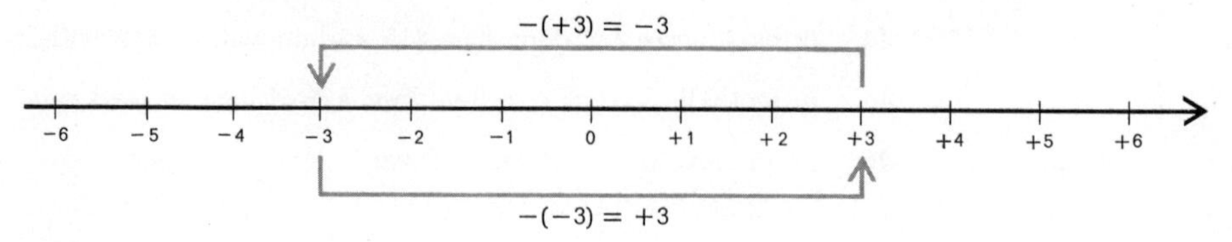

Figure 3

As a consequence of the definition of "the negative of a number," we note the following important properties:

1 The negative of a positive number is negative.
2 The negative of a negative number is positive.
3 The negative of 0 is 0.

Thus, we see that the negative of a number is not necessarily negative: $-x$ names a positive number if x is negative and a negative number if x is positive.

You should be aware by now that the minus sign "−" is used in the following three distinct ways:

1 As the operation "subtract": $7 \stackrel{\downarrow}{-} 5 = 2$
2 As the operation "the negative of": $\stackrel{\downarrow}{-}(-6) = +6$
3 As part of a number symbol: $\stackrel{\downarrow}{-}8$

The *absolute value of a number* x is an operation on x, denoted symbolically by

$$|x|$$

(not square brackets) that produces another number. What number? If x is positive or 0 it leaves it alone; if x is negative it makes it positive. Symbolically, and more formally,

$$|x| = \begin{cases} x & \text{if } x \geq 0 \\ -x & \text{if } x < 0 \end{cases}$$

Do not be afraid of this symbolic form of the definition. It represents a first exposure to more precise mathematical representations and it will take on more meaning with repeated exposure.

EXAMPLE 3

(A) $|+7| = +7$

(B) $|-7| = +7$

(C) $|0| = 0$

PROBLEM 3 Evaluate:

(A) $|+5|$ (B) $|-5|$ (C) $|0|$

ANSWER (A) $+5$ (B) $+5$ (C) 0

Thus we see that

1. The absolute value of a positive number is positive;
2. The absolute value of a negative number is positive;
3. The absolute value of 0 is 0;

and conclude that

The absolute value of a number is never negative.

"The absolute value" and "the negative of" operations are often used in combination and it is important to perform the operations in the right order—generally, from the inside out.

EXAMPLE 4

(A) $|-(-3)| = |+3| = +3$

(B) $-|-3| = -(+3) = -3$

(C) $-(|-5| - |-2|) = -[(+5) - (+2)] = -(+3) = -3$

PROBLEM 4 Evaluate:

(A) $|-(+5)|$ (B) $-|+5|$ (C) $-(|-7| + |-3|)$

ANSWER (A) $+5$ (B) -5 (C) -10

The reader may wonder if we are not leading him on. We went to all of the trouble to introduce negative numbers, and now we turn around and define an operation that makes negative numbers positive! We really are not trying to make things difficult. The absolute value operation has many uses in mathematics, and the reader will encounter it many times in this and other courses. In particular, as was noted above, it will be used in the following sections to define addition and multiplication for the integers.

Exercise 12

A *Evaluate:*

1. $-(+9)$ **2.** $-(+14)$ **3.** $-(-2)$

4. $-(-3)$ **5.** $|+4|$ **6.** $|+10|$

7. $|-6|$

8. $|-7|$

9. $-(0)$

10. $|0|$

11. The negative of a number is (*always, sometimes, never*) a negative number.

12. The negative of a number is (*always, sometimes, never*) a positive number.

13. The absolute value of a number is (*always, sometimes, never*) a negative number.

14. The absolute value of a number is (*always, sometimes, never*) a positive number.

Replace each question mark with an appropriate integer.

15. $-(+11) = ?$

16. $-(-15) = ?$

17. $-(?) = +5$

18. $-(?) = -8$

19. $|-13| = ?$

20. $|+17| = ?$

21. $|?| = +2$

22. $|?| = +8$

23. $|?| = -4$

24. $|?| = 0$

B *Evaluate:*

25. $-[-(+6)]$

26. $-[-(-11)]$

27. $|-(-5)|$

28. $|-(+7)|$

29. $-|-5|$

30. $-|+7|$

31. $(|-3| + |-2|)$

32. $(|-7| - |+3|)$

33. $-(|-12| - |-4|)$

34. $-(|-6| + |-2|)$

Evaluate for $x = +7$ and $y = -5$.

35. $-x$

36. $|x|$

37. $|y|$

38. $-y$

39. $-|x|$

40. $-|y|$

41. $-(-y)$

42. $-(-x)$

43. $|-y|$

44. $|-x|$

45. $|x| - |y|$

46. $-(|x| + |y|)$

Guess at the solution set of each equation from the set of integers.

47. $|+5| = x$

48. $|-7| = x$

49. $-x = -3$

50. $-x = +8$

51. $|x| = +6$

52. $|x| = +9$

53. $|x| = -4$

54. $-|x| = +4$

C *Describe the elements in each set.*

55. $\{x \in I \mid |x| = 0\}$

56. $\{x \in I \mid -x = 0\}$

57. $\{x \in I \mid -x = |x|\}$

58. $\{x \in I \mid x = |x|\}$

59. $\{x \in I \mid -(-x) = x\}$

60. $\{x \in I \mid |-x| = |x|\}$

61. $\{x \in I \mid |-x| = x\}$

62. $\{x \in I \mid -x = x\}$

63. $\{x \in I \mid -|x| = |x|\}$

64. $\{x \in I \mid |-x| = -|x|\}$

2.3 Addition of Integers

How should addition on the integers be defined so that we can assign numbers to each of the following sums?

$$(+2) + (+5) = ?$$

$$(+2) + (-5) = ?$$

$$(-2) + (+5) = ?$$

$$(-2) + (-5) = ?$$

$$(+7) + 0 = ?$$

$$0 + (-3) = ?$$

To give us an idea, let us think of addition of integers in terms of deposits and withdrawals in a checking account, starting with a 0 balance. If we do this, then a deposit of \$2 followed by another deposit of \$5 would provide us with a balance of \$7; thus, as we would expect from addition in the natural numbers,

$$(+2) + (+5) = +7$$

Similarly, a deposit of \$2 followed by a withdrawal of \$5 would yield an overdrawn account of \$3, and we would write

$$(+2) + (-5) = -3$$

Continuing in the same way, we can assign to each sum the value that indicates the final status of our account after the two transactions have been completed. Hence,

$$(-2) + (+5) = +3$$

$$(-2) + (-5) = -7$$

$$(+7) + 0 = +7$$

$$0 + (-3) = -3$$

We would like any formal definition of addition for integers to yield the same results as above and, also, to yield the same closure, commutative, and associative properties we had with the natural numbers. With these considerations in mind, it turns out that we have very little choice but to define addition as follows.

DEFINITION OF ADDITION OF INTEGERS

Numbers with like sign. If a and b are positive integers, add as in the set of natural numbers. If a and b are both negative, the sum is the negative of the sum of their absolute values.

Numbers with unlike signs. The sum of two integers with unlike signs is a number of the same sign as the integer with the larger absolute value. The absolute value of the sum is the difference of the absolute values of the two numbers found by subtracting the smaller absolute value from the larger. If the numbers have the same absolute values, their sum is 0.

Zero. The sum of any integer and 0 is that integer; the sum of 0 and any integer is that integer.

This definition when applied to the deposit-withdrawal illustrations above will produce the same results. But you will no doubt object to the difficulty in its use. Fortunately, we will be able to mechanize the process so that you will be able to handle addition problems without difficulty. You should not forget, however, that these mechanical rules that are to be discussed shortly are justified on the basis of the above definition and not vice versa.

The following important properties of addition are an immediate consequence of this definition.

THEOREM 1 For each integer a, b, and c

(*A*) $a + b$ is an integer closure property

(*B*) $a + b = b + a$ commutative property

(*C*) $(a + b) + c = a + (b + c)$ associative property

As a consequence of this theorem we will have essentially the same kind of freedom that we had with the natural numbers in rearranging terms and inserting or removing parentheses.

Now let us turn to the mechanics of adding signed numbers. It may relieve you to know that no one (not even the professional mathematician) in his everyday routine calculations involving the addition of signed numbers goes through the steps precisely as they are described in the formal definition of addition; mechanical shortcuts soon take over. The following process, or something close to it, is very likely used. We will restrict our attention to nonzero quantities since addition involving zero seems to offer few difficulties.

MECHANICS OF ADDING SIGNED NUMBERS

Are the signs of the two numbers alike or unlike?

alike

(*A*) Mentally block out the signs.
(*B*) Add the two numbers as if they were natural numbers.
(*C*) Prefix the common sign of the original numbers to the sum.

examples

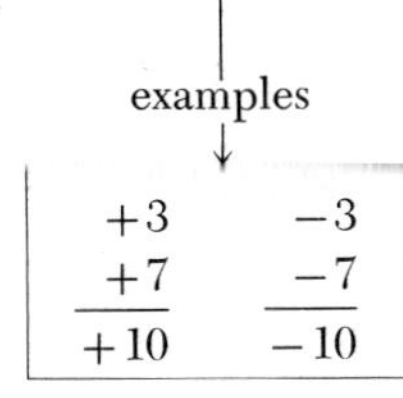

unlike

(*A*) Mentally block out the signs.
(*B*) Subtract the smaller unsigned number from the larger.
(*C*) Prefix the sign associated with the larger of the two unsigned numbers.

examples

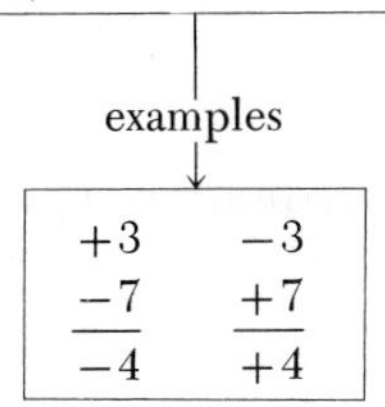

EXAMPLE 5 (*A*) Add:

$$\begin{array}{r}-8\\-3\\\hline-11\end{array}\qquad\begin{array}{r}+8\\-3\\\hline+5\end{array}\qquad\begin{array}{r}-8\\+3\\\hline-5\end{array}\qquad\begin{array}{r}+4\\0\\\hline+4\end{array}\qquad\begin{array}{r}0\\-6\\\hline-6\end{array}$$

(*B*) Add:

$$(-4) + (-6) = -10$$
$$(-4) + (+6) = +2$$
$$(+4) + (-6) = -2$$
$$0 + (-1) = -1$$

PROBLEM 5 (*A*) Add:

$$\begin{array}{r}-4\\-5\\\hline\end{array}\qquad\begin{array}{r}+4\\-5\\\hline\end{array}\qquad\begin{array}{r}-4\\+5\\\hline\end{array}\qquad\begin{array}{r}-9\\0\\\hline\end{array}$$

(*B*) Add:

$(-2) + (+7)$, $(+2) + (-7)$, $(-2) + (-7)$, $0 + (-5)$

ANSWER (*A*) $-9, -1, +1, -9$ (*B*) $+5, -5, -9, -5$

To add three or more integers, add all of the positive integers together, add all of the negative integers together (the commutative and associative properties of integers justify this procedure), and then add the two resulting sums as above.

EXAMPLE 6 Add: $(+3) + (-6) + (+8) + (-4) + (-5)$

SOLUTION

$$\begin{aligned}(+3) + (-6) + (+8) + (-4) + (-5) &= [(+3) + (+8)] + [(-6) + (-4) + (-5)] \\ &= (+11) + (-15) \\ &= -4\end{aligned}$$

or vertically,

	Done mentally or on scratchpaper		
+3	+3	−6	+11
−6	+8	−4	−15
+8	+11	−5	−4
−4		−15	
−5			
−4 → (−4)			

PROBLEM 6 Add: $(-8) + (-4) + (+6) + (-3) + (+10) + (+1)$

ANSWER $+2$

We conclude this section by stating without proof another, but less obvious, property of addition that follows from the definition of addition. We will refer to this property in future developments.

THEOREM 2 (*A*) For each integer a the sum of it and its negative is 0; that is,

$$a + (-a) = 0$$

(*B*) If the sum of two numbers is 0, then each must be the negative of the other; symbolically, if

$$a + b = 0$$

then $a = -b$ and $b = -a$.

Exercise 13

A *Add:*

1. $\begin{array}{r} +5 \\ \underline{+6} \end{array}$ **2.** $\begin{array}{r} +7 \\ \underline{-4} \end{array}$ **3.** $\begin{array}{r} -9 \\ \underline{+6} \end{array}$ **4.** $\begin{array}{r} -6 \\ \underline{+8} \end{array}$

5. $\begin{array}{r} +6 \\ \underline{-8} \end{array}$ **6.** $\begin{array}{r} -3 \\ \underline{-4} \end{array}$ **7.** $\begin{array}{r} -7 \\ \underline{-1} \end{array}$ **8.** $\begin{array}{r} +8 \\ \underline{+2} \end{array}$

9. $\begin{array}{r} 0 \\ +3 \\ \hline \end{array}$

10. $\begin{array}{r} -4 \\ 0 \\ \hline \end{array}$

11. $(+5) + (+4)$

12. $(-7) + (-3)$

13. $(-8) + (+2)$

14. $(+3) + (-7)$

15. $(-6) + (-3)$

16. $(+2) + (+3)$

17. $0 + (-9)$

18. $(+2) + 0$

19. $\begin{array}{r} +4 \\ -3 \\ -5 \\ -7 \\ +9 \\ \hline \end{array}$

20. $\begin{array}{r} -6 \\ -4 \\ +8 \\ +3 \\ -5 \\ \hline \end{array}$

21. $\begin{array}{r} -7 \\ +2 \\ -3 \\ -1 \\ +5 \\ \hline \end{array}$

22. $\begin{array}{r} +6 \\ -4 \\ -8 \\ -2 \\ +9 \\ \hline \end{array}$

23. $(+5) + (-8) + (-9) + (+7)$

24. $(-8) + (-7) + (+3) + (+9)$

25. $(-6) + 0 + (+5) + (-2) + (-1)$

26. $(+9) + (-3) + 0 + (-8)$

B *Add:*

27. $\begin{array}{r} +11 \\ -23 \\ \hline \end{array}$

28. $\begin{array}{r} -12 \\ -21 \\ \hline \end{array}$

29. $\begin{array}{r} -403 \\ -219 \\ \hline \end{array}$

30. $\begin{array}{r} -307 \\ +231 \\ \hline \end{array}$

31. $(-63) + (+25)$

32. $(-45) + (-73)$

33. $(-237) + (-431)$

34. $(-197) + (+364)$

35. $\begin{array}{r} +12 \\ -18 \\ -23 \\ +\ 4 \\ -11 \\ \hline \end{array}$

36. $\begin{array}{r} -63 \\ +45 \\ -\ 3 \\ +17 \\ +12 \\ \hline \end{array}$

37. $(+12) + (+7) + (-37) + (+14)$

38. $(-23) + (-35) + (+43) + (-33)$

Replace each question mark with an appropriate integer.

39. $(-3) + ? = -7$

40. $? + (-9) = -13$

41. $(+8) + ? = +3$

42. $(-12) + ? = +4$

43. $? + (-12) = -7$

44. $(+54) + ? = -33$

45. $(+33) + ? = -44$

46. $? + (-14) = +20$

Evaluate:

47. $|-8| + |+6|$

48. $|(-8) + (+6)|$

49. $(-|-3|) + (-|+3|)$

50. $|-5| + [-(-8)]$

Evaluate for $x = -5$, $y = +3$, and $z = -2$.

51. $x + y$ **52.** $y + z$ **53.** $|(-x) + z|$ **54.** $-(|x| + |z|)$

55. You own a stock that is traded on the New York Stock Exchange. On Monday it closed at $23 per share, it fell $3 on Tuesday and another $6 on Wednesday, it rose $2 on Thursday, and finished strongly on Friday by rising $7. Use addition of signed numbers to determine the closing price of the stock on Friday.

56. Your football team is on the opponent's 10-yd line and in 4 downs gains 8 yd, loses 4 yd, loses another 8 yd, and gains 13 yd. Use addition of signed numbers to determine if a touchdown was made.

57. A spelunker (cave explorer) had gone down 2,340 (vertical) ft into the 3,300-ft Gouffre Berger, the world's deepest pothole cave, located in the Isere province of France. On his ascent he climbed 732 ft, slipped back 25 ft and then another 60 ft, climbed 232 ft, and finally slipped back 32 ft. Use addition of signed numbers, starting with $-2{,}340$, to find his final position.

58. In a card game (such as rummy, where cards held in your hand after someone goes out are counted against you) the following scores were recorded after four hands of play. Who was ahead at this time and what was his score?

RUSS	JAN	PAUL	MEG
+35	+80	−5	+15
+45	+5	+40	−10
−15	−35	+25	+105
−5	+15	+35	−5

C *Replace each question mark with an appropriate symbol (variables represent integers).*

59. $a + (-a) = ?$ **60.** $(-x) + x = ?$

61. $m + ? = 0$ **62.** $(-x) + ? = 0$

63. Give a reason for each step:

$$\begin{aligned}[a + b] + (-a) &= (-a) + [a + b] \\ &= [(-a) + a] + b \\ &= 0 + b \\ &= b\end{aligned}$$

2.4 Subtraction of Integers

From subtraction in the natural numbers we know that

$$(+8) - (+5) = +3$$

but what can we write for the differences

$$(+8) - (-5) = ?$$

$(+5) - (+8) = ?$

$(-8) - (+5) = ?$

$(-8) - (-5) = ?$

$(-5) - (-8) = ?$

$0 - (-5) = ?$

We are going to define subtraction in such a way that all of these problems will have answers. Where do we start? We start with some of the notions you learned about subtraction in elementary school and then generalize on these ideas.

You will recall that to check the subtraction problem

$$\begin{array}{r} 8 \\ -5 \\ \hline 3 \end{array}$$

we add 3 to 5 to obtain 8. We can use this checking requirement to transform subtraction into addition: Instead of saying

Subtract 5 from 8.

we can ask

What must be added to 5 to produce 8?

Notice that the answer to each is the same, 3. The latter way of looking at subtraction is the most useful way of the two for its generalization to more involved number systems. And it motivates the following general definition that will not only apply to the integers, but also to any other number system we encounter.

DEFINITION OF SUBTRACTION

We write

$M - S = D$ if and only if $M = S + D$

Let us use the definition to find answers to the problems stated earlier. In

$(+8) - (-5) = ?$ or $\begin{array}{r} (+8) \\ -(-5) \\ \hline ? \end{array}$

we ask, "What must be added to (-5) to produce $(+8)$?" From our definition of addition we know the answer to be $(+13)$. Thus we write

$(+8) - (-5) = +13$ and $\begin{array}{r} (+8) \\ -(-5) \\ \hline +13 \end{array}$

since

$(-5) + (+13) = (+8)$

EXAMPLE 7

Similarly,

$$\begin{array}{ccccc} (+5) & (-8) & (-8) & (-5) & 0 \\ -(+8) & -(+5) & -(-5) & -(-8) & -(-5) \\ \hline -3 & -13 & -3 & +3 & +5 \end{array}$$

since each difference added to the second number (subtrahend) produces the top number (minuend).

PROBLEM 7

Subtract by proceeding as above; that is, "What must be added to the bottom number to produce the top number?"

$$\begin{array}{ccccccc} (+7) & (+3) & (+7) & (-3) & (-7) & (-3) & 0 \\ -(+3) & -(+7) & -(-3) & -(+7) & -(-3) & -(-7) & -(+3) \\ \hline ? & ? & ? & ? & ? & ? & ? \end{array}$$

ANSWER $+4, -4, +10, -10, -4, +4, -3$

You no doubt found these problems a little troublesome. Fortunately, we will be able to mechanize the process so that subtraction will be as easy as addition. Nevertheless, you should still do the problems to make sure you understand the definition of subtraction. The mechanical rule that we will formulate, as well as other properties of subtraction, is based on the above definition. In fact, the following theorem is a direct consequence of the definition and leads directly to the mechanical rule just referred to.

THEOREM 3

To subtract S from M, add the negative of S to M; symbolically,

$$M - S = M + (-S)$$

To prove the theorem, we must show that the sum of the subtrahend and the difference is equal to the minuend, that is, that $S + [M + (-S)] = M$. To this end, we make direct use of the properties of addition discussed in the last section (try to supply the reasons for each step). Thus

$$\begin{aligned} S + [M + (-S)] &= [M + (-S)] + S \\ &= M + [(-S) + S] \\ &= M + 0 \\ &= M \end{aligned}$$

If you look at this theorem very carefully and recall that when we take the negative of a number we change its sign, then you should be able to uncover a simple *mechanical rule for subtraction,* namely

> To subtract one number from another, change the sign of the number being subtracted and add.

EXAMPLE 8 Subtract:

(A)
$$\begin{array}{r} (+7) \\ \underline{-(-8)} \\ +15 \end{array}$$

(B)
$$\begin{array}{r} (-4) \\ \underline{-(+5)} \\ -9 \end{array}$$

(C) $(-9) - (-4) = (-9) + (+4) = -5$

(D) $0 - (+8) = 0 + (-8) = -8$

PROBLEM 8 Subtract:

(A)
$$\begin{array}{r} (+6) \\ \underline{-(-9)} \end{array}$$

(B)
$$\begin{array}{r} (-3) \\ \underline{-(-5)} \end{array}$$

(C) $(-7) - (-2)$

(D) $(-3) - (+8)$

ANSWER (A) $+15$ (B) $+2$ (C) -5 (D) -11

As a consequence of the definition of subtraction and Theorem 3, we can conclude that the integers are closed under subtraction; that is, the difference of *any* two integers is always an integer.

Exercise 14

A *Subtract (using the mechanical rule).*

1. $\begin{array}{r} (+9) \\ \underline{-(+4)} \end{array}$ **2.** $\begin{array}{r} (+10) \\ \underline{-(+7)} \end{array}$ **3.** $\begin{array}{r} (+9) \\ \underline{-(-4)} \end{array}$ **4.** $\begin{array}{r} (+10) \\ \underline{-(-7)} \end{array}$

5. $\begin{array}{r} (+4) \\ \underline{-(+9)} \end{array}$ **6.** $\begin{array}{r} (+7) \\ \underline{-(+10)} \end{array}$ **7.** $\begin{array}{r} (-9) \\ \underline{-(-4)} \end{array}$ **8.** $\begin{array}{r} (-10) \\ \underline{-(-7)} \end{array}$

9. $\begin{array}{r} (-4) \\ \underline{-(-9)} \end{array}$ **10.** $\begin{array}{r} (-7) \\ \underline{-(-10)} \end{array}$ **11.** $\begin{array}{r} 0 \\ \underline{-(+6)} \end{array}$ **12.** $\begin{array}{r} 0 \\ \underline{-(-4)} \end{array}$

13. $\begin{array}{r} (-2) \\ \underline{-(+6)} \end{array}$ **14.** $\begin{array}{r} (+7) \\ \underline{-(+9)} \end{array}$ **15.** $\begin{array}{r} (-3) \\ \underline{-\quad 0} \end{array}$ **16.** $\begin{array}{r} (+2) \\ \underline{-\quad 0} \end{array}$

17. $(+6) - (-8)$

18. $(-4) - (+7)$

19. $(+6) - (+10)$

20. $(+6) - (-10)$

21. $(-9) - (-3)$

22. $0 - (-7)$

23. $0 - (+5)$

24. $(-1) - (+6)$

B 25. $(-12) - (-27)$

26. $(+57) - (+92)$

27. $0 - (-87)$

28. $0 - (+101)$

29. $(-271) - (+44)$

30. $(+327) - (-73)$

31. $(-245) - 0$

32. $(+732) - 0$

Perform the indicated operations.

33. $[(-2) - (+4)] - (-7)$

34. $(-2) - [(+4) - (-7)]$

35. $(-23) - [(-7) + (-13)]$

36. $[(-23) - (-7)] + (-13)$

37. $[(+6) - (-8)] + [(-8) - (+6)]$

38. $[(+3) - (+5)] - [(-5) - (-8)]$

Evaluate for $x = +2$, $y = -5$, and $z = -3$.

39. $x - y$

40. $y - z$

41. $(x + z) - y$

42. $x - (y - z)$

43. $(-y) - (-z)$

44. $(-z) - y$

45. $-(|x| - |y|)$

46. $|(x - y) - (y - z)|$

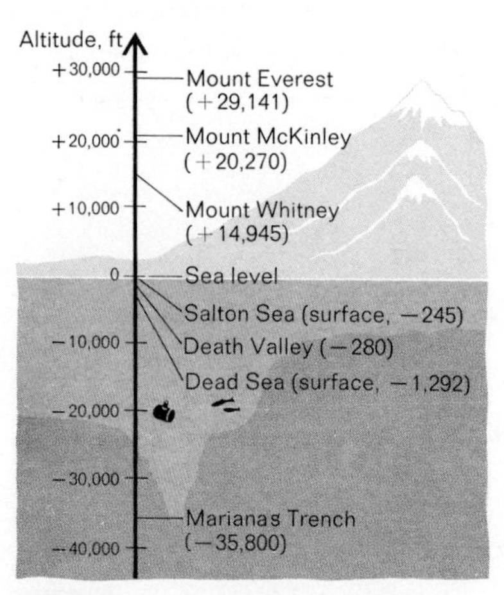

Figure 4

Use Fig. 4 and subtraction of integers, subtracting the one lower on the scale from the higher one, to find each of the following:

47. The difference in height between the highest point on earth, Mount Everest, and the deepest point in the ocean, Marianas Trench.

48. The difference in height between the highest point in the United States, Mount McKinley, and the lowest point in the United States, Death Valley.

49. The difference in height between the Salton Sea and Death Valley (both in California).

50. The difference in height between the Dead Sea, the deepest fault in the earth's crust, and Death Valley.

C *Which of the following statements hold for all integers a, b, and c? Illustrate each false statement with an example that shows that it is false.*

51. $a + b = b + a$

52. $a + (-a) = 0$

53. $a - b = b - a$

54. $a - b = a + (-b)$

55. $(a + b) + c = a + (b + c)$

56. $(a - b) - c = a - (b - c)$ false

57. $|a + b| = |a| + |b|$

58. $|a - b| = |a| - |b|$

59. Write down the reason for each step in the proof for Theorem 3.

60. Supply the reasons for each step:

$$
\begin{aligned}
(a + b) + [(-a) + (-b)] &= (b + a) + [(-a) + (-b)] \\
&= [(b + a) + (-a)] + (-b) \\
&= \{b + [a + (-a)]\} + (-b) \\
&= (b + 0) + (-b) \\
&= b + (-b) \\
&= 0
\end{aligned}
$$

Therefore,

$$-(a + b) = (-a) + (-b)$$

2.5 Multiplication of Integers

Having defined addition and subtraction for the integers, we now turn to multiplication. Again, from the natural numbers we know that

$$(+3)(+2) = +6$$

but how shall we define

$$(+3)(-2) = ?$$

$$(-3)(+2) = ?$$

$$(-3)(-2) = ?$$

$$(0)(-4) = ?$$

We would like to define multiplication of integers in such a way that closure, associative, commutative, and distributive properties hold. Let us assume for the moment that these properties do hold and investigate the consequences. Consider the following argument (assuming that $a \cdot 0 = 0$ for all integers a):

$(+3) + (-3) = 0$	Theorem 2, Sec. 2.3
$(+2)[(+3) + (-3)] = (+2)(0)$	Property of equality (see Prob. 32, Exercise 4)
$(+2)(+3) + (+2)(-3) = 0$	Assumed distributive property and $(+2)(0) = 0$
$(+2)(-3) = -[(+2)(+3)]$	Theorem 2, Sec. 2.3
$= -6$	Definition of the negative of a number

What the preceding argument indicates is that if the multiplication properties of natural numbers carry over to the integers, then $(+2)(-3)$ has no choice but to be defined to be -6. A similar argument will show that $(-2)(-3)$ will have to be defined to be $+6$.

Thus, in order to continue to use the basic properties of the natural numbers for the integers, we are led to the following formal definition of multiplication.

DEFINITION OF MULTIPLICATION

Numbers with like signs. The product of two integers with like signs is a positive number and is found by multiplying the absolute values of the two numbers.

Numbers with unlike signs. The product of two integers with unlike signs is a negative integer and is found by taking the negative of the product of the absolute values of the two integers.

Zero. The product of any integer and 0 is 0; the product of 0 and any integer is 0.

Fortunately, as in the case with addition, we will be able to mechanize the process of multiplication of integers so that you will be able to write answers with very little effort. However, as before, it is important that you realize that it is the basic definition that is the basis for any mechanical procedures as well as the many useful properties of multiplication that we will state. For example, an immediate consequence of the definition is

THEOREM 4 For each integer a, b, and c

(A) ab is an integer closure property

(B) $ab = ba$ commutative property

(C) $(ab)c = a(bc)$ associative property

(D) $a(b + c) = ab + ac$ distributive property

Because of this theorem we continue to have the same kind of freedom in rearranging factors and inserting and removing parentheses as we did with multiplication in the natural numbers. The distributive property will be seen to be a particularly useful tool.

By taking a close look at the definition, we easily formulate the following *mechanical rule* for routine calculations.

MECHANICS OF MULTIPLYING SIGNED NUMBERS

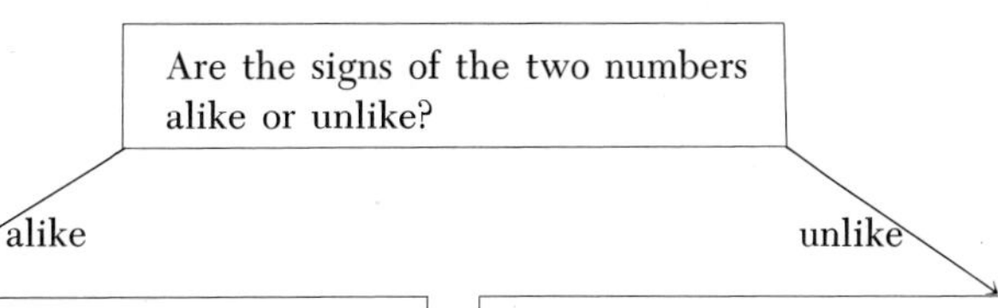

alike:
(A) Mentally block out the signs.
(B) Multiply the two numbers as if they were natural numbers.
(C) Prefix a plus sign (or no sign) to the product.

unlike:
(A) Mentally block out the signs.
(B) Multiply the two numbers as if they were natural numbers.
(C) Prefix a negative sign to the product.

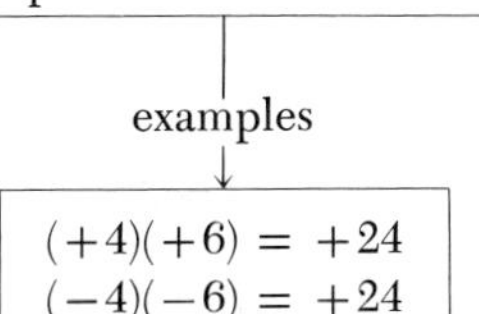

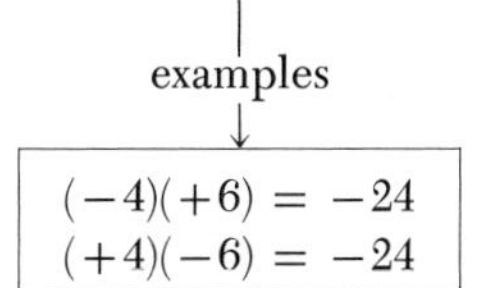

EXAMPLE 9

(A) $(+5)(+3) = +15$

(B) $(-8)(-6) = +48$

(C) $(-7)(+3) = -21$

(D) $(+9)(-4) = -36$

PROBLEM 9 Evaluate, using the mechanical rule:

(A) $(+6)(+5)$ (B) $(+7)(-6)$

(C) $(-4)(-10)$ (D) $(-9)(+8)$

ANSWER (A) $+30$ (B) -42 (C) $+40$ (D) -72

We complete this section by considering several additional properties of multiplication that are very important. To start, consider the effect of multiplying an integer by either $+1$ or -1. It is an immediate consequence of the definition of multiplication that if we multiply any integer by $+1$, we will get that integer back again, and if we multiply any integer by -1, we will get the negative of the original integer. Thus,

THEOREM 5 For any integer a

$$(+1)a = a \quad \text{and} \quad (-1)a = -a$$

Because a quantity may be substituted for its equal in any expression (substitution principle for equality), $(+1)a$ and a may be used interchangeably, and $(-1)a$ and $-a$ may be used interchangeably. We now see why we say that a has an understood coefficient of $+1$, and $-a$ has an understood coefficient of -1.

The next property of integers is often mistaken for the definition of the product of two integers; actually, it is a theorem.

THEOREM 6 For each integer a and each integer b,

(A) $(-a)b = -(ab)$

(B) $(-a)(-b) = ab$

In words this theorem states that the product of the negative of an integer and another integer is the negative of the product of the two integers, and the product of the negative of an integer and the negative of another integer is the product of the original integers.

PROOF OF PART A

$$(-a)b = [(-1)a]b \qquad \text{Theorem 5}$$

$$= (-1)(ab) \qquad \text{Associative property of multiplication}$$

$$= -(ab) \qquad \text{Theorem 5}$$

The proof of part B is left as an exercise.

EXAMPLE 10 Evaluate $(-a)b$ and $-(ab)$ for $a = -5$ and $b = +4$.

SOLUTION

$$(-a)b = [-(-5)](+4) = (+5)(+4) = +20$$

$$-(ab) = -[(-5)(+4)] = -(-20) = +20$$

PROBLEM 10 Evaluate $(-a)(-b)$ and ab for $a = -5$ and $b = +4$.

ANSWER Both are -20

Expressions of the form

$-ab$

occur frequently and at first glance are confusing to students. If you were asked to evaluate $-ab$ for $a = -3$ and $b = +2$, how would you proceed? Would you take the negative of a and then multiply it by b, or multiply a and b first and then take the negative of the product? Actually it does not matter! Because of Theorem 6 we get the same result either way since $(-a)b = -(ab)$. If, in addition, we consider other material in this section, we find that

$$-ab = \begin{cases} (-a)b \\ a(-b) \\ -(ab) \\ (-1)ab \end{cases}$$

and we are at liberty to replace any one of these five forms with another from the same group.

We conclude this section by observing two additional distributive properties that are very useful:

THEOREM 7 If all variables represent integers,

(A) $a(b + c + d + \cdots + f) = ab + ac + ad + \cdots + af$

(B) $a(b - c) = ab - ac$

Part A is easily proved for any particular case by repeated use of Theorem 4D. The proof of part B is left as an exercise.

EXAMPLE 11 (A) $2x(3x^2 + 4x + 1) = 6x^3 + 8x^2 + 2x$

(B) $3xy(x - 2y) = 3x^2y - 6xy^2$

PROBLEM 11 Multiply, using the appropriate distributive properties:

(A) $5y^2(2y^3 + y^2 + 3y + 4)$ (B) $2mn(3m - 4n)$

ANSWER (A) $10y^5 + 5y^4 + 15y^3 + 20y^2$ (B) $6m^2n - 8mn^2$

Exercise 15

All variables represent integers.

A *Multiply:*

1. $(+2)(+7)$ **2.** $(+8)(+4)$ **3.** $(-2)(-7)$

4. $(-8)(-4)$ **5.** $(-2)(+7)$ **6.** $(-8)(+4)$

7. $(+2)(-7)$ **8.** $(+8)(-4)$ **9.** $(0)(-7)$

10. $(0)(+6)$ **11.** $(-5)(0)$ **12.** $(+1)(0)$

Evaluate:

13. $(-2) + (-1)(+3)$
14. $(-3)(-2) + (+4)$
15. $(+2)[(+3) + (-2)]$
16. $(+5)[(-4) + (+6)]$
17. $(+4) - (-2)(-4)$
18. $(-7) - (+4)(-3)$
19. $(-3)[(-2) - (-4)]$
20. $(-6)[(+3) - (+8)]$
21. $(-2)^2(-3)$
22. $(-3)(-1)^2$
23. $(-3)^2 + (-2)(+1)$
24. $(-6)(+2) + (-2)^2$
25. $(-1)(-7)$ and $-(-7)$
26. $(-1)(+3)$ and $-(+3)$
27. $-(-2)(+4)$
28. $-(+3)(-4)$
29. $-(-5)(-1)$
30. $-(-2)(-7)$

Multiply:

31. $4x(2x^2 + x + 3)$
32. $3x(x^2 + 3x + 1)$
33. $2(3x - 5)$
34. $5(2y - 4)$

B *Multiply:*

35. $(-20)(+35)$
36. $(+12)(-22)$
37. $(-15)(-30)$
38. $(-34)(-12)$
39. $(-6)(-3)(+4)$
40. $(+5)(-7)(+2)$
41. $(+19)(0)(-35)$
42. $(-22)(+36)(0)$

Evaluate:

43. $(-5)^2 + (-3)^2$
44. $(+4)^2 + (-6)^2$
45. $(-6) - (-2)^2$
46. $(-6)(+7) - (-3)^2$
47. $[(-3) - (+8)][(+4) + (-2)]$
48. $[(+2) + (-7)][(+8) - (+10)]$

Evaluate for $x = -5$, $y = +2$, and $z = -7$.

49. xy
50. yz
51. xyz
52. $6yz$
53. $x(y + z)$
54. $y(x + z)$
55. $xy - xz$
56. $yx - yz$
57. $(-2)(x + y + z)$
58. $(-2)x + (-2)y + (-2)z$
59. $-x$ and $(-1)x$
60. $(-x)y$ and $x(-y)$
61. $-xy$ and $(-1)xy$
62. $-xy$
63. $(-x)^2$ and $-x^2$
64. $(-x)^3$ and $-y^3$
65. $|x|\,|y|$
66. $-(|y|\,|z|)$

67. A product made up of an even number of negative factors is (*sometimes, always, never*) negative.

68. A product made up of an odd number of negative factors is (*sometimes, always, never*) negative.

69. For all integers x, $-x$ is (*sometimes, always, never*) negative.

70. For all integers x, $(-x)^2 = -x^2$. (True or False?)

Multiply:

71. $4t^2(2t^2 + 3t + 5)$

72. $4x^3(2x^4 + 3x^2 + x + 3)$

73. $6xy(x^2 - y^2)$

74. $2yz(3y^2z - 5yz^2)$

75. $3ab^2(2a + 6b + ab + b^2)$

76. $3x^2y^2(x^3 + 2x^2y + 3xy^2 + 2y^3)$

77. $4u^3(2u^2 - 3v)$

78. $4t^3(t^5 - 2t^2)$

C *Evaluate for $x = -3$, $y = -6$, and $z = +4$.*

79. $x(x^2 - z^2)$

80. $x^3 - xy^2$

81. $x^2y(2x^2 + 3y)$

82. $3xz^2(xy^2 + 2z^3)$

83. Prove part B of Theorem 6.

2.6 Division of Integers

Three down and one to go! The last arithmetic operation is division. Because of our knowledge of natural numbers, we know that

$$(+8) \div (+4) = +2$$

but how shall we define

$$(+8) \div (-4) = ?$$

$$(-8) \div (+4) = ?$$

$$(-8) \div (-4) = ?$$

$$(0) \div (-4) = ?$$

$$(-8) \div (0) = ?$$

Division will be defined in such a way that all of these problems will have answers in the integers, except one. (Can you guess the exception?) Our approach here will parallel that used in defining subtraction in that we will investigate some elementary school notions about division to find motivation for a general definition of division for the integers.

You will recall that to check division in the problem

$$8\overline{)40}^{\;5}$$

we multiply 8 by 5 to obtain 40. We will use this checking requirement to transform division into multiplication. Instead of saying,

Divide 8 into 40.

we can ask

What must 8 be multiplied by to produce 40?

Notice that the answer to each is the same, 5. The latter way of looking at division is the most useful of the two for its generalization to more involved number systems, and it provides the motivation for the following general definition that will not only apply to the integers, but also to any other number system we will encounter.

DEFINITION OF DIVISION

$a \div b = Q$ if and only if $a = bQ$ and Q is unique

Let us use this definition to find answers to the problem stated earlier. In

$$(+8) \div (-4) = ? \qquad \text{or} \qquad -4\overline{)+8}^{\;?}$$

we ask, "what must (-4) be multiplied by to produce $(+8)$?" From the definition of multiplication we know the answer is (-2). Thus we write

$$(+8) \div (-4) = -2 \qquad \text{and} \qquad -4\overline{)+8}^{\;-2}$$

since

$$(-4)(-2) = +8$$

EXAMPLE 12 Similarly,

(A) $(-8) \div (+4) = -2$ since $(+4)(-2) = -8$

(B) $(-8) \div (-4) = +2$ since $(-4)(+2) = -8$

(C) $(0) \div (-4) = 0$ since $(-4)(0) = 0$

(D) $(-8) \div (0)$ is not defined since no number times 0 is -8

PROBLEM 12 Divide by using the definition above; that is, answer the question, "What number times the divisor b produces the dividend a in $b\overline{)a}$?"

(A) $+3\overline{)+12}$ (B) $-3\overline{)+12}$ (C) $+3\overline{)-12}$

(D) $-3\overline{)-12}$ (E) $-3\overline{)0}$ (F) $0\overline{)-12}$

ANSWER (A) $+4$ (B) -4 (C) -4 (D) $+4$ (E) 0 (F) not defined (no number times 0 is -12)

The two symbols "$\div$" and "$\overline{)\;\;}$" that you have used in arithmetic are not used a great deal in algebra and higher mathematics. The horizontal bar "–" and slash mark "/" are the symbols for division we will use most frequently. Thus a/b, $\frac{a}{b}$, $a \div b$, and $b\overline{)a}$ all name the same number (assuming the quotient is defined), and we can write

$$a/b = \frac{a}{b} = a \div b = b\overline{)a}$$

Thus, $8/2$, $\frac{8}{2}$, $8 \div 2$, and $2\overline{)8}$ all equal 4.

As with the other three arithmetic operations, we can mechanize the process for division. But once again you should realize that any mechanical rules that are formulated, as well as many useful properties of division that exist, have their basis in the above definition. In particular, the following theorem, which leads directly to a mechanical rule, is an immediate consequence of the definition.

THEOREM 8 (A) If a and b are integers with like signs and the quotient on the right is defined in the natural numbers, then

$$\frac{a}{b} = \frac{|a|}{|b|}$$

(B) If a and b are integers with unlike signs and the quotient on the right is defined in the natural numbers, then

$$\frac{a}{b} = -\left(\frac{|a|}{|b|}\right)$$

(C) If a is any nonzero integer, then

$$\frac{0}{a} = 0 \qquad \frac{a}{0} \text{ is not defined} \qquad \frac{0}{0} \text{ is not defined}$$

Zero cannot be used as a divisor—ever!

A careful look at this theorem in conjunction with Example 13 leads directly to the *simple mechanical rule:*

> If neither number is 0, use the same rule of signs as in multiplication (that is, quotients of numbers with like sign are positive and quotients of numbers with unlike signs are negative) and divide as in the natural numbers, affixing the appropriate sign to the result.
>
> Zero divided by a nonzero number is always zero.
>
> Division by 0 is not defined.

EXAMPLE 13 Divide:

(*A*) $(+27)/(+9) = +3$

(*B*) $(-27)/(-9) = +3$

(*C*) $(-27)/(+9) = -3$

(*D*) $0/(-4) = 0$

(*E*) $(-27)/0$ is not defined

(*F*) $0/0$ is not defined

PROBLEM 13 Divide:

(*A*) $(+18)/(+6)$ (*B*) $(-18)/(-6)$ (*C*) $(+18)/(-6)$

(*D*) $(-18)/(+6)$ (*E*) $0/(-6)$ (*F*) $(-18)/0$

(*G*) $0/0$

ANSWER (*A*) $+3$ (*B*) $+3$ (*C*) -3 (*D*) -3 (*E*) 0
(*F*) not defined (*G*) not defined

It is clear that division is not closed in the integers since, for example

$$\frac{+7}{+3}$$

is not an integer. In the next chapter this deficiency will be removed by extending the integers to a number system that includes fractions. You will recall, however, that the integers are closed with respect to addition, multiplication, and subtraction, while the natural numbers are only closed with respect to addition and multiplication. We have thus gained quite a bit by extending the natural numbers to the integers—more than you realize at the moment.

Exercise 16

A *Divide:*

1. $(+10) \div (+2)$

2. $(-10) \div (-2)$

3. $(+10) \div (-2)$

4. $(-10) \div (+2)$

5. $(0) \div (-2)$

6. $(-10) \div 0$

7. $(-9)/(-3)$

8. $(+9)/(+3)$

9. $(-9)/(+3)$

10. $(+9)/(-3)$

11. $(-9)/0$

12. $0/(+3)$

13. $\dfrac{+14}{+7}$

14. $\dfrac{-4}{-2}$

15. $\dfrac{-0}{+2}$

16. $\dfrac{+12}{-3}$

17. $\dfrac{0}{0}$

18. $\dfrac{-6}{0}$

B *Divide and check:*

19. $(-36)/(-9)$

20. $(-56)/(+7)$

21. $0/(-14)$

22. $(-83)/0$

23. $(+45)/(+15)$

24. $(-55)/(-11)$

25. $0/(-1)$

26. $(-54)/(+9)$

Evaluate:

27. $\dfrac{(-10)}{(+5)} + (-7)$

28. $(-12) + \dfrac{(-14)}{(-7)}$

29. $\dfrac{(-3)(+5)}{(+3)}$

30. $\dfrac{(+6)(-4)}{(-8)}$

31. $(-2)(+3) - \dfrac{(-10)}{(-5)}$

32. $\dfrac{(+22)}{(-11)} - (-4)(-3)$

33. $\dfrac{(+27)}{(-9)} - \dfrac{(-21)}{(-7)}$

34. $\dfrac{(-16)}{(+2)} - \dfrac{(+3)}{(-1)}$

Evaluate for $w = +2$, $x = -3$, $y = 0$, and $z = -24$.

35. z/x

36. z/w

37. y/x

38. $\dfrac{w}{y}$

39. $wx - \dfrac{z}{w}$

40. $\dfrac{z}{x} - wz$

41. $wxy - \dfrac{y}{z}$

42. $\dfrac{xy}{w} - xyz$

43. The number 0 divided by *any* integer is (*always, sometimes, never*) 0.

44. Any integer divided by 0 is (*always, sometimes, never*) 0.

Guess at the solution for each equation assuming x is restricted to the set of integers.

45. $\frac{(-24)}{x} = (-3)$

46. $\frac{x}{(-4)} = (+12)$

47. $\frac{x}{(+32)} = 0$

48. $\frac{(-52)}{(-13)} = x$

49. $\frac{(-1)}{(-6)} = x$

50. $\frac{(-2)}{(+8)} = x$

51. $\frac{x}{0} = (+4)$

52. $\frac{0}{x} = 0$

C *Evaluate for* $w = +2$, $x = -3$, $y = 0$, *and* $z = -24$.

53. $(w^3x)/z$

54. $w^2x^2 - x^3$

55. $xyz + \frac{y}{z} + x$

56. $wx + \frac{z}{wx} + wz$

57. $\frac{w - x}{w + x} - \frac{z}{2x}$

58. $\frac{8x}{z} - \frac{z - 6x}{wx}$

59. If the quotient x/y exists, when is it equal to $\frac{|x|}{|y|}$?

60. If the quotient x/y exists, when is it equal to $\frac{-|x|}{|y|}$?

2.7 Simplifying Algebraic Expressions

We are close to where we can use algebra to solve practical problems. First, however, we need to streamline our methods of representing algebraic expressions. For ease of reading and faster manipulation, it will be desirable to reduce the number of grouping symbols and plus signs to a minimum.

DROPPING UNNECESSARY PLUS SIGNS

To start, we drop the plus sign from numerals that name positive integers unless a particular emphasis is desired. Thus, we will write

1, 2, 3, . . . instead of +1, +2, +3, . . .

ADDITION AND SUBTRACTION WITHOUT GROUPING SYMBOLS

When three or more terms are combined by addition or subtraction and symbols of grouping are omitted, we convert (mentally) any subtraction to addition (Theorem 3) and add. Thus,

$8 - 5 + 3$ $= 8 + (-5) + 3$ *think* $= 6$

EXAMPLE 14 (*A*) $2 - 3 - 7 + 4$ $= 2 + (-3) + (-7) + 4$ *think* $= -4$

(*B*) $-4 - 8 + 2 + 9$ $= (-4) + (-8) + 2 + 9$ *think* $= -1$

PROBLEM 14 Evaluate:

(*A*) $5 - 8 + 2 - 6$ (*B*) $-6 + 12 - 2 - 1$

ANSWER (*A*) -7 (*B*) 3

Building on these ideas, we formulate the following general theorem:

THEOREM 9 When two or more terms in an algebraic expression are combined by addition or subtraction:

(*A*) Any subtraction sign may be replaced with an addition sign if the term following it is replaced by its negative [thus, $a - b - c = a + (-b) + (-c)$].

(*B*) The terms may be reordered without restriction as long as the sign preceding an involved term accompanies it in the process (thus, $a - b + c = a + c - b$).

Using Theorem 9, we extend the idea of a numerical coefficient and devise a simple mechanical rule for combining like terms in more involved algebraic expressions. The numerical coefficient of a given term in an algebraic expression is to include the sign that precedes it.

EXAMPLE 15 Since $3x^3 - 2x^2 - x + 3$ $= 3x^3 + (-2x^2) + (-1x) + 3$ *think* the coefficient of

(*A*) the first term is 3.

(*B*) the second term is -2.

(*C*) the third term is -1.

PROBLEM 15 In $2y^3 - y^2 - 2y + 4$, what is the coefficient of:

(A) the first term (B) the second term (C) the third term

ANSWER (A) 2 (B) -1 (C) -2

Using Theorem 9 again, and distributive and commutative properties, we can write an expression such as

$$3x - 2y - 5x + 7y$$

in the form

$$3x - 5x - 2y + 7y$$

or

$$3x + (-5x) + (-2y) + 7y$$

or

$$[3 + (-5)]x + [(-2) + 7]y$$

or

$$-2x + 5y$$

which leads to the same *mechanical rule* we had before for combining like terms. That is,

> Like terms are combined by simply adding their numerical coefficients.

EXAMPLE 16 Combine like terms:

(A) $2x - 3x = -x$

(B) $-4x - 7x = -11x$

(C) $3x - 5y + 6x + 2y = 9x - 3y$

(D) $2x^2 - 3x - 5 + 5x^2 - 2x + 3 = 7x^2 - 5x - 2$

PROBLEM 16 Combine like terms:

(A) $4x - 5x$ (B) $-8x - 2x$

(C) $7x + 8y - 5x - 10y$ (D) $4x^2 + 5x - 8 - 3x^2 - 7x - 2$

ANSWER (A) $-x$ (B) $-10x$ (C) $2x - 2y$ (D) $x^2 - 2x - 10$

REMOVING SYMBOLS OF GROUPING

How can we simplify expressions such as

$$2(3x - 5y) - 2(x + 3y)$$

You no doubt would guess that we could rewrite this expression as

$$6x - 10y - 2x - 6y$$

and combine like terms to obtain

$$4x - 16y$$

and your guess would be correct. Because of Theorem 9 and other properties of the integers, we find that parentheses (and other symbols of grouping) can be cleared by multiplying each term within a parentheses by the coefficient of the parentheses (including the sign that precedes it). We then combine like terms as in Example 16.

EXAMPLE 17

(A)
$$\begin{aligned} 2(x - 3y) + 3(2x - y) &= 2x - 6y + 6x - 3y \\ &= 8x - 9y \end{aligned}$$

(B)
$$\begin{aligned} 4x(2x + y) - 3x(x - 3y) &= 8x^2 + 4xy - 3x^2 + 9xy \\ &= 5x^2 + 13xy \end{aligned}$$

(C)
$$\begin{aligned} (x + 5y) + (x - 3y) &= 1(x + 5y) + 1(x - 3y) \quad \textit{think} \\ &= x + 5y + x - 3y \\ &= 2x + 2y \end{aligned}$$

(D)
$$\begin{aligned} (x + 5y) - (2x - 3y) &= 1(x + 5y) - 1(2x - 3y) \quad \textit{think} \\ &= x + 5y - 2x + 3y \\ &= -x + 8y \end{aligned}$$

PROBLEM 17 Remove parentheses and simplify:

(A) $3(2x - 3y) + 2(x - 2y)$ (B) $5x(x - 2y) - 3x(2x + y)$

(C) $(3x + 2y) + (2x - 4y)$ (D) $(3x + 2y) - (2x - 4y)$

ANSWER (A) $8x - 13y$ (B) $-x^2 - 13xy$ (C) $5x - 2y$ (D) $x + 6y$

Exercise 17

A *Evaluate:*

1. $5 + 7$

2. $3 + 6$

3. $5 - 7$

4. $3 - 6$

5. $3 - 2 + 4$

6. $3 + 4 - 2$

7. $4 - 8 - 9$

8. $-8 + 4 - 9$

9. $-4 + 7 - 6$

10. $7 - 6 - 4$

11. $2 - 3 - 6 + 5$

12. $5 + 2 - 6 - 3$

Simplify:

13. $7x + 3x$

14. $9x + 8x$

15. $7x - 3x$

16. $9x - 8x$

17. $2x + 5x + x$

18. $5x + x + 2x$

19. $2x - 5x + x$

20. $4t - 8t - 9t$

21. $-3y + 2y + 5y - 6y$

22. $2y - 3y - 6y + 5y$

23. $2x + 3y + 5x$

24. $4y + 3x + y$

25. $2x - 3y - 5x$

26. $4y - 3x - y$

27. $2x + 8y - 7x - 5y$

28. $5m + 3n - m - 9n$

29. $2(m + 3n) + 4(m - 2n)$

30. $3(u - 2v) + 2(3u + v)$

31. $2(x - y) - 3(3x - 2y)$

32. $4(m - 3n) - 3(2m + 4n)$

33. $(x + 3y) + (2x - 5y)$

34. $(2u - v) + (3u - 5v)$

35. $x - (2x - y)$

36. $m - (3m + n)$

37. $(x + 3y) - (2x - 5y)$

38. $(2u - v) - (3u - 5v)$

B *Evaluate:*

39. $-7 + 1 + 6 + 2 - 1$

40. $9 - 5 - 4 + 7 - 6 + 10$

41. $-5 - 3 - 8 + 15 - 1$

42. $1 - 12 + 5 + 7 - 1 + 6 - 8$

Simplify:

43. $3xy + 4xy - xy$

44. $3xy - xy + 4xy$

45. $-x^2y + 3x^2y - 5x^2y$

46. $-4r^3t^3 - 7r^3t^3 + 9r^3t^3$

47. $3x^2 - 2x + 5 - x^2 + 4x - 8$

48. $y^3 + 4y^2 - 10 + 2y^3 - y + 7$

49. $2x^2y + 3xy^2 - 5xy + 2xy^2 - xy - 4x^2y$

50. $a^2 - 3ab + b^2 + 2a^2 + 3ab - 2b^2$

51. $a + b - 2(a - b)$

52. $x - 3y - 4(2x - 3y)$

53. $x - 3(x + 2y) + 5y$

54. $y - 2(x - y) - 3x$

55. $-3(-t + 7) - (t - 1)$

56. $-2(-3x + 1) - (2x + 4)$

57. $-2(y - 7) - 3(2y + 1) - (-5y + 7)$

58. $2(x - 1) - 3(2x - 3) - (4x - 5)$

59. $3x(2x^2 - 4) - 2(3x^2 - x)$

60. $5y(2y - 3) + 3y(-2y + 4)$

61. $3x - 2[2x - (x - 7)]$

62. $2t - 3t[4 - 2(t - 1)]$

Replace each question mark with an appropriate algebraic expression.

63. $2 + 3x - y = 2 + (?)$

64. $5 + m - 2n = 5 + (?)$

65. $2 + 3x - y = 2 - (?)$

66. $5 + m - 2n = 5 - (?)$

67. $x - 4y - 8z = x - 4(?)$

68. $2x - 3a + 12b = 2x - 3(?)$

69. $w^2 - x + y - z = w^2 - (?)$

70. $w^2 - x + y - z = w^2 + (?)$

71. The width of a rectangle is 5 in. less than its length. If x is the length of the rectangle, write an algebraic expression that represents the perimeter of the rectangle and simplify the expression.

72. The length of a rectangle is 8 ft more than its width. If y is the width of the rectangle, write an algebraic expression that represents its area. Change the expression to a form without parentheses.

C *Simplify:*

73. $x - \{x - [x - (x - 1)]\}$

74. $2t - 3\{t + 2[t - (t + 5)] + 1\}$

75. $2x[3x - 2(2x + 1)] - 3x[8 + (2x - 4)]$

76. $-2t\{-2t(-t - 3) - [t^2 - t(2t + 3)]\}$

77. $3x^2 - 2\{x - x[x + 4(x - 3)] - 5\}$

78. $w - \{x - [z - (w - x) - z] - (x - w)\} + x$

79. A coin purse contains dimes and quarters only. There are 4 more dimes than quarters. If x equals the number of dimes, write an algebraic expression that represents the value of the money in the purse. Simplify the expression. HINT: If x represents the value of dimes, then what does $x - 4$ represent?

80. A pile of coins consists of nickels, dimes, and quarters. There are 5 less dimes than nickels and 2 more quarters than dimes. If x equals the number of nickels, write an algebraic expression that represents the value of the pile in cents. Simplify the expression. HINT: If x represents the number of nickels, then what do $x - 5$ and $(x - 5) - 2$ represent?

2.8 Equations

We have reached the place where we can discuss methods of solving equations other than by guessing. For example, you would not likely guess the solution to

$$2x + 2(x - 6) = 52$$

an equation related to a practical problem we will consider later.

A *solution* or *root* of an equation is a replacement of x that makes the left side equal to the right. The set of all solutions is called the *solution set.* To *solve an equation* is to find its solution set.

Knowing what we mean by the solution set of an equation is one thing, finding it is another. Our objective now is to develop a systematic method of solving equations that is free from guess work. We start by introducing the idea of equivalent equations. We say that two equations are *equivalent* if they both have the same solution set.

The basic idea in solving equations is to perform operations on equations that produce simpler equivalent equations and to continue the process until we reach an equation whose solution is obvious—generally, an equation such as

$$x = -3$$

With a little practice you will find the methods that we are going to develop very easy to use and very powerful. Before proceeding further, it is recommended that you briefly review the properties of equality discussed in Sec. 1.4. The following important theorem is a direct consequence of these properties.

THEOREM 10 (Further properties of equality.) For a, b, and c integers,

(A) If $a = b$, then $a + c = b + c$ addition property

(B) If $a = b$, then $a - c = b - c$ subtraction property

(C) If $a = b$, then $ca = cb$ multiplication property

(D) If $a = b$ and $c \neq 0$, then $\frac{a}{c} = \frac{b}{c}$ division property

The proofs of these properties are very easy. We will prove part A and leave the others as exercises:

$a + c = a + c$ identity property of equality

$a = b$ given

$a + c = b + c$ substitution principle

The next theorem, which we will freely use but not prove, provides us with our final instructions for solving simple equations.

THEOREM 11 An equivalent equation will result if

(*A*) An equation is changed in any way by use of Theorem 10, except for multiplication or division by 0.

(*B*) Any algebraic expression in an equation is replaced by its equal (substitution principle).

We are now ready to solve equations! Several examples will illustrate the process.

EXAMPLE 18 Solve $x - 5 = -2$ and check.

SOLUTION	COMMENTS
$x - 5 = -2$	How can we eliminate the -5 from the left side?
$x - 5 + 5 = -2 + 5$	Add 5 to each side (addition property).
$x = 3$	Solution is obvious.

CHECK

$$3 - 5 \stackrel{?}{=} -2$$

$$-2 \stackrel{\checkmark}{=} -2$$

PROBLEM 18 Solve $x + 8 = -6$ and check.

ANSWER $x = -14$

EXAMPLE 19 Solve $-3x = 15$ and check.

SOLUTION	COMMENTS
$-3x = 15$	How can we make the coefficient of x plus 1?
$\dfrac{-3x}{-3} = \dfrac{15}{-3}$	Divide each side by -3 (division property).
$x = -5$	Solution is obvious.

CHECK

$$(-3)(-5) \stackrel{?}{=} 15$$
$$15 \stackrel{\checkmark}{=} 15$$

PROBLEM 19 Solve $5x = -20$ and check.

ANSWER $x = -4$

The following examples are a little more difficult, but each can be converted into one of the above types simply by following a two-step process:

1. Simplify the left- and right-hand sides of the equation.
2. Perform operations on the resulting equations that will get all of the variable terms on one side (usually the left) and all of the constant terms on the other side (usually the right), then solve as in Examples 18 or 19.

EXAMPLE 20 Solve $2x - 8 = 5x + 4$ and check.

SOLUTION

$$2x - 8 = 5x + 4$$

$$\boxed{2x - 8 + 8 = 5x + 4 + 8}$$

$$2x = 5x + 12$$

$$\boxed{2x - 5x = 5x + 12 - 5x}$$

$$-3x = 12$$

$$\boxed{\frac{-3x}{-3} = \frac{12}{-3}}$$

$$x = -4$$

CHECK

$$2(-4) - 8 \stackrel{?}{=} 5(-4) + 4$$

$$-8 - 8 \stackrel{?}{=} -20 + 4$$

$$-16 \stackrel{\checkmark}{=} -16$$

PROBLEM 20 Solve $3x - 9 = 7x + 3$ and check.

ANSWER $x = -3$

EXAMPLE 21 Solve $3x - 2(2x - 5) = 2(x + 3) - 8$ and check.

SOLUTION This equation is not as difficult as it might at first appear; simplify the expressions on each side of the equal sign first, and then proceed as in the preceding example. (Note that some steps in the following solution are done mentally.)

$$3x - 2(2x - 5) = 2(x + 3) - 8$$

$$3x - 4x + 10 = 2x + 6 - 8$$

$$-x + 10 = 2x - 2$$

$$-x = 2x - 12$$

$$-3x = -12$$

$$x = 4$$

CHECK

$$3(4) - 2[2(4) - 5] \stackrel{?}{=} 2[(4) + 3] - 8$$

$$12 - 2(8 - 5) \stackrel{?}{=} 2(7) - 8$$

$$12 - 2(3) \stackrel{?}{=} 14 - 8$$

$$12 - 6 \stackrel{?}{=} 6$$

$$6 \stackrel{\checkmark}{=} 6$$

PROBLEM 21 Solve $8x - 3(x - 4) = 3(x - 4) + 6$ and check.

ANSWER $x = -9$

We will show in Chap. 4 that equations of the above types have at most one solution. Other equations that we will encounter later can have more than one solution. For example

$$x^2 = 4$$

has two solutions. Can you guess them?

Exercise 18

A *Solve and check.*

1. $x + 5 = 8$

2. $x + 2 = 7$

3. $x + 8 = 5$

4. $x + 7 = 2$

5. $x + 9 = -3$

6. $x + 4 = -6$

7. $x - 3 = 2$

8. $x - 4 = 3$

9. $x - 5 = -8$

10. $x - 7 = -9$

11. $y + 13 = 0$

12. $x - 5 = 0$

13. $4x = 32$

14. $9x = 36$

15. $6x = -24$

16. $7x = -21$

17. $-3x = 12$

18. $-2x = 18$

19. $-8x = -24$

20. $-9x = -27$

21. $3y = 0$

22. $-5m = 0$

23. $4x - 7 = 5$

24. $3y - 8 = 4$

25. $2y + 5 = 9$

26. $4x + 3 = 19$

27. $2y + 5 = -1$

28. $2w + 18 = -2$

29. $-3t + 8 = -13$

30. $-4m + 3 = -9$

31. $4m = 2m + 8$

32. $3x = x + 6$

33. $2x = 8 - 2x$

34. $3x = 10 - 7x$

35. $2n = 5n + 12$

36. $3y = 7y + 8$

37. $2x - 7 = x + 1$

38. $4x - 9 = 3x + 2$

39. $3x - 8 = x + 6$

40. $4y + 8 = 2y - 6$

41. $2t + 9 = 5t - 6$

42. $3x - 4 = 6x - 19$

B

43. $x - 3 = x + 7$

44. $2y + 8 = 2y - 6$

45. $2x + 2(x - 6) = 52$

46. $5x + 10(x + 7) = 100$

47. $x + (x + 2) + (x + 4) = 54$

48. $10x + 25(x - 3) = 275$

49. $2(x + 7) - 2 = x - 3$

50. $5 + 4(t - 2) = 2(t - 7) + 1$

51. $-3(4 - t) = 5 - (t + 1)$

52. $5x - (7x - 4) - 2 = 5 - (3x + 2)$

53. $x(x + 2) = x(x + 4) - 12$

54. $x(x - 1) + 5 = x^2 + x - 3$

55. $t(t - 6) + 8 = t^2 - 6t - 3$

56. $x(x - 4) - 2 = x^2 - 4(x + 3)$

57. Which of the following are equivalent to $3x - 6 = 6$: $3x = 12$, $3x = 0$, $x = 4$, $x = 0$?

58. Which of the following are equivalent to $2x + 5 = x - 3$: $2x = x - 8$, $2x = x + 2$, $3x = -8$, $x = -8$?

C

Which of the following are true (I is the set of integers)?

59. $\{x \in I | 3x = 5\} = \emptyset$

60. $\{t \in I | 2t - 1 = 19\} = \{9, 10\}$

61. $\{t \in I | 2t - 1 = 19\} = \{9\}$

62. $\{t \in I | 2t - 1 = 19\} = \{10\}$

63. $\{y \in I | -2y = 8, y > 0\} = \emptyset$

64. $\{u \in I | 2u - 5 = 11\} = \{8\}$

65. $\{x \in I | 3x + 11 = 5, x > 0\} = \emptyset$

66. $\{x \in I | x^2 = 4\} = \{-2, 2\}$

67. Prove the subtraction property in Theorem 10.

68. Prove the multiplication property in Theorem 10.

2.9 Applications

At this time we will consider only a few applications. A large variety of applications from many different fields will be considered in detail in following chapters.

To start, we will solve a fairly simple problem dealing with numbers. Through this problem you will learn a method of attack that can be applied to many other problems.

EXAMPLE 22 Find 3 consecutive integers whose sum is 66.

SOLUTION

Let

$x =$ the first integer

then

$x + 1 =$ the next integer

and

$x + 2 =$ the third integer

COMMENTS: Identify one of the unknowns with a letter, and then write other unknowns in terms of this letter.

$$x + (x + 1) + (x + 2) = 66$$

Write an equation that relates the unknown quantities with other facts in the problem.

$$\begin{aligned} x + x + 1 + x + 2 &= 66 \\ 3x + 3 &= 66 \\ 3x &= 63 \\ x &= 21 \\ x + 1 &= 22 \\ x + 2 &= 23 \end{aligned}$$

Solve the equation.

Write all answers requested.

CHECK

$$\begin{array}{r} 21 \\ 22 \\ \underline{23} \\ 66 \end{array}$$

Thus we have found three consecutive integers whose sum is 66.

Checking back in the equation is not enough since you might have made a mistake in setting up the equation; a final check is provided only if the conditions in the original problem are satisfied.

PROBLEM 22 Find 3 consecutive integers whose sum is 54.

ANSWER 17, 18, 19

EXAMPLE 23 Find 3 consecutive even numbers such that twice the second plus 3 times the third is 7 times the first.

SOLUTION Let

$x =$ the first even number

then

$x + 2 =$ the second even number

and

$x + 4 =$ the third even number

twice the second even number	+	three times the third even number	=	seven times the first even number

$$\begin{aligned} 2(x+2) + 3(x+4) &= 7x \\ 2x + 4 + 3x + 12 &= 7x \\ 5x + 16 &= 7x \\ -2x &= -16 \\ x &= 8 \\ x + 2 &= 10 \\ x + 4 &= 12 \end{aligned}$$

CHECK

8, 10, and 12 are three consecutive even numbers

$$\begin{aligned} 2 \cdot 10 + 3 \cdot 12 &\stackrel{?}{=} 7 \cdot 8 \\ 20 + 36 &\stackrel{?}{=} 56 \\ 56 &\stackrel{\checkmark}{=} 56 \end{aligned}$$

PROBLEM 23 Find 3 consecutive even numbers such that the second plus twice the third is 4 times the first.

ANSWER 10, 12, 14

EXAMPLE 24 Find the dimensions of a rectangle with a perimeter of 52 in. if its length is 5 in. more than twice its width.

SOLUTION

2(length) + 2(width) = perimeter

$$\begin{aligned} 2(2x+5) + 2x &= 52 \\ 4x + 10 + 2x &= 52 \\ 6x &= 42 \\ x &= 7 \quad \text{width} \\ 2x + 5 &= 19 \quad \text{length} \end{aligned}$$

CHECK

19 is 5 more than twice 7

$$\begin{aligned} 2 \cdot 19 + 2 \cdot 7 &\stackrel{?}{=} 52 \\ 38 + 14 &\stackrel{?}{=} 52 \\ 52 &\stackrel{\checkmark}{=} 52 \end{aligned}$$

PROBLEM 24 Find the dimensions of a rectangle with a perimeter of 30 ft if its length is 7 ft more than its width.

ANSWER 4 ft by 11 ft

EXAMPLE 25 In a pile of coins containing only dimes and nickels, there are 7 more dimes than nickels. If the total value of all of the coins in the pile is $1, how many of each type of coin is in the pile?

SOLUTION Let

$x =$ the number of nickels in the pile

then

$x + 7 =$ the number of dimes in the pile

$$\begin{array}{c}\text{value of nickels}\\\text{in cents}\end{array} + \begin{array}{c}\text{value of dimes}\\\text{in cents}\end{array} = \begin{array}{c}\text{value of pile}\\\text{in cents}\end{array}$$

$$\begin{aligned} 5x + 10(x + 7) &= 100 \\ 5x + 10x + 70 &= 100 \\ 15x &= 30 \\ x &= 2 \qquad \text{nickels} \\ x + 7 &= 9 \qquad \text{dimes} \end{aligned}$$

CHECK

9 dimes is seven more than 2 nickels

value of nickels in cents = 10
value of dimes in cents = 90
total value 100

PROBLEM 25 A person has dimes and quarters worth $1.80 in his pocket. If there are twice as many dimes as quarters, how many of each does he have?

ANSWER 4 quarters and 8 dimes

EXAMPLE 26 An airplane flew out to an island from the mainland and back in 5 hr. How far is the island from the mainland if the pilot averaged 600 mph going to the island and 400 mph returning?

SOLUTION In this problem we will find it convenient to find the time out to the island first. The formula $d = rt$, of course, will be of great use to us here. Let

$x =$ the time it took to get to the island

then

$5 - x =$ the time to return (since round trip time is 5 hr)

distance out = distance back

$$\begin{aligned} 600x &= 400(5 - x) \\ 600x &= 2{,}000 - 400x \\ 1{,}000x &= 2{,}000 \\ x &= 2 \text{ hr} \qquad \text{time going} \\ 5 - x &= 3 \text{ hr} \qquad \text{time returning} \end{aligned}$$

Since distance = (rate) · (time), the distance to the island from the mainland is $600 \cdot 2 = 1{,}200$ miles.

CHECK

time going + time returning = 5

$$\frac{1{,}200}{600} + \frac{1{,}200}{400} \stackrel{?}{=} 5$$

$$2 + 3 \stackrel{\checkmark}{=} 5$$

PROBLEM 26 An airplane flew from San Francisco to a distressed ship out at sea and back in 7 hr. How far was the ship from San Francisco if the pilot averaged 400 mph going and 300 mph returning?

ANSWER 1,200 miles.

You are now beginning to see the power of algebra. It was an historic occasion when it was realized that a solution to a problem that was difficult to obtain by arithmetic computation could be obtained instead by a deductive process involving conditions that the solution was required to satisfy.

There are many different types of algebraic applications, so many, in fact, that no single approach will apply to all. The following suggestions, however, may be of help to you:

1 Read the problem very carefully—a second and third time if necessary.
2 Write down important facts and relationships on a piece of scratch paper.
3 Identify unknown quantities in terms of a single letter if possible.
4 Write an equation that relates these unknown quantities and the facts in the problem.
5 Solve the equation.
6 Write down all of the solutions asked for in the original problem
7 Check the solution(s) in the original problem.

Remember, mathematics is not a spectator sport! Just reading examples is not enough; you must set up and solve problems yourself.

Exercise 19

A

1. Find 3 consecutive numbers whose sum is 78.

2. Find 3 consecutive numbers whose sum is 96.

3. Find 3 consecutive even numbers whose sum is 54.

4. Find 3 consecutive even numbers whose sum is 42.

5. How long would it take you to drive from San Francisco to Los Angeles, a distance of about 424 miles, if you could average 53 mph? (Use $d = rt$.)

6. If you drove from Berkeley to Lake Tahoe, a distance of 200 miles, in 4 hr, what is your average speed?

7. About 8 times as much of an iceberg is under water as is above the water. If the total height of an iceberg from bottom to top is 117 ft, how much is above and how much is below the surface?

8. A chord called an octave can be produced by dividing a stretched string into two parts so that one part is twice as long as the other part. How long will each part of the string be if the total length of the string is 57 in.?

9. The sun is about 390 times as far from the earth as the moon. If the sun is approximately 93,210,000 miles from the earth, how far is the moon from the earth?

10. You are asked to construct a triangle with two equal angles so that the third angle is twice the size of either of the two equal ones. How large should each angle be? NOTE: The sum of the three angles in any triangle is $180°$.

B

11. Find three consecutive odd numbers such that the sum of the first and third is twice the second.

12. Find 3 consecutive odd numbers such that the sum of the second and third is 1 more than 3 times the first.

13. Find the dimensions of a rectangle with perimeter 66 ft if its length is 3 ft more than twice the width.

14. Find the dimensions of a rectangle with perimeter 128 in. if its length is 6 in. less than 4 times the width.

15. In a pile of coins containing only quarters and dimes, there are 3 less quarters than dimes. If the total value of the pile is $2.75, how many of each type of coin is in the pile?

16. If you have 20 dimes and nickels in your pocket worth $1.40, how many of each do you have?

17. A toy rocket shot vertically upward with an initial velocity of 160 fps, has at time t a velocity given by the equation $v = 160 - 32t$, where air resistance is neglected. In how many seconds will the rocket reach its highest point? HINT: Find t when $v = 0$.

18. In the preceding problem, when will the rocket's velocity be 32 fps?

19. Air temperature drops approximately 5°F per 1,000 ft in altitude above the surface of the Earth up to 30,000 ft. If T represents temperature and A represents altitude in thousands of feet, and if the temperature on the ground is 60°F, then we can write

$$T = 60 - 5A \qquad 0 \le A \le 30$$

If you were in a balloon, how high would you be if the thermometer registered −50°F?

20. A mechanic charges $6 per hr for his labor and $4 per hr for his assistant. On a repair job his bill was $190 with $92 for labor and $98 for parts. If the assistant worked 2 hr less than the mechanic, how many hours did each work?

21. In a recent election involving five candidates, the winner beat his opponents by 805, 413, 135, and 52, respectively. If the total number of votes cast was 10,250, how many votes did each receive?

22. If an adult with pure brown eyes marries an adult with blue eyes, their children, because of the dominance of brown, will all have brown eyes but will be carriers of the gene for blue. If the children marry others with the same type of parents, then according to Mendel's laws of heredity, we would expect the third generation (the children's children) to include 3 times as many with brown eyes as with blue. Out of a sample of 1,748 third-generation children with second-generation parents as described, how many brown-eyed children and blue-eyed children would you expect?

C **23.** A man in a canoe went up a river and back in 6 hr. If his rate up the river was 2 mph and back 4 mph, how far did he go up the river?

24. You are at a river resort and rent a motor boat for 5 hr at 7 A.M. You are told that the boat will travel at 8 mph upstream and 12 mph returning. You decide that you would like to go as far up the river as you can and still be back at noon. At what time should you turn back, and how far from the resort will you be at that time?

25. One ship leaves England and another leaves the United States at the same time. The distance between the two ports is 3,150 miles. The ship from the United States averages 25 mph and the one from England 20 mph. If they both travel the same route, how long will it take the ships to reach a rendezvous point, and how far from the United States will they be at that time?

26. At 8 A.M. your father left by car on a long trip. An hour later you find that he has left his wallet behind. You decide to take another car to try to catch up with him. From past experience you know that he averages about 48 mph. If you can average 60 mph, how long will it take you to catch him?

27. In a computer center two electronic card sorters are used to sort 52,000 IBM cards. If the first sorter operates at 225 cards per min and the second sorter operates at 175 cards per min, how long will it take both sorters together to sort all of the cards?

28. Find 4 consecutive even numbers so the sum of the first and last is the same as the sum of the second and third. (Be careful!)

2.10 Inequalities and Line Graphs

Earlier when we discussed inequality symbols relative to the natural numbers, we appealed directly to your common-sense notions about these ideas. Now, with the negative integers and 0, we will have to proceed more carefully. It no doubt seems obvious that

$$2 < 3$$

but does it seem equally obvious that

$$-3 < -2$$

$$-10 < 3$$

$$-5 < 0$$

$$-1 > -1{,}000$$

The above inequalities may seem more reasonable if we think of the quantities as temperature readings or altitudes above and below sea level. (Which number indicates a warmer temperature or higher altitude?) To free the inequality relation from any particular application, making it available for many different applications, we present a mathematical definition of the concept.

DEFINITION OF THE INEQUALITY RELATION

If a and b are integers, then we write

$$a < b$$

if there exists a positive integer p such that $a + p = b$. We write

$$a > b$$

if $b < a$.

Certainly, one would expect that if a positive number were added to *any* number, the sum would be larger than the original. That is essentially what the definition states.

EXAMPLE 27

(A) $2 < 3$ since $2 + 1 = 3$

(B) $-3 < -2$ since $-3 + 1 = -2$

(C) $-10 < 3$ since $-10 + 13 = 3$

(D) $-5 < 0$ since $-5 + 5 = 0$

(E) $-1 > -1{,}000$ since $-1{,}000 < -1$, Why?

PROBLEM 27 Replace each question mark with either $<$ or $>$:

(A) $4 \, ? \, 6$ (B) $6 \, ? \, 4$ (C) $-6 \, ? \, -4$

(D) $-8 \, ? \, 8$ (E) $3 \, ? \, -9$ (F) $0 \, ? \, -4$

ANSWER (A) $4 < 6$ (B) $6 > 4$ (C) $-6 < -4$ (D) $-8 < 8$

(E) $3 > -9$ (F) $0 > -4$

The same simple *geometric interpretation* applys to inequalities in the integers as well as to the natural numbers. That is, any number to the right of another number on a number line is larger than that number.

EXAMPLE 28 If x is an integer find and graph the solution set for

(A) $-3 < x \leq 2$

SOLUTION

Solution set $= \{-2, -1, 0, 1, 2\}$

Graph:

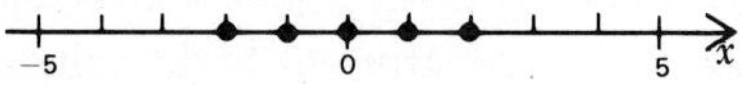

(B) $-7 \leq x < 0$

SOLUTION

Solution set $= \{-7, -6, -5, -4, -3, -2, -1\}$

Graph:

PROBLEM 28 If x is an integer find and graph the solution set for

(A) $-4 < x \leq 3$ (B) $-6 < x \leq 0$

ANSWER (A) $\{-3, -2, -1, 0, 1, 2, 3\}$

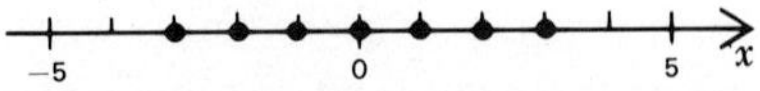

(B) $\{-5, -4, -3, -2, -1, 0\}$

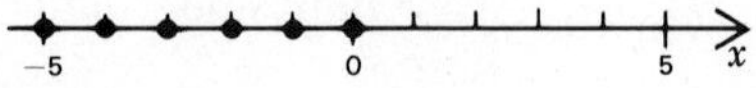

Exercise 20

A *Replace each question mark with < or >.*

1. $7 ? 5$ **2.** $3 ? 6$ **3.** $5 ? 7$

4. $6 ? 3$ **5.** $-7 ? -5$ **6.** $-3 ? -6$

7. $-5 ? -7$ **8.** $-6 ? -3$ **9.** $0 ? 8$

10. $5 ? 0$ **11.** $0 ? -8$ **12.** $-5 ? 0$

13. $-7 ? 5$ **14.** $-6 ? 3$ **15.** $-7 ? -5$

16. $-6 ? -3$

Referring to

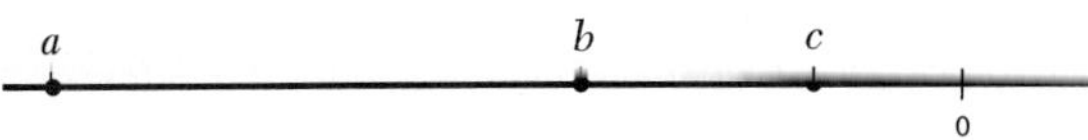

replace each question mark with either < or >.

17. $a ? d$ **18.** $e ? a$ **19.** $b ? a$

20. $0 ? d$ **21.** $e ? f$ **22.** $d ? e$

23. $e ? c$ **24.** $0 ? a$

For x an integer, find the solution set and graph.

25. $-1 \leq x < 3$ **26.** $-4 < x \leq 1$

27. $-6 < x \leq -1$ **28.** $-5 \leq x < 0$

29. If we add a positive integer to any integer, will the sum be to the right or left of the original number on a number line?

30. If we add a positive integer to any integer, will the sum be greater than or less than the original number?

B *Replace each question mark with < or >.*

31. $(-3) + 5 \quad ? \quad 0$ **32.** $5 - (-3) \quad ? \quad 3$

33. $\frac{-6}{2} - 4 \quad ? \quad -2$ **34.** $5 + \frac{-18}{3} \quad ? \quad 0$

35. $(-7)(2) - \frac{-12}{3} \quad ? \quad 9$ **36.** $\frac{24}{-3} - (4)(-2) \quad ? \quad -6$

Find the solution set for each inequality for x restricted to $\{-6, -4, -2, 0, 2, 4, 6\}$.

37. $x \leq 2$ **38.** $x > -4$ **39.** $x > 10$

40. $x \leq -8$ **41.** $3x > 0$ **42.** $2x < -2$

43. $2 - x \geq 0$ **44.** $4 - x > 6$ **45.** $\frac{x}{-2} + x \leq 0$

46. $-2x - \frac{12}{x} \leq 10$

C *Graph each set on a number line where I is the set of integers.*

47. $\{x \in I \mid -6 < x \leq 3\}$ **48.** $\{x \in I \mid -8 \leq x \leq 0\}$

49. $\{x \in I \mid -3 < x < 3\}$ **50.** $\{x \in I \mid -10 < x \leq 5\}$

51. Show that if $a < b$, then $b - a$ is a positive integer.

52. Show that if $b - a$ is a positive integer, than $a < b$.

53. Show that if $a < b$ and $b < c$, then $a < c$.

54. Show that if $a < b$ and c is any integer, then $a + c < b + c$.

Exercise 21 Chapter Review

All variables represent integers.

A **1.** Graph $\{-5, -2, 0, +3\}$ on a number line.

2. Evaluate: (A) $-(+4)$, (B) $|-8|$, (C) $-(-2)$, (D) $|+9|$

3. Express each quantity by means of an appropriate integer. (A) Salton Sea's surface at 245 ft below sea level. (B) Mount Whitney's height of 14,495 ft above sea level.

4. Add: (A) $\begin{array}{r}(-8)\\ \underline{(+3)}\end{array}$ (B) $\begin{array}{r}(-9)\\ \underline{(-4)}\end{array}$

5. Add: (A) $(-6) + (-3) + (+7) + (-1)$, (B) $0 + (-3)$

6. Subtract: (A) $\begin{array}{r}(-3)\\ \underline{-(-9)}\end{array}$ (B) $(+4) - (+7)$

7. Multiply: (A) $(-7)(-4)$, (B) $(+3)(-6)$

8. Divide: (A) $(-16) \div (+4)$, (B) $(-12)/(-2)$

9. Divide: (A) $(-6)/0$, (B) $0/(+2)$

10. Multiply: $3m(m - 4n)$

11. Simplify: (A) $-3 + 8 - 1$, (B) $4x - 8y - 3x + 4y$

12. Simplify: $3(m + 2n) - (m - 3n)$

13. Solve and check: $4x - 9 = x - 15$

14. Find 3 consecutive numbers whose sum is 159.

15. Replace each question mark with either $<$ or $>$: (A) $7 ? 2$, (B) $-7 ? -2$, (C) $0 ? -5$, (D) $-125 ? -5$.

B **16.** If $P =$ set of positive integers
$M =$ set of negative integers
$I =$ set of integers

indicate true (T) or false (F).

(A) $+7$ is in I (B) 0 is in M (C) -4 is in I
(D) 0 is in P (E) M is a subset of I.

17. Evaluate for $x = -8$ and $y = +3$: (A) $-x$, (B) $-(x + y)$.

18. Evaluate: (A) $(-54) + (+44)$, (B) $(-62) + (-18) + (0) + (+20)$

19. Evaluate: $[-(-4)] + [-|3|]$

20. Guess at all solutions: (A) $|x| = 5$, (B) $-x = +7$

21. Evaluate: $[(-3) - (-3)] - (-4)$

22. Evaluate for $x = +6$, $y = -8$, and $z = +4$: $(x + y) - z$

23. Evaluate: $(-2)(-4) - (-3)^2$

24. Evaluate for $w = -10$, $x = -2$, and $z = 0$: $\left(wx - \frac{z}{x}\right) - \frac{w}{x}$

25. Multiply: $2y^4(3y^3 + y + 5)$

26. Simplify: $3x^2y^2 - xy - 5x^2y^2 - 4xy$

27. Simplify: $7x - 3[(x + 7y) - (2x - y)]$

28. Solve and check: $2x + 3(x - 1) = 5 - (x - 4)$

29. A pile of coins consists of nickels and quarters. How many of each kind are there if the value of the pile is \$1.75 and there are 5 more nickels than quarters?

30. Graph $-4 \leq x < -1$ for x an integer.

C **31.** Express the net gain by means of an appropriate integer: A 15° rise in temperature followed by a 30° drop, another 15° drop, a 25° gain, and finally a 40° drop.

32. Describe the elements in the set $\{x \in I \mid |x| = x\}$

33. For a an integer, (A) $-a + ? = 0$, (B) $a + (-a) = ?$

34. Show that subtraction is not associative by evaluating: (A) $(x - y) - z$ and (B) $x - (y - z)$ for $x = +7$, $y = -3$, and $z = -5$.

35. Show that division is not associative by evaluating: (A) $(x \div y) \div z$ and (B) $x \div (y \div z)$ for $x = +16$, $y = -8$, and $z = -2$.

36. Evaluate for $x = -3$ and $y = -5$: $xy^2 - x^3$.

37. Evaluate for $x = -6$, $y = 0$, and $z = -3$: $[(xyz + xz) - z]/z$.

38. $3x - 6y + 9 = 3x - 3(\ ?\ )$

39. Simplify: $3x - 2\{x - 2[x - (4x + 2)]\}$.

40. $\{x \in I \mid 2(x - 5) = 4x + 8\} = ?$

41. An unmanned space capsule passes over Cape Canaveral at 8 A.M. traveling at 17,000 mph. A manned capsule, attempting a rendezvous, passes over the same spot at 9 A.M. traveling at 18,000 mph. How long will it take the second capsule to catch up with the first?

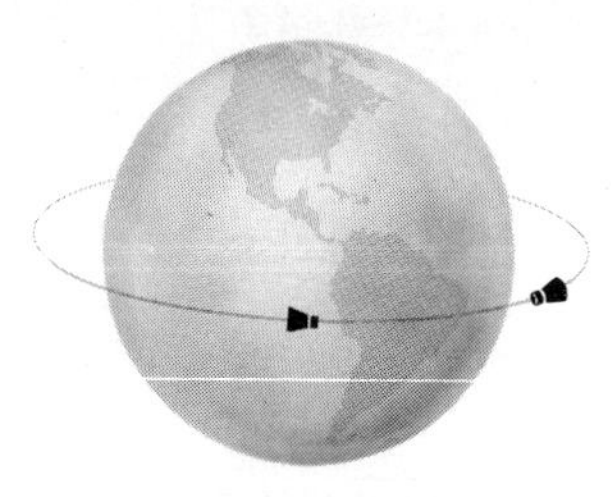

42. If $a + p = b$ for some positive number b, then a is (*greater than, less than*) b.

CHAPTER 3 Rational and Real Numbers

3.1 The Set of Rational Numbers

In the last chapter we formed the set of integers by extending the natural numbers to include 0 and the negative integers. With this extension came more power to perform arithmetic operations and more power to solve equations. In spite of this added power, however, we are still not able to solve the simple equation

$$2x = 3$$

or find

$$-2 \div 5$$

You have no doubt guessed the direction of the next extension of the number system. We need fractions!

You will recall in the last chapter that we said that we would use

$$a \div b \qquad a/b \qquad \frac{a}{b}$$

interchangeably; hence,

$$4 \div 2 \qquad 4/2 \qquad \frac{4}{2}$$

are different names for the number 2. However, what does

$$\frac{3}{2}$$

name? Certainly not an integer. We are going to extend the set of integers so that $\frac{3}{2}$ will name a number and division will always be defined (except by 0). The extended number system will be called the set of rational numbers.

We will not attempt to give a precise definition of a rational number. We will, instead, accept the intuitive notion of a rational number that you acquired through your experience with fractions in arithmetic. We will postulate their existence and name them with appropriate symbols. Actually, it is how we work with symbols that name numbers that interests us most at this point rather than what we really mean by number.

POSTULATE There exists a set of numbers called the *rational numbers*. Every fractional form a/b where a and b are integers with $b \neq 0$ names a rational number; every rational number has a fractional form of the type described as a name.

Thus

$$\frac{1}{3} \qquad \frac{3}{5} \qquad \frac{8}{1} \qquad \frac{-2}{7} \qquad \frac{10}{-5} \qquad \frac{-3}{-2}$$

all name rational numbers. It is important to note that the integers are a subset of the rational numbers since any integer can be expressed as the quotient of two integers. For example,

$$9 = \frac{9}{1} = \frac{18}{2}$$

$$23 = \frac{23}{1} = \frac{-46}{-2}$$

and so on

Thus, every integer is a rational number, but not every rational number is an integer—for example, $\frac{3}{2}$ is not an integer.

It would seem reasonable, from our experience with multiplying and dividing signed quantities in the preceding chapter, to define the quotient of any two integers with like signs as a *positive rational number* and the quotient of any two integers with unlike signs as a *negative rational number*. Thus,

$$\frac{+2}{+3} = \frac{2}{3} \qquad \frac{-2}{-3} = \frac{2}{3} \qquad \frac{-2}{+3} = -\frac{2}{3} \qquad \frac{+2}{-3} = -\frac{2}{3}$$

Identifying the rational numbers with points on a number line proceeds as one would expect; the positive numbers to the right of the origin and the negative numbers to the left. Where do we locate $\frac{7}{4}$? We divide the interval between 1 and 2 into 4 equal line segments and take the end of the third segment. Where is $-\frac{3}{2}$ located? Halfway between -1 and -2.

Proceeding as described, every rational number can be associated with a point on a number line.

EXAMPLE 1 Locate $\frac{1}{2}$, $-\frac{3}{4}$, $\frac{5}{2}$, $-\frac{9}{4}$ on a number line.

SOLUTION

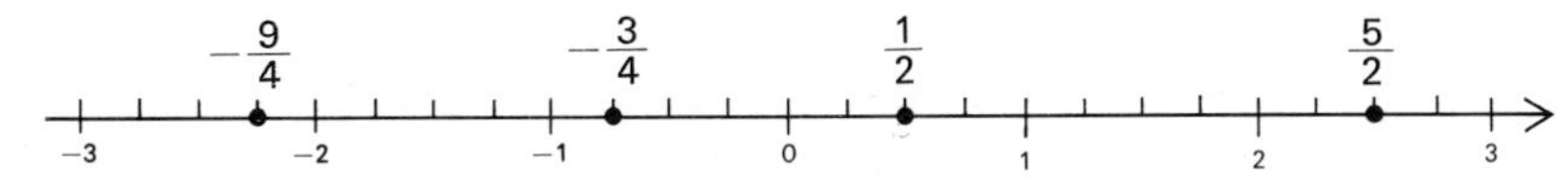

PROBLEM 1 Locate $\frac{3}{4}$, $-\frac{1}{2}$, $\frac{7}{4}$, $-\frac{5}{2}$ on a number line.

ANSWER

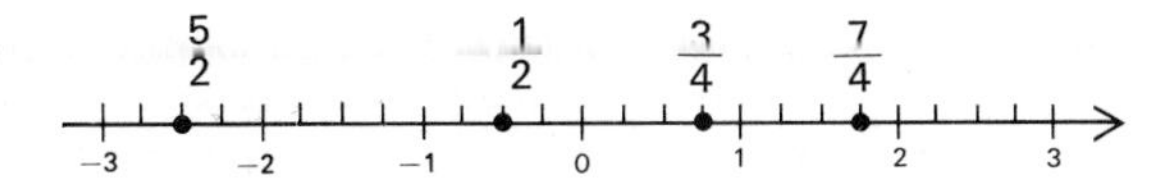

The *negative of a rational number* and the *absolute value of a rational number* are defined as in the integers. Thus

$$-\left(\frac{2}{3}\right) = -\frac{2}{3} \qquad -\left(-\frac{2}{3}\right) = \frac{2}{3} \qquad \left|\frac{2}{3}\right| = \frac{2}{3} \qquad \left|-\frac{2}{3}\right| = \frac{2}{3}$$

3.2 Multiplication and Fundamental Properties

In arithmetic you learned to multiply fractions by multiplying their numerators and their denominators. This is exactly what we do with rational numbers in general.

DEFINITION OF MULTIPLICATION OF RATIONAL NUMBERS

If a, b, c, and d are integers with b and d different from 0, then

$$\frac{a}{b} \cdot \frac{c}{d} = \frac{ac}{bd}$$

EXAMPLE 2 (A) $\dfrac{2}{5} \cdot \dfrac{3}{7} = \dfrac{2 \cdot 3}{5 \cdot 7} = \dfrac{6}{35}$

(B) $(-8) \cdot \dfrac{9}{5} = \dfrac{-8}{1} \cdot \dfrac{9}{5} = \dfrac{(-8)(9)}{(1)(5)} = \dfrac{-72}{5} = -\dfrac{72}{5}$

(C) $\dfrac{2x}{3y^2} \cdot \dfrac{x^2}{5y} = \dfrac{(2x)(x^2)}{(3y^2)(5y)} = \dfrac{2x^3}{15y^3}$

PROBLEM 2 Multiply:

(A) $\frac{3}{4} \cdot \frac{3}{5}$ (B) $(-5) \cdot \frac{3}{4}$ (C) $\frac{3x^2}{2y} \cdot \frac{x}{4y^2}$

ANSWER (A) $\frac{9}{20}$ (B) $-\frac{15}{4}$ (C) $\frac{3x^3}{8y^3}$

The following important properties of the rational numbers are a direct consequence of the above definition.

THEOREM 1 The rational numbers are closed, associative, and commutative relative to the operation of multiplication

Thus, relative to multiplication, we may continue to regroup and reorder factors at will.

A single rational number may have many different names, in fact, infinitely many. We will discuss several properties of the rational numbers that will enable us to change expressions that represent rational numbers to different (but equivalent) forms.

We start by stating two rather obvious properties. The first property follows from the definition of the quotient of two integers. The establishment of the second property is left as an exercise.

THEOREM 2 For any nonzero integers a and c, and any integer b,

(A) $$\frac{a}{a} = 1$$

(B) $$1 \cdot \frac{b}{c} = \frac{b}{c} \cdot 1 = \frac{b}{c}$$

Now to one of the basic theorems for rational numbers:

THEOREM 3 (Fundamental principle of fractions) For each integer a and nonzero integers b and k,

$$\frac{ak}{bk} = \frac{a}{b}$$

Can you supply the reasons in the following proof?

$$\frac{ak}{bk} = \frac{a}{b} \cdot \frac{k}{k} = \frac{a}{b} \cdot 1 = \frac{a}{b}$$

This theorem is the basis for reducing fractions to lower terms and raising fractions to higher terms.

EXAMPLE 3

(A) $\dfrac{27}{18} = \dfrac{3}{2}$ lower terms

(B) $\dfrac{6x}{9x} = \dfrac{2}{3}$ lower terms

(C) $\dfrac{3}{4} = \dfrac{9}{12}$ higher terms

(D) $\dfrac{3}{5} = \dfrac{6x^2}{10x^2}$ higher terms

PROBLEM 3

Replace question marks with appropriate symbols:

(A) $\dfrac{24}{32} = \dfrac{?}{4}$ (B) $\dfrac{8m}{12m} = \dfrac{2}{?}$

(C) $\dfrac{2}{3} = \dfrac{?}{12y}$ (D) $\dfrac{7}{4} = \dfrac{14y^2}{?}$

ANSWER (A) 3 (B) 3 (C) $8y$ (D) $8y^2$

The next and last theorem in this section, pertaining to rational numbers and signs, is used with great frequency in mathematics; its misuse is a major contributor to errors.

THEOREM 4

For each integer a and each nonzero integer b,

(A) $\dfrac{-a}{-b} = \dfrac{a}{b}$

(B) $\dfrac{-a}{b} = \dfrac{a}{-b} = -\dfrac{a}{b}$

(C) $(-1)\dfrac{a}{b} = -\dfrac{a}{b}$

Proofs of parts of this theorem are left to the exercises. We conclude this section by considering examples that illustrate most of what we have been talking about.

EXAMPLE 4

(A) $$\frac{5x^2}{9y^2} \cdot \frac{6y}{10x} = \frac{\overset{x}{\cancel{(5x^2)}}\overset{2}{\cancel{(6y)}}}{\underset{3y}{\cancel{(9y^2)}}\underset{\underset{1}{\cancel{2}}}{\cancel{(10x)}}} = \frac{x}{3y}$$

(B) $$\frac{-3x}{2y} \cdot \frac{6y^2}{9x^2} = \frac{\overset{-1}{\cancel{(-3x)}}\overset{\overset{y}{\cancel{3y}}}{\cancel{(6y^2)}}}{\underset{1}{\cancel{(2y)}}\underset{\underset{x}{\cancel{3x}}}{\cancel{(9x^2)}}} = \frac{-y}{x} = -\frac{y}{x}$$

PROBLEM 4 Multiply and write answer in lowest terms:

(A) $\dfrac{10x}{6y^2} \cdot \dfrac{12y}{5x^2}$ (B) $\dfrac{-7x^2}{3y^2} \cdot \dfrac{12y}{14x}$

ANSWER (A) $\dfrac{4}{xy}$ (B) $-\dfrac{2x}{y}$

Exercise 22

In answer, do not change improper fractions to mixed fractions; that is, write $\frac{7}{2}$, not $3\frac{1}{2}$.

A **1.** What are the coordinates of points a, b, and c?

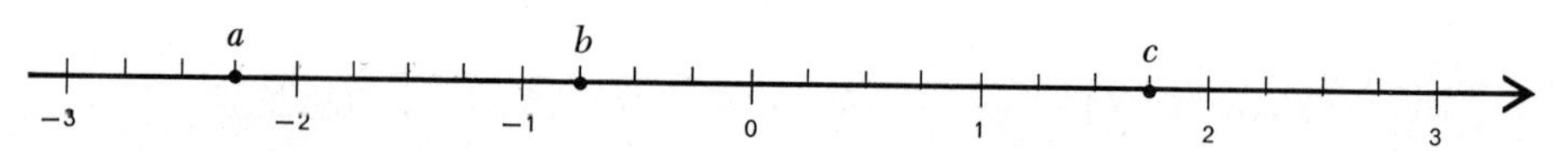

2. What are the coordinates of points c, d, and e?

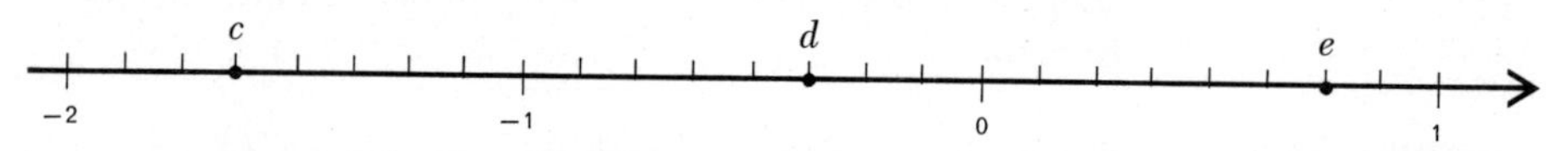

3. Graph $\left\{\dfrac{5}{4}, \dfrac{-5}{4}, \dfrac{2}{-1}, \dfrac{-7}{4}\right\}$ on a number line.

4. Graph $\left\{\dfrac{3}{2}, \dfrac{-3}{2}, \dfrac{-1}{-2}, -\dfrac{1}{2}\right\}$ on a number line.

Multiply:

5. $\dfrac{2}{5} \cdot \dfrac{3}{7}$ **6.** $\dfrac{3}{8} \cdot \dfrac{3}{5}$ **7.** $\dfrac{-2}{5} \cdot \dfrac{4}{3}$

8. $\dfrac{3}{7} \cdot \dfrac{-2}{11}$ **9.** $\dfrac{2}{-5} \cdot \dfrac{-3}{7}$ **10.** $\dfrac{-5}{3} \cdot \dfrac{2}{-7}$

11. $\dfrac{5}{7} \cdot \dfrac{2x}{3y}$ **12.** $\dfrac{4}{5} \cdot \dfrac{7x}{3y}$ **13.** $\dfrac{x}{2y} \cdot \dfrac{3x}{y^2}$

14. $\dfrac{2m}{n^2} \cdot \dfrac{3m^2}{5n^2}$

Replace question marks with appropriate symbols.

15. $\dfrac{8}{12} = \dfrac{?}{3}$ **16.** $\dfrac{12}{16} = \dfrac{?}{4}$ **17.** $\dfrac{1}{5} = \dfrac{3}{?}$

18. $\dfrac{3}{4} = \dfrac{?}{20}$ **19.** $\dfrac{21x}{28x} = \dfrac{?}{4}$ **20.** $\dfrac{36y}{54y} = \dfrac{2}{?}$

21. $\frac{3}{7} = \frac{?}{21x^2}$ **22.** $\frac{4}{5} = \frac{28m^3}{?}$ **23.** $\frac{6x^3}{4xy} = \frac{?}{2y}$

24. $\frac{9xy}{12y^2} = \frac{3x}{?}$

B *Reduce to lowest terms.*

25. $\frac{9x}{6x}$ **26.** $\frac{27y}{15y}$ **27.** $\frac{-3}{12}$

28. $\frac{18}{-8}$ **29.** $\frac{2y^2}{8y^3}$ **30.** $\frac{6x^3}{15x}$

31. $\frac{12a^2b}{3ab^2}$ **32.** $\frac{21x^2y^3}{35x^3y}$

33. $\frac{-2xy^2}{8x^2}$ **34.** $\frac{25mn^3}{-15m^2n^2}$

Multiply and reduce to lowest terms.

35. $\frac{3}{7} \cdot \frac{14}{9}$ **36.** $\frac{10}{9} \cdot \frac{12}{15}$ **37.** $\frac{4}{-5} \cdot \frac{15}{16}$

38. $\frac{8}{3} \cdot \frac{-12}{24}$ **39.** $\frac{2x}{3yz} \cdot \frac{6y}{4x}$ **40.** $\frac{2a}{3bc} \cdot \frac{9c}{a}$

41. $\frac{2x^2}{3y^2} \cdot \frac{9y}{4x}$ **42.** $\frac{3x^2}{4} \cdot \frac{16y}{12x^3}$ **43.** $\frac{6a^2}{7c} \cdot \frac{21cd}{12ac}$

44. $\frac{8x^2}{3xy} \cdot \frac{12y^3}{6y}$

C *Multiply and reduce to lowest terms.*

45. $\frac{-21}{16} \cdot \frac{12}{-14} \cdot \frac{8}{9}$ **46.** $\frac{18}{15} \cdot \frac{-10}{21} \cdot \frac{3}{-1}$

47. $\frac{2x^2}{3y^2} \cdot \frac{-6yz}{2x} \cdot \frac{y}{-xz}$ **48.** $\frac{-a}{-b} \cdot \frac{12b^2}{15ac} \cdot \frac{-10}{4b}$

49. Supply the reasons for the following proof of Theorem 2*B*.

PROOF

	STATEMENT	REASON
1	$1 \cdot \frac{b}{c} = \frac{1}{1} \cdot \frac{b}{c}$	*1*
2	$= \frac{1b}{1c}$	2
3	$= \frac{b}{c}$	3

50. Supply the reasons for the proof of Theorem 3 given in the text.

51. Prove Theorem 4A.

52. Prove Theorem 4B.

3.3 Division

In arithmetic courses you were probably told: "To divide one fraction by another, invert the divisor and multiply." It is not difficult to see why this mechanical rule is valid. To start, we define division for rational numbers as in the integers.

DEFINITION OF DIVISION OF RATIONAL NUMBERS

If a/b and c/d are any two rational numbers, then

$$\frac{a}{b} \div \frac{c}{d} = Q \quad \text{if and only if} \quad \frac{a}{b} = \frac{c}{d} \cdot Q \text{ and } Q \text{ is unique}$$

As a result of this definition, we can convert division into multiplication by means of the following theorem:

THEOREM 5 If a, b, c, and d are integers with b, c, and d different from 0, then

$$\frac{a}{b} \div \frac{c}{d} = \frac{a}{b} \cdot \frac{d}{c}$$

To prove this theorem, all we have to do is to show that the product of the divisor (c/d) and the quotient $[(a/b) \cdot (d/c)]$ is equal to the dividend (a/b)—a task we leave to the exercises.

It should be clear that this theorem is simply a symbolic statement of the *mechanical rule:*

To divide one rational number by another, invert the divisor and multiply.

EXAMPLE 5 (A)

$$\frac{6}{14} \div \frac{21}{2} = \frac{\overset{2}{\cancel{6}}}{\underset{7}{\cancel{14}}} \cdot \frac{\overset{1}{\cancel{2}}}{\underset{7}{\cancel{21}}} = \frac{2}{49}$$

(B)

$$\frac{12x}{5y} \div \frac{9y}{8x} = \frac{\overset{4x}{\cancel{12x}}}{5y} \cdot \frac{8x}{\underset{3y}{\cancel{9y}}} = \frac{32x^2}{15y^2}$$

$$(C)\quad \frac{18a^2b}{15c} \div \frac{12ab^2}{5c} = \frac{\overset{a}{\overset{\cancel{3}a}{\cancel{18a^2b}}}}{\underset{\cancel{3}}{\cancel{15c}}} \cdot \frac{\overset{1}{\cancel{5c}}}{\underset{2b}{\cancel{12ab^2}}} = \frac{a}{2b}$$

$$(D)\quad \frac{-3x}{yz} \div 12x = \frac{-3x}{yz} \div \frac{12x}{1} = \frac{\overset{-1}{\cancel{-3x}}}{yz} \cdot \frac{1}{\underset{4}{\cancel{12x}}} = \frac{-1}{4yz} \text{ or } -\frac{1}{4yz}$$

PROBLEM 5 Divide and reduce to lowest terms:

(A) $\frac{8}{9} \div \frac{4}{3}$ (B) $\frac{8x}{3y} \div \frac{6x}{9y}$

(C) $\frac{15mn^2}{12x} \div \frac{9m^2n}{8x}$ (D) $\frac{6x}{wz} \div (-3x)$

ANSWER (A) $\frac{2}{3}$ (B) 4 (C) $10n/9m$ (D) $-2/wz$

The following important property of the rational numbers follows directly from Theorems 1 and 5 above:

THEOREM 6 The rational numbers are closed with respect to division, except for division by 0.

We now turn to solutions of simple equations that involve rational numbers. We will find that by having extended the integers to the rational numbers we now can solve many more equations. A few examples should make the process clear. (NOTE: The properties of equality discussed in Chap. 2 apply to rational numbers as well as integers. In fact, they apply to any number system!)

First we define the reciprocal of a number. When one inverts a number other than 0, one obtains the reciprocal of the number. In general, two numbers are *reciprocals* of each other if their product is 1.

EXAMPLE 6 (A) The reciprocal of $\frac{2}{3}$ is $\frac{3}{2}$ since $\frac{2}{3} \cdot \frac{3}{2} = 1$.

(B) The reciprocal of 5 is $\frac{1}{5}$ since $(5) \cdot \frac{1}{5} = 1$.

PROBLEM 6 Find the reciprocal of:

(A) $\frac{5}{7}$ (B) 9 (C) 0

ANSWER (A) $\frac{7}{5}$ (B) $\frac{1}{9}$ (C) does not exist

EXAMPLE 7 Solve:

(A) $2x = 3$ (B) $\frac{x}{3} = -2$ (C) $-\frac{2}{3}x = \frac{4}{9}$

SOLUTION

In each problem we operate on both members of the equation so as to make the coefficient of x a positive 1. (Note that since $x/3 = \frac{1}{3}x$, the coefficient of $x/3$ is $\frac{1}{3}$.) This can be done in two ways: (1) Divide both members by the coefficient of x or (2) multiply both members by the reciprocal of the coefficient of x. Choose the easier method for any particular problem.

(*A*) $2x = 3$ or $2x = 3$ CHECK

$$\frac{2x}{2} = \frac{3}{2} \qquad \tfrac{1}{2}(2x) = \tfrac{1}{2}(3) \qquad 2(\tfrac{3}{2}) \stackrel{?}{=} 3$$

$$x = \tfrac{3}{2} \qquad x = \tfrac{3}{2} \qquad 3 \stackrel{\checkmark}{=} 3$$

(*B*) CHECK

$$\frac{x}{3} = -2 \qquad \frac{-6}{3} \stackrel{?}{=} -2$$

$$3\left(\frac{x}{3}\right) = (3)(-2) \qquad -2 \stackrel{\checkmark}{=} -2$$

$$x = -6$$

(*C*) CHECK

$$-\tfrac{2}{3}x = \tfrac{4}{9} \qquad (-\tfrac{2}{3})(-\tfrac{2}{3}) \stackrel{?}{=} \tfrac{4}{9}$$

$$(-\tfrac{3}{2})(-\tfrac{2}{3}x) = (-\tfrac{3}{2})(\tfrac{4}{9}) \qquad \tfrac{4}{9} \stackrel{\checkmark}{=} \tfrac{4}{9}$$

$$x = -\tfrac{2}{3}$$

PROBLEM 7 Solve and check:

(*A*) $5x = 4$ (*B*) $x/2 = -7$ (*C*) $-\frac{3}{4}x = \frac{9}{20}$

ANSWER (*A*) $\frac{4}{5}$ (*B*) -14 (*C*) $\frac{-3}{5}$

Exercise 23

Each variable represents a nonzero integer.

A *Divide and reduce to lowest terms.*

1. $\frac{2}{3} \div \frac{4}{9}$ **2.** $\frac{5}{11} \div \frac{55}{44}$ **3.** $\frac{7}{3} \div \frac{2}{3}$

4. $\frac{1}{25} \div \frac{15}{4}$ **5.** $7 \div \frac{7}{5}$ **6.** $\frac{5}{3} \div 3$

7. $\frac{16}{15} \div (-32)$

8. $7 \div \frac{21}{-2}$

9. $\frac{6x}{5y} \div \frac{3x}{10y}$

10. $\frac{9m}{8n} \div \frac{3m}{4n}$

11. $\frac{2x}{3y} \div \frac{4x}{6y^2}$

12. $\frac{a}{4c} \div \frac{a^2}{12c^2}$

13. $2xy \div \frac{x}{y}$

14. $\frac{x}{3y} \div 3y$

Write the reciprocal of each number.

15. $\frac{7}{8}$

16. $\frac{5}{9}$

17. 3

18. 6

19. $-\frac{2}{3}$

20. -7

Solve and check:

21. $3x = 2$

22. $4x = 3$

23. $\frac{x}{2} = 5$

24. $\frac{x}{3} = 4$

25. $\frac{2}{3}x = 4$

26. $\frac{3}{4}x = 9$

27. $\frac{3}{5}x = \frac{4}{5}$

28. $\frac{2}{7}x = \frac{4}{21}$

B *Perform the indicated operations and reduce to lowest terms.*

29. $\frac{-8}{9} \div \frac{24}{15}$

30. $\frac{36}{21} \div \frac{20}{-32}$

31. $\frac{3uv^2}{5w} \div \frac{6u^2v}{15w}$

32. $\frac{21x^2y^2}{12cd} \div \frac{14xy}{9d}$

33. $\frac{-6x^3}{5y^2} \div \frac{18x}{10y}$

34. $\frac{9u^4}{4v^3} \div \frac{-12u^2}{15v}$

35. $\left(\frac{9}{10} \div \frac{4}{6}\right) \cdot \frac{3}{5}$

36. $\frac{9}{10} \div \left(\frac{4}{5} \cdot \frac{3}{5}\right)$

37. $\left(\frac{a}{b} \div \frac{c}{d}\right) \cdot \frac{e}{f}$

38. $\frac{a}{b} \div \left(\frac{c}{d} \cdot \frac{e}{f}\right)$

Solve and check.

39. $-4x = 6$

40. $15x = -21$

41. $\frac{x}{6} = -\frac{1}{4}$

42. $-\frac{x}{3} = \frac{2}{9}$

43. $\frac{2}{3}x = \frac{4}{9}$

44. $\frac{1}{5}x = -\frac{3}{5}$

45. $\frac{5x}{-7} = -\frac{15}{28}$

46. $-\frac{3}{8}x = \frac{27}{16}$

47. $\frac{1}{3}$ of what number is $\frac{1}{5}$?

48. $\frac{2}{3}$ of what number is $-\frac{8}{9}$?

49. If you walk at 5 mph, how long will it take you to walk 3 miles? (Use $d = rt$.)

50. About $\frac{1}{9}$ of an iceberg is above water. If 50 ft are observed above water, what is the total height of the iceberg?

C **51.** An electronic card sorter sorts IBM cards at 320 cards per min. How long will it take the sorter to sort 2,800 cards? Leave the answer as the quotient of two integers reduced to lowest terms.

52. An arrow is shot vertically upward with an initial velocity of 200 fps. Neglecting air resistance, its velocity at time t is given by the formula $v = 200 - 32t$. In how many seconds will the arrow reach its highest point? Leave the answer as a quotient of two integers reduced to lowest terms.

Which of the following statements are true for all rational numbers a, b, c (division by 0 excluded)? For each statement that is false, find replacements for the variables to show that it is false.

53. ab is a rational number

54. a/b is a rational number

55. $a/b = b/a$

56. $ab = ba$

57. $(ab)c = a(bc)$

58. $(a/b)c = a/(b/c)$

59. Supply the reasons for the following proof of Theorem 5:

PROOF

$$\frac{c}{d}\cdot\left[\frac{a}{b}\cdot\frac{d}{c}\right] = \frac{c}{d}\cdot\left[\frac{d}{c}\cdot\frac{a}{b}\right]$$

$$= \left[\frac{c}{d}\cdot\frac{d}{c}\right]\cdot\frac{a}{b}$$

$$= \left(\frac{cd}{dc}\right)\cdot\frac{a}{b}$$

$$= 1\cdot\frac{a}{b}$$

$$= \frac{a}{b}$$

3.4 Addition and Subtraction

As in the preceding sections, we will again generalize from arithmetic. In adding $\frac{1}{2}$ and $\frac{2}{3}$ you probably proceed somewhat as follows:

$$\begin{array}{r} \frac{1}{2} = \frac{3}{6} \\ \frac{2}{3} = \frac{4}{6} \\ \hline \frac{7}{6} = 1\frac{1}{6} \end{array}$$

That is, you changed each fraction to an equivalent form having a common denominator, then added the numerators and placed the sum over the common denominator. We will proceed in the same way with rational numbers, but we will find it more convenient to work horizontally.

In changing fractions to equivalent forms having a common denominator, we make direct use of the fundamental principle of fractions

$$\frac{a}{b} = \frac{ak}{bk}$$

This principle coupled with the following definition takes care of addition for all rational numbers.

DEFINITION OF ADDITION OF RATIONAL NUMBERS

If a, b, and c are integers with $b \neq 0$, then

$$\frac{a}{b} + \frac{c}{b} = \frac{a+c}{b}$$

EXAMPLE 8

(A) $\frac{2}{3} + \frac{5}{3} = \frac{2+5}{3} = \frac{7}{3}$

(B) $\frac{3}{5x} + \frac{-4}{5x} = \frac{3+(-4)}{5x} = \frac{-1}{5x}$ or $-\frac{1}{5x}$

(C) $\frac{5}{6} + \frac{3}{4} = \frac{2(5)}{2(6)} + \frac{3(3)}{3(4)} = \frac{10}{12} + \frac{9}{12} = \frac{10+9}{12} = \frac{19}{12}$

(D) $\frac{2}{3x} + \frac{1}{2} = \frac{2(2)}{2(3x)} + \frac{3x(1)}{3x(2)} = \frac{4}{6x} + \frac{3x}{6x} = \frac{4+3x}{6x}$ (We cannot cancel the x's.)

PROBLEM 8

Add and reduce to lowest terms:

(A) $\frac{3}{5} + \frac{8}{5}$ (B) $\frac{-2}{5x} + \frac{4}{5x}$

(C) $\frac{2}{3} + \frac{3}{4}$ (D) $\frac{5}{6x} + \frac{3}{4}$

ANSWER (A) $\frac{11}{5}$ (B) $\frac{2}{5x}$ (C) $\frac{17}{12}$ (D) $\frac{10+9x}{12x}$

Even though any common denominator will do, using the least common denominator (LCD) will generally result in less work. (Recall that the least common denominator is the "smallest" expression exactly divisible by each denominator.)

NOTE: Generally we will not change improper fractions to mixed fractions (except for some answers to word problems). An expression such as $3\frac{4}{5}$ is too easily mistaken for the product $(3)(\frac{4}{5})$ when actually it means $3 + \frac{4}{5}$.

Now that we have defined addition of rational numbers, we can define subtraction of rational numbers just as we defined subtraction in the integers. That is, $M - S = D$ if and only if $M = S + D$. As with the integers, we immediately have the following theorem: $M - S = M + (-S)$ (that is, to subtract a quantity add its negative). From this definition and theorem we can easily prove the following useful theorem on subtraction of rational numbers.

THEOREM 7 If a, b, and c are integers with $b \neq 0$, then

$$\frac{a}{b} - \frac{c}{b} = \frac{a - c}{b}$$

Try to supply the reasons for each step in the following proof:

$$\frac{a}{b} - \frac{c}{b} = \frac{a}{b} + \left(-\frac{c}{b}\right) = \frac{a}{b} + \frac{-c}{b} = \frac{a + (-c)}{b} = \frac{a - c}{b}$$

EXAMPLE 9

(A) $$\frac{3}{4} - \frac{1}{4} = \frac{3 - 1}{4} = \frac{2}{4} = \frac{1}{2}$$

(B) $$\frac{3}{7} - \frac{-5}{7} = \frac{3 - (-5)}{7} = \frac{3 + 5}{7} = \frac{8}{7}$$

(C) $$\frac{5x}{6} - \frac{-3x}{2} = \frac{5x}{6} - \frac{3(-3x)}{3(2)} = \frac{5x}{6} - \frac{-9x}{6} = \frac{5x - (-9x)}{6}$$

$$= \frac{5x + 9x}{6} = \frac{14x}{6} = \frac{7x}{3}$$

(D) $$\frac{7x}{4y} - \frac{-2y}{6x} = \frac{3x(7x)}{3x(4y)} - \frac{2y(-2y)}{2y(6x)} = \frac{21x^2}{12xy} - \frac{-4y^2}{12xy}$$

$$= \frac{21x^2 - (-4y^2)}{12xy} = \frac{21x^2 + 4y^2}{12xy}$$

PROBLEM 9 Subtract and reduce to lowest terms:

(A) $\frac{4}{5} - \frac{2}{5}$

(B) $\frac{5}{6} - \frac{-7}{6}$

(C) $\frac{3x}{4} - \frac{-2x}{3}$

(D) $\frac{4x}{3y} - \frac{-3y}{5x}$

ANSWER (A) $\frac{2}{5}$ (B) 2 (C) $\frac{17x}{12}$ (D) $\frac{20x^2 + 9y^2}{15xy}$

Because of the definition of addition, Theorem 8, the fact that $a - b = a + (-b)$, and because the integers are closed under addition, we can conclude that the rational numbers are closed under addition and subtraction. We summarize this result with others listed in preceding sections under one general theorem for convenient reference.

THEOREM 8 The set of rational numbers is closed relative to the operations of addition, subtraction, multiplication, and division (except for division by 0); the rational numbers are associative and commutative relative to addition and multiplication; and multiplication distributes over addition.

To combine three or more rational numbers, or terms representing rational numbers, by addition or subtraction, we use the above results in combination with the material covered in Sec. 2.7. A couple of examples should make the process clear.

EXAMPLE 10 (A)

$$\frac{-3}{4} - \frac{-1}{3} + \frac{5}{6} = \frac{3(-3)}{3(4)} - \frac{4(-1)}{4(3)} + \frac{2(5)}{2(6)} = \frac{-9}{12} - \frac{-4}{12} + \frac{10}{12}$$

$$= \frac{-9 - (-4) + 10}{12}$$

NOTE: LCD = 12

$$= \frac{-9 + 4 + 10}{12}$$

$$= \frac{5}{12}$$

(B)

$$\frac{3}{2x^2} - \frac{-5}{x} + 1 = \frac{3}{2x^2} - \frac{2x(-5)}{2x(x)} + \frac{2x^2}{2x^2} = \frac{3}{2x^2} - \frac{-10x}{2x^2} + \frac{2x^2}{2x^2}$$

$$= \frac{3 - (-10x) + 2x^2}{2x^2}$$

NOTE: LCD = $2x^2$

$$= \frac{3 + 10x + 2x^2}{2x^2}$$

PROBLEM 10 Combine into one fraction:

(A) $\frac{-1}{2} - \frac{-3}{4} + \frac{2}{3}$ (B) $2 - \frac{-3}{x} + \frac{4}{3x^2}$

ANSWER (A) $\frac{11}{12}$ (B) $\frac{6x^2 + 9x + 4}{3x^2}$

Exercise 24

A *Combine into single fractions and reduce to lowest terms (work horizontally).*

1. $\frac{2}{3}+\frac{4}{3}$ **2.** $\frac{3}{4}+\frac{5}{4}$ **3.** $\frac{-3}{5}+\frac{7}{5}$

4. $\frac{2}{3}+\frac{-5}{3}$ **5.** $\frac{3}{8}+\frac{1}{2}$ **6.** $\frac{2}{5}+\frac{3}{10}$

7. $\frac{2}{3}+\frac{3}{5}$ **8.** $\frac{1}{2}+\frac{4}{7}$ **9.** $\frac{7}{11}-\frac{3}{11}$

10. $\frac{5}{3}-\frac{2}{3}$ **11.** $\frac{7}{11}-\frac{-3}{11}$ **12.** $\frac{5}{3}-\frac{-2}{3}$

13. $\frac{1}{2}-\frac{3}{8}$ **14.** $\frac{2}{5}-\frac{3}{10}$ **15.** $\frac{3}{5}-\frac{2}{3}$

16. $\frac{1}{2}-\frac{4}{7}$ **17.** $\frac{3}{5xy}+\frac{-6}{5xy}$ **18.** $\frac{-6}{5x^2}+\frac{4}{5x^2}$

19. $\frac{3y}{x}+\frac{2y}{x}$ **20.** $\frac{x}{5y}+\frac{2x}{5y}$ **21.** $\frac{3}{7y}-\frac{-3}{7y}$

22. $\frac{-2}{3x}-\frac{2}{3x}$ **23.** $\frac{1}{2x}+\frac{2}{3x}$ **24.** $\frac{3}{5m}+\frac{5}{2m}$

25. $\frac{3x}{2}+\frac{2x}{3}$ **26.** $\frac{4m}{3}+\frac{m}{7}$ **27.** $\frac{3}{5x}-\frac{2}{3}$

28. $\frac{2}{3}-\frac{3}{4y}$

B **29.** $\frac{x}{y}-\frac{y}{x}$ **30.** $\frac{a}{b}+\frac{b}{a}$ **31.** $\frac{x}{y}-2$

32. $1-\frac{1}{x}$ **33.** $5-\frac{-3}{x}$ **34.** $\frac{-2}{m}-4$

35. $\frac{1}{xy}-\frac{3}{y}$ **36.** $\frac{2a}{b}+\frac{-1}{ab}$ **37.** $\frac{3}{2x^2}+\frac{4}{3x}$

38. $\frac{5}{3y}+\frac{3}{4y^2}$ **39.** $\frac{5}{8m^3}-\frac{1}{12m}$ **40.** $\frac{2}{9n^2}-\frac{5}{12n^4}$

41. $\frac{1}{3}-\frac{-1}{2}+\frac{5}{6}$ **42.** $\frac{-3}{4}+\frac{2}{5}-\frac{-3}{2}$

43. $\frac{x^2}{4}-\frac{x}{3}+\frac{-1}{2}$ **44.** $\frac{2}{5}-\frac{x}{2}-\frac{-x^2}{3}$

45. $\frac{3}{4x}-\frac{2}{3y}+\frac{1}{8xy}$ **46.** $\frac{1}{xy}-\frac{1}{yz}+\frac{-1}{xz}$

47. $\frac{3}{y^3}-\frac{-2}{3y^2}+\frac{1}{2y}-3$ **48.** $\frac{1}{5x^3}+\frac{-3}{2x^2}-\frac{-2}{3x}-1$

C NOTE: *When the LCD is not obvious, as for example in* $\frac{1}{8} + \frac{1}{15} + \frac{1}{24}$, *we can use the fact that the LCD must contain each different prime factor present in each denominator to the highest power it occurs in any one denominator to find the LCD. Thus, since*

$$8 = 2^3$$
$$15 = 3 \cdot 5$$
$$24 = 2^3 \cdot 3$$

the LCD $= 2^3 \cdot 3 \cdot 5 = 120$. *Use this method to find LCD's in the following problems.*

49. $\frac{2y}{18} - \frac{-1}{28} - \frac{y}{42}$

50. $\frac{5x}{6} - \frac{3}{8} + \frac{x}{15} - \frac{3}{20}$

51. $\frac{x^2}{12} + \frac{x}{18} - \frac{1}{30}$

52. $\frac{3x}{50} - \frac{x}{15} - \frac{-2}{6}$

53. Supply the reasons for the proof of Theorem 7 in the text.

3.5 Equations and Applications

Except for a few problems, we have avoided equations with rational number coefficients. In practical applications rational number coefficients occur more frequently than integers. We are now at the place where we can easily convert an equation with rational coefficients into one with integer coefficients, thus placing it in a position to be solved by previously known methods.

EXAMPLE 11 What operation can we perform on the equation

$$\frac{x+1}{3} - \frac{x}{4} = \frac{1}{2}$$

to eliminate the denominators? If we could find a number that was exactly divisible by each denominator, then we would be able to use the multiplication property of equality to clear the denominators. The least common denominator of the fractions is exactly what we are looking for! Thus, we multiply both members of the equation by 12:

$$12\left(\frac{x+1}{3} - \frac{x}{4}\right) = 12\left(\frac{1}{2}\right)$$

$$\overset{4}{\cancel{12}} \cdot \frac{(x+1)}{\underset{1}{\cancel{3}}} - \overset{3}{\cancel{12}} \cdot \frac{x}{\underset{1}{\cancel{4}}} = \overset{6}{\cancel{12}} \cdot \frac{1}{\underset{1}{\cancel{2}}}$$

$$4(x+1) - 3x = 6$$
$$4x + 4 - 3x = 6$$
$$x = 2$$

PROBLEM 11 Solve

$$\frac{x+2}{2} - \frac{x}{3} = 5$$

ANSWER $x = 24$

EXAMPLE 12 Equations will often have rational coefficients written as decimal fractions. Some equations of this type are more easily solved if they are first cleared of decimals. The following is a case in point:

$$\begin{aligned} 0.2x + 0.3(x - 5) &= 13 \\ 10(0.2x) + 10[0.3(x - 5)] &= 10 \cdot 13 \\ 2x + 3(x - 5) &= 130 \\ 5x - 15 &= 130 \\ 5x &= 145 \\ x &= 29 \end{aligned}$$

PROBLEM 12 Solve:

$$0.3(x + 2) + 0.5x = 3$$

ANSWER $x = 3$

EXAMPLE 13 Five individuals formed a glider club and decided to share the cost of a glider equally. They found, however, that if they let three more join the club, the share for each of the original five would be reduced by \$120. What was the total cost of the glider?

SOLUTION Let $C =$ the total cost of the glider.

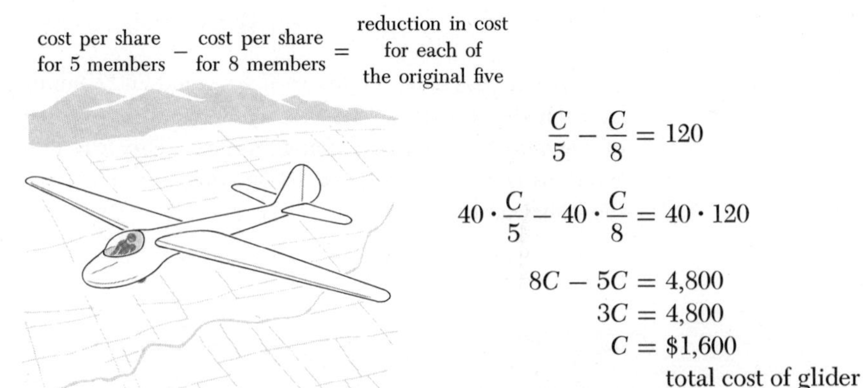

cost per share for 5 members − cost per share for 8 members = reduction in cost for each of the original five

$$\begin{aligned} \frac{C}{5} - \frac{C}{8} &= 120 \\ 40 \cdot \frac{C}{5} - 40 \cdot \frac{C}{8} &= 40 \cdot 120 \\ 8C - 5C &= 4{,}800 \\ 3C &= 4{,}800 \\ C &= \$1{,}600 \end{aligned}$$

total cost of glider

PROBLEM 13 Three individuals bought a sail boat together. If they had taken in a fourth person, the cost for each would have been reduced \$200. What was the total cost of the boat?

ANSWER \$2,400

A very common error that occurs at this stage is to confuse algebraic expressions involving fractions with equations involving fractions. Compare the following two problems:

ADD: $\frac{x}{2} + \frac{x}{3} + 10 = \frac{3x}{6} + \frac{2x}{6} + \frac{60}{6}$

$= \frac{3x + 2x + 60}{6}$

$= \frac{5x + 60}{6}$

SOLVE: $\frac{x}{2} + \frac{x}{3} = 10$

$6 \cdot \frac{x}{2} + 6 \cdot \frac{x}{2} = 6 \cdot 10$

$3x + 2x = 60$

$5x = 60$

$x = 12$

In the addition problem, we cannot multiply "through" by 6. There is no "through," we are dealing with an algebraic expression, not an equation!

Exercise 25

A *Solve:*

1. $\frac{x}{7} - 1 = \frac{1}{7}$ **2.** $\frac{x}{5} - 2 = \frac{3}{5}$ **3.** $\frac{y}{4} + \frac{y}{2} = 9$

4. $\frac{x}{3} + \frac{x}{6} = 4$ **5.** $\frac{x}{2} + \frac{x}{3} = 5$ **6.** $\frac{y}{4} + \frac{y}{3} = 7$

7. $\frac{x}{3} - \frac{1}{4} = \frac{3x}{8}$ **8.** $\frac{y}{5} - \frac{1}{3} = \frac{2y}{15}$ **9.** $\frac{n}{5} - \frac{n}{6} = \frac{6}{5}$

10. $\frac{m}{4} - \frac{m}{3} = \frac{1}{2}$ **11.** $\frac{2}{3} - \frac{x}{8} = \frac{5}{6}$ **12.** $\frac{5}{12} - \frac{m}{3} = \frac{4}{9}$

13. $0.8x = 16$ **14.** $0.5x = 35$

15. $0.3x + 0.5x = 24$ **16.** $0.7x + 0.9x = 32$

B **17.** $\frac{x+3}{2} - \frac{x}{3} = 4$ **18.** $\frac{x-2}{3} + 1 = \frac{x}{7}$

19. $\frac{2x+1}{4} = \frac{3x+2}{3}$ **20.** $\frac{4x+3}{9} = \frac{3x+5}{7}$

21. $3 - \frac{x-1}{2} = \frac{x-3}{3}$ **22.** $4 - \frac{x-3}{4} = \frac{x-1}{8}$

23. $3 - \frac{2x-3}{3} = \frac{5-x}{2}$ **24.** $1 - \frac{3x-1}{6} = \frac{2-x}{3}$

25. $0.4(x + 5) - 0.3x = 17$ **26.** $0.1(x - 7) + 0.05x = 0.8$

27. Four students share an apartment. If they take in two more, the cost for each will be reduced \$20. What is the monthly cost for the apartment?

28. A travel agent, arranging a round-trip charter flight from New York to Europe, tells 100 signed-up members of a group that they can save \$120 each if they can get 50 more people to sign up for the trip to fill the plane. What is the total round-trip charter cost of the jet transport? What is the cost per person if 100 go? What is the cost per person if 150 go?

29. A pole is located in a pond. One-fifth of the pole is in the sand, 10 ft of it is in the water, and two-thirds of it is in the air. How long is the pole?

30. Two-thirds of a bar of silver balances exactly with one-half of a bar of silver and a $\frac{1}{4}$-lb weight (Fig. 1). How much does one whole bar of silver weigh?

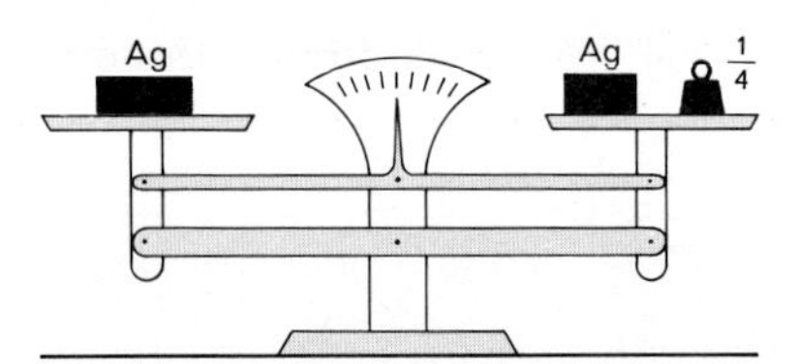

Figure 1

31. The retail price of a record is \$2.80. The markup on the cost is 40 percent. What did the store pay for the record? HINT: Cost + markup = retail.

32. A taxi company charges a flat fee of 75 cents and then 60 cents per mile traveled. (*A*) Write a formula relating cost c and the number of miles traveled m, using a decimal representation for cents. (*B*) How long a ride can you get for \$3.15 (not including tip)?

C *Solve:*

33. $\frac{3x-1}{8} - \frac{2x+1}{3} = \frac{1-x}{12} - 1$

34. $\frac{2x-3}{9} - \frac{x+5}{6} = \frac{3-x}{2} + 1$

35. A skydiver free falls (because of air resistance) at about 176 fps or 120 mph (Fig. 2); with his parachute open he falls at about 22 fps or 15 mph. If the skydiver opened his chute halfway down and the total time for the descent was 6 min, how high was the plane when he jumped?

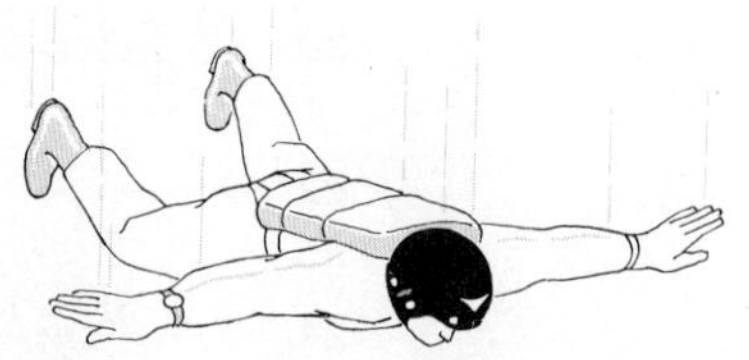

Figure 2

36. It is known that a carton contains 100 packages and that some of the packages weigh $\frac{1}{2}$ lb each and the rest weigh $\frac{1}{3}$ lb each. To save time counting each type of package in the carton, you can weigh the whole contents of the box (45 lb) and determine the number of each kind of package by use of algebra. How many are there of each kind?

37. Diophantus, an early Greek algebraist (A.D. 280), was the subject for a famous ancient puzzle. See if you can find Diophantus' age at death from the following information: Diophantus was a boy for one-sixth of his life; after one-twelfth more he grew a beard; after one-seventh more he married, and after 5 years of marriage he was granted a son; the son lived one-half as long as his father; and Diophantus died 4 years after his son's death.

3.6 Ratio and Proportion

The comparison of quantities by use of ratios is very common. In fact, Pythagoras (569? B.C.) practically founded a whole religion on the ratio of whole numbers. You have no doubt been using ratios for many years and will recall that the *ratio of two quantities* is simply their quotient—the first divided by the second.

EXAMPLE 14 If there are 10 boys and 20 girls in a class, then the ratio of boys to girls is $\frac{10}{20}$ or $\frac{1}{2}$, which is often written

$1:2$ (read 1 to 2)

The ratio of girls to boys is $\frac{20}{10}$ or $\frac{2}{1}$ or

$2:1$ (read 2 to 1)

PROBLEM 14 If there are 500 boys in a school and 400 girls:

(*A*) what is the ratio of boys to girls (*B*) of girls to boys?

ANSWER (*A*) $5:4$ (*B*) $4:5$.

In addition to comparing known quantities, another reason why we want to know something about ratios is that they often lead to a simple way of finding unknown quantities.

EXAMPLE 15 Suppose you are told that in a school the ratio of girls to boys is $3:5$ and that there are 1,450 boys. How many girls are in the school?

SOLUTION Let $x =$ the number of girls in the school, then

$$\frac{x}{1{,}450} = \frac{3}{5}$$

$$x = 1{,}450 \cdot \tfrac{3}{5}$$

$$= 870 \text{ girls}$$

PROBLEM 15 If in a school the ratio of boys to girls is 2:3, and there are 1,200 girls, how many boys are in the school?

ANSWER 800

The statement of equality between two ratios is called a *proportion*. Knowing that various pairs of quantities are proportional leads to simple solutions of many types of problems. For example, to find the height of a tree, the distance across a lake, or many other inaccessible distances, one can use the proportional property of similar triangles (two triangles are similar if their corresponding angles are equal): If two triangles are similar, the ratios of corresponding sides are equal. Thus, in the similar triangles in the figure

$$\frac{a}{d} = \frac{b}{e} \qquad \frac{a}{d} = \frac{c}{f} \qquad \frac{e}{b} = \frac{f}{c}$$

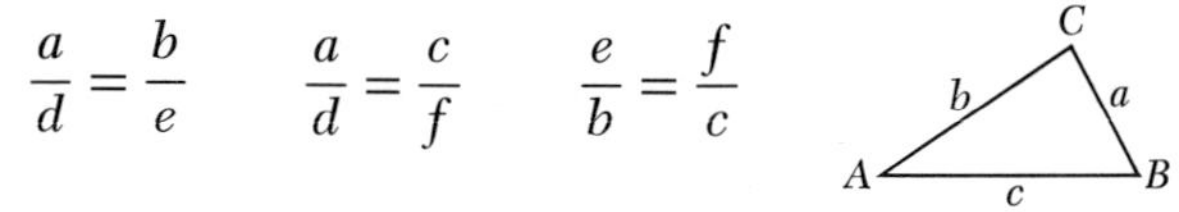

EXAMPLE 16 To measure the height of the tree in Fig. 3, you and a friend could place two stakes vertically in the ground in line with the tree at any convenient distance apart and distance from the tree. You then place your eye at any point on the farthest stake and your friend marks A and B as indicated. Triangles EAB and ECD are similar. Measure EA, AB, and EC, and determine the height h from the proportion

$$\frac{h}{AB} = \frac{EC}{EA}$$

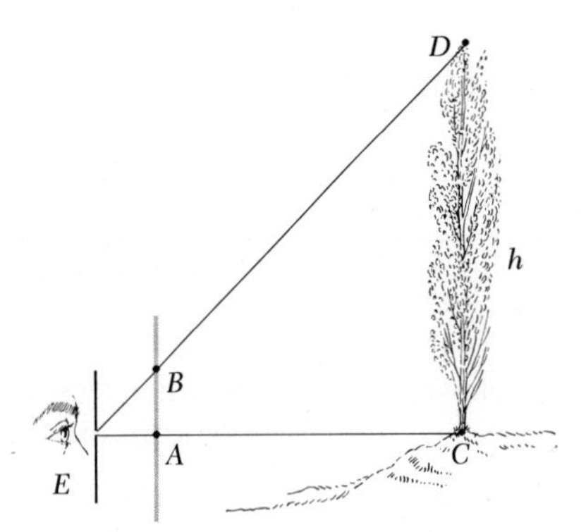

Figure 3

Suppose that $AB = 1.5$ ft, $EC = 200$ ft, and $EA = 5$ ft, then

$$\frac{h}{1.5} = \frac{200}{5}$$

$$h = (1.5) \cdot \tfrac{200}{5}$$

$$= 60 \text{ ft} \qquad \text{height of tree}$$

PROBLEM 16 Find the height of the tree in Example 16 if $AB = 2$ ft, $EC = 150$ ft, and $EA = 4$ ft.

ANSWER 75 ft

Exercise 26

A *Write as a ratio.*

1. 16 boys to 64 girls

2. 64 girls to 16 boys

3. 25 ft to 5 ft

4. 30 in. to 10 in.

5. 30 sq in. to 90 sq in.

6. 25 sq in. to 100 sq in.

Solve each proportion.

7. $\frac{x}{12} = \frac{2}{3}$

8. $\frac{y}{16} = \frac{5}{4}$

9. $\frac{d}{12} = \frac{27}{18}$

10. $\frac{y}{13} = \frac{21}{39}$

11. $\frac{18}{27} = \frac{h}{6}$

12. $\frac{35}{56} = \frac{x}{32}$

13. If in a school the ratio of boys to girls is 5:7 and there are 840 girls, how many boys are there?

14. If the ratio of girls to boys is 7:9 and there are 630 boys, how many girls are there?

15. Find the height of the tree in Fig. 3 if $AB = 1$ ft, $EC = 120$ ft, and $EA = 2$ ft.

16. Find the height of the tree in Fig. 3 if $AB = 1.4$ ft, $EC = 100$ ft, and $EA = 0.7$ ft.

B **17.** SCALE DRAWINGS. An architect wishes to make a scale drawing of a 48- by 30-ft rectangular building. If his drawing of the building is 6 in. long, how wide is it?

18. PHOTOGRAPHY. If you enlarge a 6- by 3-in. picture so that the longer side is 8 in., how wide will the enlargement be?

19. OPTICS (MAGNIFICATION). In Fig. 4 triangles ABC and EDC are similar; hence, corresponding parts are proportional. If the object is 0.4 in., $AC = 1.4$ in., and $CD = 4.9$ in., what is the size of the image?

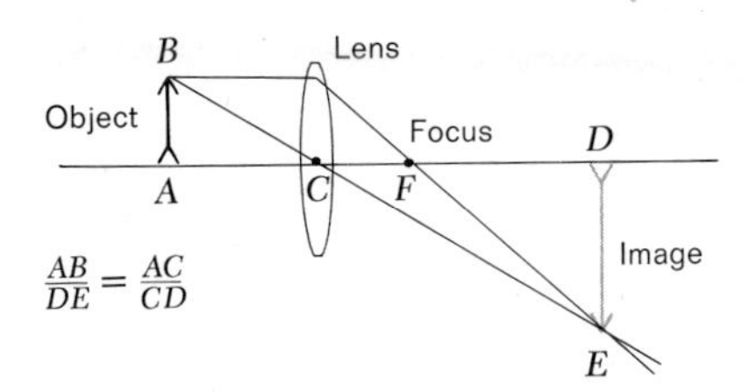

Figure 4

20. COMPUTERS. If an IBM electronic card sorter can sort 1,250 cards in 5 min, how long will it take the sorter to sort 11,250 cards?

21. COMMISSIONS. If you were charged a commission of $57 on the purchase of 300 shares of stock, what would be the proportionate commission on 500 shares?

22. INTELLIGENCE. The IQ (Intelligence Quotient) is found by dividing the mental age, as indicated by standard tests, by the chronological age, and then multiplying by 100. For example, if a child has a mental age of 12 and has a chronological age of 10, his IQ is 120. If an 11-year-old has an IQ of 132 (superior intelligence), compute his mental age.

C **23.** HYDRAULIC LIFTS. If in Fig. 5 the diameter of the smaller pipe is $\frac{1}{2}$ in. and the diameter of the larger pipe is 10 in., how much force would be required to lift a 3,000-lb car? (Neglect weight of lift equipment.)

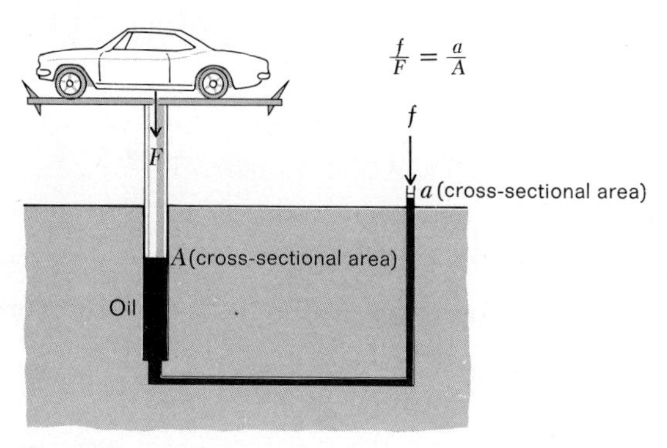

Figure 5

24. POPULATION SAMPLING. Zoologists Green and Evans (1940) estimated the total population of snowshoe hares in the Lake Alexander area of Minnesota as follows: They captured and banded 948 hares, and then released them. After an appropriate period for mixing, they again captured a sample of 421 and found 167 of these marked. Set up an appropriate proportion and estimate the total hare population in the region.

25. ASTRONOMY. Do you have any idea how one might measure the circumference of the earth? In 240 B.C. Eratosthenes measured the size of the earth from its curvature. At Syene, Egypt (lying on the Tropic of Cancer) the sun was directly overhead at noon

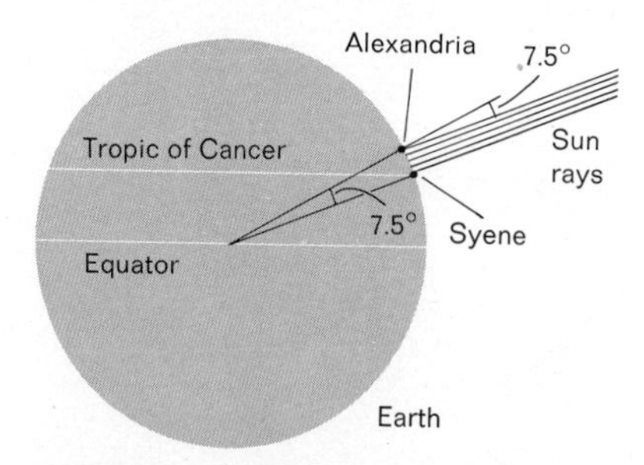

Figure 6

on June 21. At the same time in Alexandria, a town 500 miles directly north, the sun's rays fell at an angle of 7.5° to the vertical. Using this information and a little knowledge of geometry (see Fig. 6), Eratosthenes was able to approximate the circumference of the earth using the following proportion: 7.5 is to 360 as 500 is to the circumference of the earth. Compute Eratosthenes' estimate.

3.7 Real Numbers and Line Graphs

By extending the integers to the rational numbers, we significantly increased our ability to perform certain operations, to solve equations, and to attack practical problems. From a certain point of view it may appear that the rational numbers are capable of satisfying all of our number needs. Can we stop with the rational numbers or do we need to go further?

Let us take a closer look at the relationship between the rational numbers and the points on a number line. If we take any two points on a number line with rational number coordinates, add the rational numbers, and divide by 2, the result (because of closure properties) will be a rational number. In addition, it will name a point between the other two (see Fig. 7).

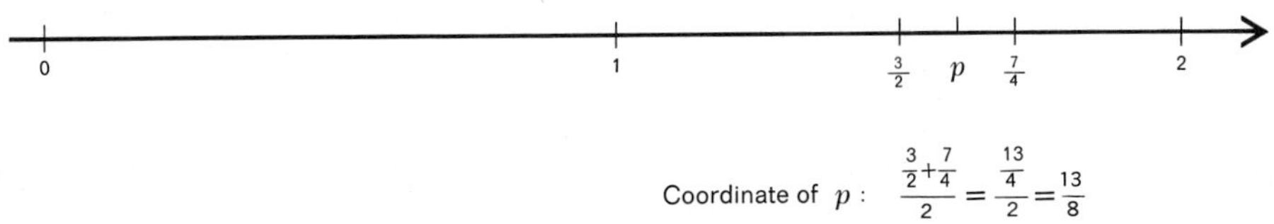

Figure 7

In general, it can be shown that between any two points with rational numbers as coordinates there always exists another point with a rational number coordinate. It would appear that points with rational number coordinates are very "close" together. And that if we used all of the rational numbers to label points on a number line, all of the points would be used up. It may startle you to know that in a certain sense there are more points that have not been named by rational numbers than have been named. The line is like a sieve!

It is not difficult to find a point that cannot be named by a rational number. As a matter of fact, Pythagoras (569? B.C.) managed on the basis of this discovery to destroy the very foundations of the Pythagorean sect which he had founded. This group of religious zealots believed that all natural and supernatural phenomena was built on whole number and ratio of whole number relationships.

Using his famous Pythagorean theorem,* he found that he could find a point that

*The square of the hypotenuse of a right triangle is equal to the sum of the squares of the other two sides.

$c^2 = a^2 + b^2$

c b a

must be named by a number whose square is 2 (Fig. 8). Now one naturally asks if there is a rational number whose square is 2. Pythagoras found that if one answered this question in the affirmative, then one could show that certain odd numbers are even, which, of course, is absurd. The original assumption must be false! That is, if the point is to be named by a number, it must be a number other than a rational number. This discovery, though devastating to Pythagoras at the time, is of utmost importance to modern-day mathematics and has assured Pythagoras a permanent position in the "mathematics hall of fame."

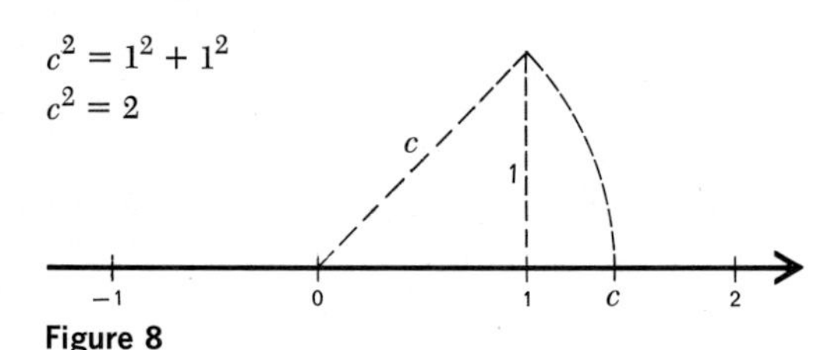

Figure 8

It turns out that there are infinitely many other points on the number line that do not have rational number coordinates. All of these points can be associated with line segments of given lengths, and many can be associated with solutions to simple equations such as

$$x^2 - 7 = 0$$

Once again we must extend our number system. We will discuss this extension informally at this time. In later chapters and future courses the extension will be treated in greater detail.

What are the numbers called that take up the points not named by rational numbers? If you guessed the *irrational numbers*, you are right. The irrational numbers include numbers such as

$\sqrt{2}$ a number whose square is 2

$\sqrt[3]{7}$ a number whose cube is 7

π the ratio of the circumference of any circle to its diameter

In fact, irrational numbers include such numbers as the square root of any positive integer that is not the square of an integer, the cube root of any integer that is not the cube of an integer, and so on.

If we combine the set of rational numbers with the set of irrational numbers we obtain the set of *real numbers*, the number system that will meet most of your number needs for some time to come.

It is an amazing and extremely useful fact that there exists a perfect matching (a one-to-one correspondence) between the set of real numbers and the points on a straight line. This fact is often referred to as the fundamental theorem of analytic geometry. An immediate consequence of this theorem is that graphs of inequality statements such as

$$-3 \leq x \leq 5$$

where x is any real number, take up *all* of the points between -3 and 5. We indicate this by using a solid line between the two points as follows:

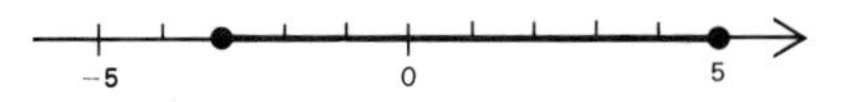

EXAMPLE 17 Graph:

(A) $-3 < x \leq 5$ (B) $-3 \leq x < 5$ (C) $-3 < x < 5$

SOLUTION

(A) -5 0 5

(B) -5 0 5

(C) -5 0 5

PROBLEM 17 Graph:

(A) $-4 \leq x \leq 2$ (B) $-4 \leq x < 2$ (C) $-4 < x < 2$

ANSWER

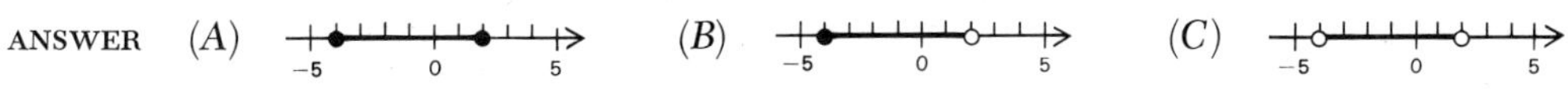

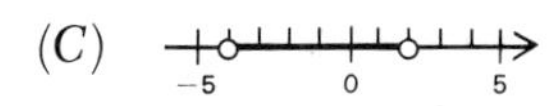

We now take a brief look back and observe that the number systems we have studied are related to each other as follows:

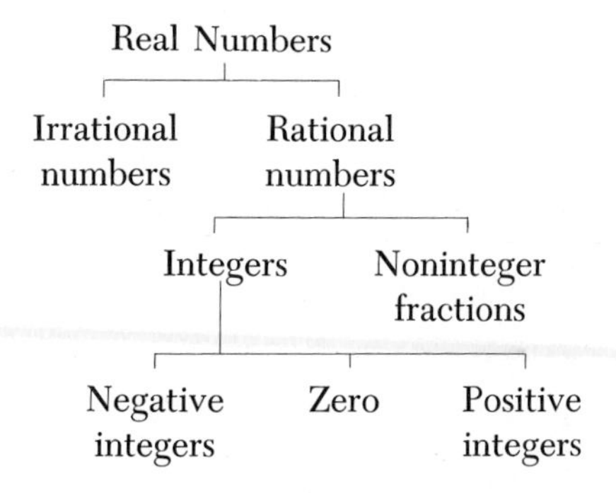

Starting with the positive integers (natural numbers) we continued adding numbers to our system, never throwing anything away, until now we have the very important real number system.

Of course, with this brief introduction to the real numbers you are not expected to be an expert on the subject. All that is expected at this time is that you be aware of the existence of irrational numbers, be able to give a few examples of these numbers, and to graph inequalities of the type found in Example 17.

For completeness and for convenient reference, some of the important properties of the real numbers are summarized in the following theorem, which is stated without proof:

THEOREM 9 **IMPORTANT PROPERTIES OF THE REAL NUMBERS**

(*A*) The real numbers are closed with respect to addition, subtraction, multiplication, and division (division by 0 excluded).

(*B*) The real numbers are associative and commutative relative to the operations of addition and multiplication.

(*C*) The real numbers have a unique additive identity called 0; that is, for each real number a, $a + 0 = a$, and 0 is the only real number with this property.

(*D*) The real numbers have a unique multiplicative identity called 1; that is, for each real number a, $a \cdot 1 = a$, and 1 is the only real number with this property.

(*E*) Each real number has a unique additive inverse; that is, for each real number a there exists a unique real number a such that $a + (-a) = 0$.

(*F*) Each real number, except 0, has a unique multiplicative inverse; that is, for each real number a, with $a \neq 0$, there exists a unique real number $1/a$, such that $a \cdot (1/a) = 1$.

(*G*) For all real numbers a, b, and c, $a(b + c) = ab + ac$ and $a(b - c) = ab - ac$.

(*H*) For all real numbers a and b, $(-1)a = -a$, $a(-b) = (-a)b = -ab$, $(-a)(-b) = ab$, $(-a)/b = a/(-b) = -(a/b)$, $(-a)/(-b) = a/b$.

Exercise 27

A *Indicate whether true (T) or false (F).*

1. 5 is a natural number

2. $\frac{2}{3}$ is a rational number

3. -3 is an integer

4. $\frac{3}{4}$ is a natural number

5. 0 is an integer

6. $-\frac{2}{3}$ is an integer

7. $\sqrt{2}$ is an irrational number

8. π is an irrational number

9. $\sqrt[3]{5}$ is a real number

10. 7 is a real number

11. -5 is a real number

12. $\frac{3}{7}$ is a real number

Graph on a number line for x a real number.

13. $-2 \leq x \leq 3$

14. $-2 < x \leq 3$

15. $-2 \leq x < 3$

16. $-2 < x < 3$

B *Given the sets* R = *the set of real numbers*
Q = *the set of rational numbers*
I = *the set of integers*
N = *the set of natural numbers*

Indicate true (T) or false (F).

17. 3 is in N

18. 5 is in I

19. -7 is in Q

20. -3 is in R

21. $\frac{2}{3}$ is in N

22. $-\frac{2}{3}$ is in I

23. $\frac{2}{3}$ is in Q

24. $-\frac{2}{3}$ is in R

25. $\sqrt{2}$ is in N

26. $\sqrt{2}$ is in I

27. $\sqrt{2}$ is in Q

28. $\sqrt{2}$ is in R

29. 3.14 is in Q

30. $\frac{22}{7}$ is in Q

31. π is in Q

32. π is in R

33. N is a subset of I

34. N is a subset of Q

35. N is a subset of R

36. I is a subset of N

37. I is a subset of Q

38. I is a subset of R

39. Q is a subset of N

40. Q is a subset of I

41. Q is a subset of R

42. R is a subset of Q

Graph on a number line for x a real number.

43. $-\frac{9}{4} \leq x < \frac{3}{2}$

44. $-\frac{5}{4} < x \leq \frac{3}{4}$

45. $-\frac{15}{8} < x < -\frac{1}{8}$

46. $-\frac{14}{3} < x < \frac{8}{3}$

47. $x \geq -\frac{7}{4}$

48. $x < \frac{17}{4}$

Graph problems 49 to 54 on a number line. (Use $\sqrt{2} \approx 1.4$ and $\pi \approx 3.1$.)

49. $-4 \leq x \leq 6$ for x a real number

50. $-7 \leq x \leq 3$ for x a real number

51. $-4 \leq x \leq 6$ for x an integer

52. $-7 \leq x \leq 3$ for x an integer

C **53.** $\{x \in R \mid -\sqrt{2} < x \leq \pi\}$

54. $\{x \in I \mid -\sqrt{2} < x \leq \pi\}$

55. *Every rational number has an infinite repeating-decimal representation.* For example,

$$\frac{1}{4} = 0.25\overline{00}$$

$$\frac{1}{3} = 0.33\overline{33}$$

$$\frac{15}{7} = 2.142857\overline{142857}$$

(The bar indicates the block of numbers which continues to repeat indefinitely.) Express each of the following rational numbers in repeating-decimal form:

(*A*) $\frac{3}{8}$ (*B*) $\frac{23}{9}$ (*C*) $\frac{7}{13}$

56. Repeat the preceding problem for (*A*) $\frac{37}{6}$ and (*B*) $\frac{15}{21}$

57. *Every infinitely repeating decimal is a rational number.* To find two integers whose quotient is a repeating decimal, say $2.135\overline{35}$, proceed as follows:

$$\begin{aligned} n &= 2.135\overline{35} \\ 10n &= 21.35\overline{35} \\ 1{,}000n &= 2135.35\overline{35} \\ 1{,}000n - 10n &= 2135.35\overline{35} - 21.35\overline{35} \\ 990n &= 2{,}114 \end{aligned}$$

$$n = \frac{2{,}114}{990} = \frac{1{,}057}{495}$$

Convert each of the following repeating decimals to a quotient of two integer forms:
(*A*) $0.27\overline{27}$ (*B*) $3.21\overline{21}$

58. Proceed as in the preceding problem to convert each of the following repeating decimals to a quotient of two integer forms:
(*A*) $2.17\overline{17}$ (*B*) $0.472\overline{72}$

Exercise 28 Chapter Review

A **1.** Graph $\left\{-\frac{7}{4}, -\frac{3}{4}, \frac{3}{2}\right\}$

2. Multiply: $\frac{3}{2y} \cdot \frac{5x}{4}$

3. Divide: $\frac{3}{2y} \div \frac{5x}{4}$

4. Solve and check: $6x = 5$

5. Add: $\frac{y}{2} + \frac{y}{3}$

6. Solve: $\frac{x}{4} - 3 = \frac{x}{5}$

7. Solve: $0.4x + 0.3x = 6.3$

8. If the ratio of men to women in a small town is 2:3 and there are 990 women, how many men are there?

9. True (T) or false (F)?: $-\frac{3}{5}$ is an
(*A*) integer (*B*) rational number (*C*) real number

10. Graph for x a real number: $-3 \leq x \leq 4$

B **11.** Multiply and reduce to lowest terms: $\frac{-3y}{5xz} \cdot \frac{-10z}{15xy}$

12. Divide and reduce to lowest terms: $\frac{-3y}{5xz} \div \frac{-10z}{15xy}$

13. Solve and check: $-\frac{3}{5}y = \frac{2}{3}$

14. Simplify: $\frac{-4}{9} - \frac{35}{18} - \frac{-10}{3}$

15. Combine into a single fraction: $\frac{1}{4y} + \frac{3}{2z} - \frac{1}{3x} - 2$

16. Solve: $\frac{x}{4} - \frac{x-3}{3} = 2$

17. Solve: $0.4x - 0.3(x - 3) = 5$

18. A store has a camera on sale at 20 percent off list price. If the sale price is \$64, what is the list price?

19. Given the sets R = the set of real numbers
Q = the set of rational numbers
I = the set of integers
N = the set of natural numbers

Indicate true (T) or false (F): (*A*) $\sqrt{2}$ is in R, (*B*) $\sqrt{2}$ is in Q, (*C*) -3 is in Q, (*D*) -3 is in N.

20. Graph $-1 \leq x < 3$ for x a real number.

C **21.** Simplify: $\dfrac{10x}{9y} \div \left(\dfrac{15xy}{2z} \cdot \dfrac{-z}{3y^2}\right)$

22. One-eighth of what number is $-\frac{3}{2}$? Write an equation and solve.

23. Simplify: $\dfrac{10x}{9y} + \dfrac{15xy}{2z} - \dfrac{-z}{3y^2}$

24. Solve: $\dfrac{x-3}{12} - \dfrac{(x+2)}{9} = \dfrac{1-x}{6} - 1$

25. Three-fourths of a bar of gold balances exactly with one-third of a bar and a $\frac{1}{2}$-lb weight. How much does one whole bar of gold weigh?

26. If there are 180 Fahrenheit degrees and 100 Celsius degrees between freezing and boiling for water, how many Celsius degrees correspond to 9 Fahrenheit degrees? How many Fahrenheit degrees correspond to 20 Celsius degrees?

27. Graph: $\{x \in I \mid -\pi < x < \sqrt{2}\}$

CHAPTER 4 Polynomials

4.1 Polynomials in One or More Variables

Practically all of the algebraic expressions with which we have dealt in this course so far are called polynomials. A *polynomial* is any algebraic expression that can be built up using only the operations of addition, subtraction, and multiplication on variables and real number constants. We may have polynomials in one variable, two variables, and so on; the constants may be any real number, rational or irrational. For example

$$x \qquad 2x^3 \qquad 2x - 1 \qquad 2x^2 - 5x + 8$$

$$x^2 - 2xy + 3y^2 \qquad 2x^3 - 3x^2y + xy^2 - 4y^2$$

are polynomials in one and two variables. The following are not polynomials since these algebraic expressions cannot be formed by using the operations referred to above (a variable cannot appear in a denominator, as an exponent, or within a radical sign):

$$\frac{3x - y}{x + y} \qquad 2^x \qquad x^3 - \frac{2\sqrt{x}}{y^2} + 3y^4$$

The polynomial form, particularly in one and two variables, is encountered with great frequency at all levels in mathematics and science. As a consequence, it is a form that receives a great deal of attention in beginning and intermediate algebra.

It is convenient to identify certain types of polynomials for more efficient study. The concept of degree is used for this purpose. The *degree of a term* in a polynomial is the power of the variable present in the term. If more than one variable is present as a factor, then the sum of the powers of the variables in the term is the degree of the term. The *degree of a polynomial* is the degree of the term with the highest degree in the polynomial.

EXAMPLE 1

(*A*) $3x^5$ is of degree 5

(*B*) $5x^2y^3$ is of degree 5

(*C*) In $4x^7 - 3x^5 + 2x^3 - 1$, the highest degree term is the first, with degree 7; thus, the degree of the polynomial is 7.

(*D*) The degree of each of the first three terms in $x^2 - 2xy + y^2 + 2x - 3y + 2$ is 2; the fourth and fifth terms each have degree 1; thus, this is a second-degree polynomial.

PROBLEM 1

(*A*) What is the degree of $5x^3$? Of $2x^3y^4$?

(*B*) What is the degree of the polynomial $6x^3 - 2x^2 + x - 1$? Of $2x^2 - 3xy + y^2 + x - y + 1$?

ANSWER (*A*) 3, 7 (*B*) 3, 2

NOTE: A constant is often defined as a polynomial of degree 0 for reasons that will become clear later on.

The Table of Contents should make more sense to you now. Turn to it and see how polynomial classifications are used for chapter and section headings.

4.2 Addition, Subtraction, and Multiplication

In the preceding chapters you worked many problems involving addition, subtraction, and multiplication of polynomials. Considerable use was made of the associative and commutative properties of real numbers, as well as the distributive property. The procedures will be reviewed and extended in this section.

EXAMPLE 2

Add:

$$x^4 - 3x^3 + x^2, \qquad -x^3 - 2x^2 + 3x, \qquad \text{and} \qquad 3x^2 - 4x - 5$$

SOLUTION Add horizontally:

$$\begin{aligned}(x^4 - 3x^3 + x^2) + (-x^3 - 2x^2 + 3x) + (3x^2 - 4x - 5)\\ = x^4 - 3x^3 + x^2 - x^3 - 2x^2 + 3x + 3x^2 - 4x - 5\\ = x^4 - 4x^3 + 2x^2 - x - 5\end{aligned}$$

or vertically by lining up like terms and adding their coefficients:

$$\begin{array}{rrrrr} x^4 & - 3x^3 & + x^2 & & \\ & - x^3 & - 2x^2 & + 3x & \\ & & 3x^2 & - 4x & - 5 \\ \hline x^4 & - 4x^3 & + 2x^2 & - x & - 5 \end{array}$$

PROBLEM 2 Add horizontally and vertically:

$$3x^4 - 2x^3 - 4x^2, \qquad x^3 - 2x^2 - 5x, \qquad \text{and} \qquad x^2 + 7x - 2$$

ANSWER $3x^4 - x^3 - 5x^2 + 2x - 2$

EXAMPLE 3 Subtract:

$$4x^2 - 3x + 5 \qquad \text{from} \qquad x^2 - 8$$

SOLUTION

$$\begin{aligned}&(x^2 - 8) - (4x^2 - 3x + 5)\\ &= x^2 - 8 - 4x^2 + 3x - 5\\ &= -3x^2 + 3x - 13\end{aligned}$$

or

$$\begin{array}{rrr} x^2 & & -8 \\ 4x^2 & - 3x & + 5 \\ \hline -3x^2 & + 3x & - 13 \end{array}$$

PROBLEM 3 Subtract: $2x^2 - 5x + 4$ from $5x^2 - 6$

ANSWER $3x^2 + 5x - 10$

Sometimes the horizontal arrangement is more convenient, and sometimes the vertical arrangement is more convenient. You should be able to work either way, letting the situation dictate the choice. In multiplying polynomials with several terms, a vertical arrangement is usually more convenient. The distributive property is the important principle behind multiplying polynomials, and it leads directly to the mechanical rule:

> To multiply two polynomials, multiply each term of one by each term of the other, and add like terms.

EXAMPLE 4 Multiply: $(x - 3)(x^2 - 2x + 3)$

SOLUTION

$$\begin{aligned}&(x - 3)(x^2 - 2x + 3)\\ &= x(x^2 - 2x + 3) - 3(x^2 - 2x + 3)\\ &= x^3 - 2x^2 + 3x - 3x^2 + 6x - 9\\ &= x^3 - 5x^2 + 9x - 9\end{aligned}$$

or

$$\begin{array}{rrrr} & x^2 & - 2x & + 3 \\ & x & - 3 & \\ \hline x^3 & - 2x^2 & + 3x & \\ & - 3x^2 & + 6x & - 9 \\ \hline x^3 & - 5x^2 & + 9x & - 9 \end{array}$$

PROBLEM 4 Multiply: $(2x - 4)(3x^2 - 2x + 1)$

ANSWER $6x^3 - 16x^2 + 10x - 4$

Exercise 29

A *Add:*

1. $3x - 5$ and $2x + 3$

2. $6x + 5$ and $3x - 8$

3. $2x + 3$, $-4x - 2$, and $7x - 4$

4. $7x - 5$, $-x + 3$, and $-8x - 2$

5. $2x^2 - 3x + 1$, $2x - 3$, and $4x^2 + 5$

6. $5x^2 + 2x - 7$, $2x^2 + 3$, and $-3x - 8$

Subtract:

7. $2x + 3$ from $5x + 7$

8. $5x + 2$ from $6x + 4$

9. $4x - 9$ from $2x + 3$

10. $3x - 8$ from $2x - 7$

11. $x^2 - 3x - 5$ from $2x^2 - 6x - 5$

12. $2y^2 - 6y + 1$ from $y^2 - 6y - 1$

Multiply:

13. $(2x - 3)(x + 2)$

14. $(3x - 5)(2x + 1)$

15. $(2x - 1)(x^2 - 3x + 5)$

16. $(3y + 2)(2y^2 + 5y - 3)$

17. $(x - 3y)(x^2 - 3xy + y^2)$

18. $(m + 2n)(m^2 - 4mn - n^2)$

B *Name the degree of each polynomial.*

19. $2x - 3$

20. $4x^2 - 2x + 3$

21. $3x^3 - x + 7$

22. $2x - y$

23. $x^2 - 3xy + y^2$

24. $x^3 - 2x^2y + xy^2 - 3y^3$

25. $2x^6 - 3x^5 + x^2 - x + 1$

26. $x^5 - 2x^2 + 5$

Add:

27. $3x^3 - 2x^2 + 5$, $3x^2 - x - 3$, and $2x + 4$.

28. $2x^4 - x^2 - 7$, $3x^3 + 7x^2 + 2x$, and $x^2 - 3x - 1$.

Subtract:

29. $3x^3 - 2x^2 - 5$ from $2x^3 - 3x + 2$

30. $5x^3 - 3x + 1$ from $2x^3 + x^2 - 1$

31. Subtract the sum of the last two polynomials from the sum of the first two: $2x^2 - 4xy + y^2$, $3xy - y^2$, $x^2 - 2xy - y^2$, and $-x^2 + 3xy - 2y^2$.

32. Subtract the sum of the first two polynomials from the sum of the last two: $3m^3 - 2m + 5$, $4m^2 - m$, $3m^2 - 3m - 2$, and $m^3 + m^2 + 2$.

Multiply:

33. $(a + b)(a^2 - ab + b^2)$

34. $(a - b)(a^2 + ab + b^2)$

35. $(x + 2y)^3$

36. $(2m - n)^3$

37. $(x^2 - 3x + 5)(2x^2 + x - 2)$

38. $(2m^2 + 2m - 1)(3m^2 - 2m + 1)$

39. $(2x^2 - 3xy + y^2)(x^2 + 2xy - y^2)$

40. $(a^2 - 2ab + b^2)(a^2 + 2ab + b^2)$

C *Simplify:*

41. $(3x - 1)(x + 2) - (2x - 3)^2$

42. $(2x + 3)(x - 5) - (3x - 1)^2$

43. $2(x - 2)^3 - (x - 2)^2 - 3(x - 2) - 4$

44. $(2x - 1)^3 - 2(2x - 1)^2 + 3(2x - 1) + 7$

4.3 Special Products of the Form (ax + b)(cx + d)

For reasons that will become clear shortly, it is essential that you learn to multiply first-degree factors of the type $(3x + 2)(2x - 1)$ and $(2x - y)(x + 3y)$ mentally. To discover relationships that will make this possible, let us first multiply $(3x + 2)$ and $(2x - 1)$ using a vertical arrangement.

$$\begin{array}{r} 3x + 2 \\ 2x - 1 \\ \hline 6x^2 + 4x \qquad \\ - 3x - 2 \\ \hline 6x^2 + x - 2 \end{array}$$

Notice that the two first-degree terms, $4x$ and $-3x$, combine into the single term x. Also notice that the product of the two first-degree polynomials is a second-degree polynomial. Both of these observations generalize for products of any first-degree polynomials of the same type.

A simple three-step process for carrying out this type of multiplication mentally is illustrated in the following example.

EXAMPLE 5 (*A*)

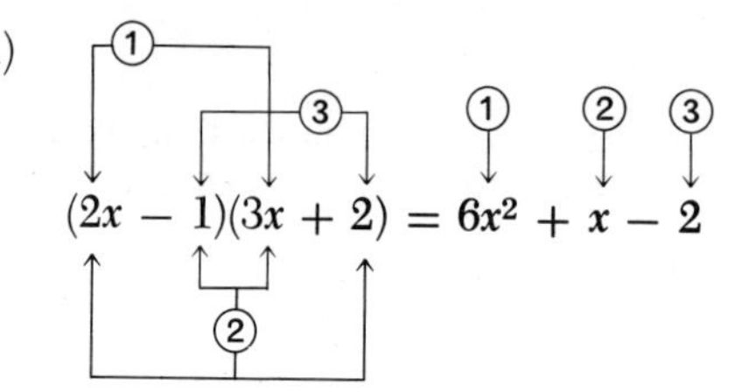

The like terms are picked up in step 2 and are combined in your head.

(*B*)

$$(2a - b)(a + 3b) = 2a^2 + 5ab - 3b^2$$

$$(2x - 3y)(2x + 3y) = 4x^2 - 9y^2$$

Notice that the middle term dropped out since its coefficient is 0.

PROBLEM 5 Multiply mentally:

(*A*) $(2x + 3)(x - 1)$ (*B*) $(a - 2b)(2a + 3b)$ (*C*) $(x - 2y)(x + 2y)$

ANSWER (*A*) $2x^2 + x - 3$ (*B*) $2a^2 - ab - 6b^2$ (*C*) $x^2 - 4y^2$

In the next section we will consider the reverse problem: given a second-degree polynomial, such as $2x^2 - 5x - 3$ or $3m^2 - 7mn + 2n^2$, find first-degree factors with integer coefficients that will produce these second-degree polynomials as products. To be able to factor second-degree polynomial forms with any degree of efficiency, it is important that you know how to mentally multiply first-degree factors of the types illustrated in this section quickly and accurately.

Exercise 30

Multiply mentally:

A

1. $(x + 1)(x + 2)$

2. $(y + 3)(y + 1)$

3. $(y + 3)(y + 4)$

4. $(x + 3)(x + 2)$

5. $(x - 5)(x - 4)$

6. $(m - 2)(m - 3)$

7. $(n - 4)(n - 3)$

8. $(u - 5)(u - 3)$

9. $(s + 7)(s - 2)$

10. $(t - 6)(t + 4)$

11. $(m - 12)(m + 5)$

12. $(a + 8)(a - 4)$

13. $(u - 3)(u + 3)$

14. $(t + 4)(t - 4)$

15. $(x + 8)(x - 8)$

16. $(m - 7)(m + 7)$

17. $(y + 7)(y + 9)$

18. $(x + 8)(x + 11)$

19. $(c - 9)(c - 6)$

20. $(u - 8)(u - 7)$

21. $(x - 12)(x + 4)$

22. $(y - 11)(y + 7)$

23. $(a + b)(a - b)$

24. $(m - n)(m + n)$

25. $(x + y)(x + 3y)$

26. $(m + 2n)(m + n)$

B

27. $(x + 2)(3x + 1)$

28. $(x + 3)(2x + 3)$

29. $(4t - 3)(t - 2)$

30. $(2x - 1)(x - 4)$

31. $(3y + 7)(y - 3)$

32. $(t + 4)(2t - 3)$

33. $(2x - 3y)(x + 2y)$

34. $(3x + 2y)(x - 3y)$

35. $(2x - 1)(3x + 2)$

36. $(3y - 2)(3y - 1)$

37. $(3y + 2)(3y - 2)$

38. $(2m - 7)(2m + 7)$

39. $(5s - 1)(s + 7)$

40. $(a - 6)(5a + 6)$

41. $(3m + 7n)(2m - 5n)$

42. $(6x - 4y)(5x + 3y)$

43. $(4n - 7)(3n + 2)$

44. $(5x - 6)(2x + 7)$

45. $(2x - 3y)(3x - 2y)$

46. $(2s - 3t)(3s - t)$

Since $(a + b)^2 = a^2 + 2ab + b^2$ *and* $(a - b)^2 = a^2 - 2ab + b^2$, *we can formulate a simple mechanical rule for squaring any binomial: The first and last terms in the product are the squares of the first and second terms respectively in the binomial being squared; the middle term in the product is twice the product of the two terms in the binomial. Use this rule to find the following squares.*

47. $(x + 3)^2$

48. $(x - 4)^2$

49. $(2x - 3)^2$

50. $(3x + 2)^2$

51. $(2x - 5y)^2$

52. $(3x + 4y)^2$

53. $(4a + 3b)^2$

54. $(5m - 3n)^2$

C *Figure out a pattern and use it to mentally multiply the following:*

55. $(x - 1)(x^2 + 2x - 1)$

56. $(x + 2)(x^2 - 3x + 2)$

57. $(2x - 3)(3x^2 - 2x + 4)$

58. $(3x + 2)(2x^2 - 3x - 2)$

59. $(x^2 - 3x + 2)(x^2 + x - 3)$

60. $(x^2 + 2x - 3)(x^2 - 3x - 2)$

4.4 Factoring Second-Degree Polynomials

It should now be very easy for you to obtain the products

$$(x - 3)(x + 2) = x^2 - x - 6$$

$$(x - 3y)(x + 2y) = x^2 - xy - 6y^2$$

but can you reverse the process? Can you, for example, find integers a, b, c, and d so that

$$2x^2 - 5x - 3 = (ax + b)(cx + d)$$

Representing a second-degree polynomial with integer coefficients as the product of two first-degree polynomials with integer coefficients is not as easy as multiplying first-degree polynomials. There is, however, a systematic approach to the problem that will enable you to find first-degree factors of special second-degree polynomials if they exist.

Let us start with a very simple polynomial whose factors you will likely guess right away:

$$x^2 + 6x + 8$$

Our problem is to find first-degree factors, if they exist, with integers as coefficients. To start, we write

$$x^2 + 6x + 8 = (\ x \qquad)(\ x \qquad)$$

leaving a space for the coefficients of x and the constant terms. First, we put in what we know for certain:

The coefficients of x are both 1.
The signs of both constant terms are +. (Why?)

Thus, we are now able to write:

$$x^2 + 6x + 8 = (x + \quad\quad)(x + \quad\quad)$$

What are the constant terms? They must be factors of $+8$. The only positive integer factors of $+8$ are

$(+1)(+8)$

$(+2)(+4)$

If we try the first pair, we do not get the middle term, $6x$, however, the second pair does give us the middle term. Thus,

$$x^2 + 6x + 8 = (x + 2)(x + 4)$$

Because of the commutative property of real numbers, we can reverse the factors if we wish.

Let us try another polynomial. Find factors with integers as coefficients for

$$2x^2 - 7x + 6$$

Again, write

$$2x^2 - 7x + 6 = (\quad x \quad\quad)(\quad x \quad\quad)$$

leaving space for the coefficients of x and the constant terms, and then insert what you know for certain:

The coefficient of one x is 2 and the other is 1.

The signs of both constant terms are $-$. (Why?)

Thus,

$$2x^2 - 7x + 6 = (2x - \quad\quad)(x - \quad\quad)$$

Now the constant terms both must be negative integer factors of 6. The possibilities are

$(-1)(-6)$

$(-6)(-1)$

$(-2)(-3)$

$(-3)(-2)$

We test each pair and find that the last pair works. Hence,

$$2x^2 - 7x + 6 = (2x - 3)(x - 2)$$

Before you get the impression that all second-degree polynomials with integers as coefficients have first-degree factors with integers as coefficients, consider the following simple polynomial:

$$x^2 + x + 2$$

Proceeding as above, we write

$$x^2 + x + 2 = (x + \quad)(x + \quad)$$

but find that no integer factors of 2 work. Hence, we conclude that $x^2 + x + 2$ cannot be factored in the integers; that is, $x^2 + x + 2$ does not have first-degree factors with integer coefficients.

EXAMPLE 6 Factor each polynomial in the integers.

(*A*) $x^2 + 7x + 12$

SOLUTION $x^2 + 7x + 12 = (x + \quad)(x + \quad)$

Positive factors of 12:

(1)(12)

(2)(6)

(3)(4)

The last pair works; thus

$$x^2 + 7x + 12 = (x + 3)(x + 4)$$

(*B*) $2x^2 + 3xy - 2y^2$

SOLUTION $2x^2 + 3xy - 2y^2 = (2x \quad y)(x \quad y)$

The signs of the coefficients of y must be opposite. (Why?) Possible factors of -2 are:

$(+1)(-2)$

$(-1)(+2)$

The second pair produces the middle term; thus

$$2x^2 + 3xy - 2y^2 = (2x - y)(x + 2y)$$

(*C*) $x^2 - 3x + 4$

SOLUTION $x^2 - 3x + 4 = (x - \quad)(x - \quad)$

Possible factors of 4:

$(-1)(-4)$

$(-2)(-2)$

Neither pair works; hence, $x^2 - 3x + 4$ is not factorable in the integers.

PROBLEM 6 Factor each polynomial in the integers. If it is not factorable, say so.

(A) $x^2 - 8x + 12$ (B) $x^2 + 2x + 5$ (C) $2x^2 + 7xy - 4y^2$

ANSWER (*A*) $(x - 2)(x - 6)$ (*B*) Not factorable in the integers
(*C*) $(2x - y)(x + 4y)$

In the second-degree polynomials

$$ax^2 + bx + c$$

$$ax^2 + bxy + cy^2$$

as a and c get larger with the possibility of more factors, the trial-and-error procedure outlined above gets more tedious. It would be helpful to know at the start if factors exist. It can be proved that these polynomials are factorable in the integers if and only if ac can be written as the product of two integers whose sum is b; that is, if and only if there exist integers m and n such that

$$ac = mn$$

and

$$m + n = b$$

We will call this *the ac test.* Use it if it helps you. Avoid it and come back to it later if it tends to confuse you.

EXAMPLE 7 Use the ac test to test for factorability and factor if possible:

(*A*) $2x^2 - 3x + 2$

SOLUTION $ac = (2)(2) = 4$

Possible integer factors of 4:

$(1)(4)$

$(-1)(-4)$

$(2)(2)$

$(-2)(-2)$

None of these pairs add up to $-3 = b$; hence, the polynomial is not factorable in the integers.

(*B*) $6x^2 + 5xy - 4y^2$

SOLUTION $ac = (6)(-4) = -24$

Out of the possible two-integer factors of -24 it is easy to see that

$$(+8)(-3) = -24$$

and

$$(+8) + (-3) = +5 = b$$

Hence, by the *ac* test, the polynomial is factorable, and with a little trial and error, we find

$$6x^2 + 5xy - 4y^2 = (2x - y)(3x + 4y)$$

PROBLEM 7 Use the *ac* test to test for factorability and factor if possible:

(A) $4x^2 - 3x + 2$ (B) $4x^2 - 4xy - 3y^2$

ANSWER (A) Not factorable in the integers (B) $(2x - 3y)(2x + y)$

In conclusion, we point out that if a, b, and c are selected at random out of the integers, the probability that

$$ax^2 + bx + c$$

is not factorable in the integers is much greater than the probability that it is. But even being able to factor some second-degree polynomials leads to marked simplification of some algebraic expressions and an easy way to solve some second-degree equations, as will be seen later.

Exercise 31

A *Factor in the integers, if possible. If not factorable, say so.*

1. $x^2 + 5x + 4$ **2.** $x^2 + 4x + 3$

3. $x^2 + 5x + 6$ **4.** $x^2 + 7x + 10$

5. $x^2 - 4x + 3$ **6.** $x^2 - 5x + 4$

7. $x^2 - 7x + 10$ **8.** $x^2 - 5x + 6$

9. $y^2 + 3y + 3$ **10.** $y^2 + 2y + 2$

11. $y^2 - 2y + 6$ **12.** $x^2 - 3x + 5$

13. $x^2 + 8xy + 15y^2$ **14.** $x^2 + 9xy + 20y^2$

15. $x^2 - 10xy + 21y^2$ **16.** $x^2 - 10xy + 16y^2$

17. $u^2 + 4uv + v^2$ **18.** $u^2 + 5uv + 3v^2$

19. $3x^2 + 7x + 2$ **20.** $2x^2 + 7x + 3$

21. $3x^2 - 7x + 4$ **22.** $2x^2 - 7x + 6$

B **23.** $3x^2 - 11xy + 6y^2$ **24.** $2x^2 - 7xy + 6y^2$

25. $n^2 - 2n - 8$ **26.** $n^2 + 2n - 8$

27. $x^2 + 4xy - 12y^2$ **28.** $x^2 - 4xy - 12y^2$

29. $x^2 + 4x - 6$

30. $x^2 - 3x - 8$

31. $3s^2 - 5s - 2$

32. $2s^2 + 5s - 3$

33. $12x^2 + 16x - 3$

34. $8x^2 + 6x - 9$

35. $6u^2 + 7uv - 3v^2$

36. $6u^2 - uv - 12v^2$

37. $3x^2 + 2xy - 3y^2$

38. $2x^2 - 3xy - 4y^2$

C

39. $12x^2 - 40xy - 7y^2$

40. $15x^2 + 17xy - 4y^2$

41. $12x^2 + 19xy - 10y^2$

42. $24x^2 - 31xy - 15y^2$

43. Find all integers p such that $x^2 + px + 12$ can be factored.

44. Find all positive integers p under 15 so that $x^2 - 7x + p$ can be factored.

4.5 More Factoring

COMMON MONOMIAL FACTORS

In the preceding chapter you had quite a bit of experience factoring out common monomial (single-term) factors. This kind of factoring is often associated with more general types of factoring. In fact, one should always take out common monomial factors first, if they exist, before using other methods.

EXAMPLE 8

(A) $4x^3 - 14x^2 + 6x = 2x(2x^2 - 7x + 3) = 2x(x - 3)(2x - 1)$

(B) $8x^3y + 20x^2y^2 - 12xy^3 = 4xy(2x^2 + 5xy - 3y^2) = 4xy(2x - y)(x + 3y)$

PROBLEM 8

Factor as far as possible in the integers:

(A) $3x^3 - 15x^2y + 18xy^2$

(B) $3x^3y + 3x^2y - 36xy$

ANSWER (A) $3x(x - 2y)(x - 3y)$ (B) $3xy(x - 3)(x + 4)$

DIFFERENCE OF TWO SQUARES

If we multiply $(A - B)$ and $(A + B)$ we obtain

$$(A - B)(A + B) = A^2 - B^2$$

a difference of two squares. Writing this result from right to left, we obtain the very useful factoring formula

$$A^2 - B^2 = (A - B)(A + B)$$

which finds far wider use than you might expect. The sum of two squares

$$A^2 + B^2$$

does not factor in the integers. Try it to see why. The difference-of-two-squares formula should be used any time you encounter a difference-of-two-squares form.

EXAMPLE 9

(A) $x^2 - y^2 = (x - y)(x + y)$

(B) $4x^2 - 9 = (2x)^2 - 3^2 = (2x - 3)(2x + 3)$

(C) $18x^3 - 8x = 2x(9x^2 - 4) = 2x(3x - 2)(3x + 2)$

(D) $x^2 + y^2$ is not factorable in the integers

PROBLEM 9

Factor as far as possible in the integers:

(A) $x^2 - 9$

(B) $4x^2 - 25y^2$

(C) $3x^3 - 48x$

(D) $4x^2 + 9$

ANSWER (A) $(x - 3)(x + 3)$ (B) $(2x - 5y)(2x + 5y)$ (C) $3x(x - 4)(x + 4)$ (D) not factorable

FACTORING BY GROUPING

Occasionally certain polynomial forms can be factored by appropriate grouping of terms. The following example illustrates the process:

EXAMPLE 10

(A)
$$\begin{aligned} x^2 + xy + 2x + 2y &= (x^2 + xy) + (2x + 2y) \\ &= x(x + y) + 2(x + y) \\ &= (x + y)(x + 2) \end{aligned}$$

(B)
$$\begin{aligned} x^2 - 2x - xy + 2y &= (x^2 - 2x) - (xy - 2y) \\ &= x(x - 2) - y(x - 2) \\ &= (x - 2)(x - y) \end{aligned}$$

PROBLEM 10

Factor by grouping terms:

(A) $x^2 - xy + 5x - 5y$

(B) $x^2 + 4x - xy - 4y$

ANSWER (A) $(x - y)(x + 5)$ (B) $(x + 4)(x - y)$

Exercise 32

A *Factor as far as possible in the integers.*

1. $6x^3 + 9x^2$

2. $8x^2 + 2x$

3. $u^4 + 6u^3 + 8u^2$

4. $m^5 + 8m^4 + 15m^3$

5. $x^3 - 5x^2 + 6x$

6. $x^3 - 7x^2 + 12x$

7. $x^2 - 4$

8. $x^2 - 1$

9. $4x^2 - 1$ **10.** $9x^2 - 4$

11. $u^2 + v^2$ **12.** $m^2 + 64$

13. $2x^2 - 8$ **14.** $3x^2 - 3$

B **15.** $9x^2 - 16y^2$ **16.** $25x^2 - 1$

17. $6u^2v^2 - 3uv^3$ **18.** $2x^3y - 6x^2y^3$

19. $4x^3y - xy^3$ **20.** $x^3y - 9xy^3$

21. $3x^4 + 27x^2$ **22.** $2x^3 + 8x$

23. $6x^2 + 36x + 48$ **24.** $4x^2 - 28x + 48$

25. $3x^3 - 6x^2 + 15x$ **26.** $2x^3 - 2x^2 + 8x$

27. $9u^2 + 4v^2$ **28.** $x^2 + 16y^2$

29. $12x^3 + 16x^2y - 16xy^2$ **30.** $9x^2y + 3xy^2 - 30y^3$

31. $x^2 + 3x + xy + 3y$ **32.** $xy + 2x + y^2 + 2y$

33. $x^2 - 3x - xy + 3y$ **34.** $x^2 - 5x + xy - 5y$

C **35.** $4x^3y + 14x^2y^2 + 6xy^3$ **36.** $3x^3y - 15x^2y^2 + 18xy^3$

37. $60x^2y^2 - 200xy^3 - 35y^4$ **38.** $60x^4 + 68x^3y - 16x^2y^2$

By noting that

$$(A - B)(A^2 + AB + B^2) = A^3 - B^3$$

$$(A + B)(A^2 - AB + B^2) = A^3 + B^3$$

we have factoring formulas for the sum and differences of two cubes. Use these formulas to factor the following polynomials.

39. $x^3 - 8$ **40.** $x^3 + 1$

41. $x^3 + 27$ **42.** $8y^3 - 1$

4.6 Reducing to Lowest Terms

If the numerator and denominator in a quotient of two polynomials contain a common factor, it may be "canceled out" according to the property of real numbers

$$\frac{ak}{bk} = \frac{a}{b}$$

The right-hand member is sometimes called the quotient of ak divided by bk, since

$$bk \cdot \frac{a}{b} = ak$$

ELIMINATION OF COMMON FACTORS FROM NUMERATORS AND DENOMINATORS

EXAMPLE 11

(A) $\dfrac{3x^3y}{9x^2y^2} = \dfrac{x}{3y}$

(B) $\dfrac{2x^2(x+3)}{6x(x+3)^2} = \dfrac{x}{3(x+3)}$

(C) $\dfrac{x^2y - xy^2}{x^2 - xy} = \dfrac{xy(x-y)}{x(x-y)} = y$

(D) $\dfrac{2x^2 + 5x - 3}{x^2 - 9} = \dfrac{(2x-1)(x+3)}{(x-3)(x+3)} = \dfrac{2x-1}{x-3}$

PROBLEM 11 Eliminate common factors from numerator and denominator:

(A) $\dfrac{2xy^3}{8x^2y}$ (B) $\dfrac{4x(x-2)^2}{12x^2(x-2)}$

(C) $\dfrac{x^2 - 3x}{x^2y - 3xy}$ (D) $\dfrac{2x^2 - 7x + 6}{x^2 - 4}$

ANSWER (A) $\dfrac{y^2}{4x}$ (B) $\dfrac{x-2}{3x}$ (C) $\dfrac{1}{y}$ (D) $\dfrac{2x-3}{x+2}$

Exercise 33

A *Eliminate common factors from numerator and denominator.*

1. $\dfrac{2x^2}{6x}$

2. $\dfrac{9y}{3y^3}$

3. $\dfrac{A}{A^2}$

4. $\dfrac{B^2}{B}$

5. $\dfrac{x+3}{(x+3)^2}$

6. $\dfrac{(y-1)^2}{y-1}$

7. $\dfrac{8(y-5)^2}{2(y-5)}$

8. $\dfrac{4(x-1)}{12(x-1)^3}$

9. $\dfrac{2x^2(x+7)}{6x(x+7)^3}$

10. $\dfrac{15y^3(x-9)^3}{5y^4(x-9)^2}$

11. $\dfrac{x^2 - 2x}{2x - 4}$

12. $\dfrac{2x^2 - 10}{4x - 20}$

13. $\dfrac{9y - 3y^2}{3y}$

14. $\dfrac{2x^2 - 4x}{2x}$

15. $\dfrac{m^2 - mn}{m^2n - mn^2}$

16. $\dfrac{a^2b + ab^2}{ab + b^2}$

17. $\dfrac{(2x-1)(2x+1)}{3x(2x+1)}$

18. $\dfrac{(x+3)(2x+5)}{2x^2(2x+5)}$

19. $\dfrac{(2x+1)(x-5)}{(3x-7)(x-5)}$

20. $\dfrac{(3x+2)(x+9)}{(2x-5)(x+9)}$

B 21. $\dfrac{x^2+5x+6}{2x^2+6x}$

22. $\dfrac{x^2+6x+8}{3x^2+12x}$

23. $\dfrac{x^2-9}{x^2+6x+9}$

24. $\dfrac{x^2-4}{x^2+4x+4}$

25. $\dfrac{x^2-4x+4}{x^2-5x+6}$

26. $\dfrac{x^2-6x+9}{x^2-5x+6}$

27. $\dfrac{2x^2+5x-3}{4x^2-1}$

28. $\dfrac{9x^2-4}{3x^2+7x-6}$

29. $\dfrac{9x^2-3x+6}{3}$

30. $\dfrac{2-6x-4x^2}{2}$

31. $\dfrac{10+5m-15m^2}{5m}$

32. $\dfrac{12t^2+4t-8}{4t}$

33. $\dfrac{4m^3n-2m^2n^2+6mn^3}{2mn}$

34. $\dfrac{6x^3y-12x^2y^2-9xy^3}{3xy}$

35. $\dfrac{x^2-x-6}{x-3}$

36. $\dfrac{x^2+2x-8}{x-2}$

37. $\dfrac{4x^2-9y^2}{4x^2y+6xy^2}$

38. $\dfrac{a^2-16b^2}{4ab-16b^2}$

39. $\dfrac{x^2-xy+2x-2y}{x^2-y^2}$

40. $\dfrac{u^2+uv-2u-2y}{u^2+2uv+v^2}$

41. $\dfrac{x^2y-8xy+15y}{xy-3y}$

42. $\dfrac{m^3+7m^2+10m}{m^2+5m}$

C 43. $\dfrac{6x^3+28x^2-10x}{12x^3-4x^2}$

44. $\dfrac{12x^3-78x^2-42x}{16x^4+8x^3}$

45. $\dfrac{x^3-8}{x^2-4}$

46. $\dfrac{y^3+27}{2y^3-6y^2+18y}$

4.7 Algebraic Long Division

There are times when it is useful to find quotients of polynomials by a long-division process similar to that used in arithmetic. Several examples will illustrate the process.

EXAMPLE 12 (*A*) Divide: $2x^2 + 5x - 12$ by $x + 4$

SOLUTION

$$x + 4\overline{)2x^2 + 5x - 12}$$

COMMENTS: Both polynomials are arranged in descending powers of the variable if this is not already done.

$$\begin{array}{r} 2x \\ x + 4\overline{)2x^2 + 5x - 12} \end{array}$$

Divide the first term of the divisor into the first term of the dividend, i.e., what must x be multiplied by so that the product is exactly $2x^2$?

$$\begin{array}{r} 2x \\ x + 4\overline{)2x^2 + 5x - 12} \\ \underline{2x^2 + 8x} \\ -3x - 12 \end{array}$$

Multiply the divisor by $2x$, line up like terms, subtract as in arithmetic, and bring down -12.

$$\begin{array}{r} 2x - 3 \\ x + 4\overline{)2x^2 + 5x - 12} \\ \underline{2x^2 + 8x} \\ -3x - 12 \\ \underline{-3x - 12} \\ 0 \end{array}$$

Repeat the process above until the degree of the remainder is less than that of the divisor.

CHECK: $(x + 4)(2x - 3) = 2x^2 + 5x - 12.$

(*B*) Divide: $x^3 + 8$ by $x + 2$

SOLUTION

$$\begin{array}{r} x^2 - 2x + 4 \\ x + 2\overline{)x^3 + 0x^2 + 0x + 8} \\ \underline{x^3 + 2x^2} \\ -2x^2 + 0x \\ \underline{-2x^2 - 4x} \\ 4x + 8 \\ \underline{4x + 8} \\ 0 \end{array}$$

COMMENT: Insert, with 0 coefficients, any missing terms of lower degree than 3, and proceed as in part *A*.

Can you check this problem?

(*C*) Divide: $3 - 7x + 6x^2$ by $3x + 1$

SOLUTION

$$\begin{array}{r} 2x - 3 \\ 3x + 1 \overline{\big) 6x^2 - 7x + 3} \\ 6x^2 + 2x \\ \hline -9x + 3 \\ -9x - 3 \\ \hline 6 \end{array} = R \quad \text{(remainder)}$$

COMMENT

Arrange $3 - 7x + 6x^2$ in descending powers of x, then proceed as above until the degree of the remainder is less than the degree of the divisor.

CHECK: Just as in arithmetic, when there is a remainder we check by adding the remainder to the product of the divisor and quotient. Thus

$$(3x + 1)(2x - 3) + 6 \stackrel{?}{=} 6x^2 - 7x + 3$$

$$6x^2 - 7x - 3 + 6 \stackrel{?}{=} 6x^2 - 7x + 3$$

$$6x^2 - 7x + 3 \stackrel{\checkmark}{=} 6x^2 - 7x + 3$$

PROBLEM 12 Divide, using the long-division process, and check:

(A) $(2x^2 + 7x + 3)/(x + 3)$ (B) $(x^3 - 8)/(x - 2)$
(C) $(2 - x + 6x^2)/(3x - 2)$

ANSWER (A) $2x + 1$ (B) $x^2 + 2x + 4$ (C) $2x + 1$, $R = 4$

Exercise 34

A *Divide, using the long-division process. Check the answers.*

1. $(x^2 + 5x + 6)/(x + 3)$
2. $(x^2 + 6x + 8)/(x + 4)$
3. $(2x^2 + x - 6)/(x + 2)$
4. $(3x^2 - 5x - 2)/(x - 2)$
5. $(2x^2 - 3x - 4)/(x - 3)$
6. $(3x^2 - 11x - 1)/(x - 4)$
7. $(2m^2 + m - 10)/(2m + 5)$
8. $(3y^2 + 5y - 12)/(3y - 4)$
9. $(6x^2 + 5x - 6)/(3x - 2)$
10. $(8x^2 - 14x + 3)/(2x - 3)$
11. $(6x^2 + 11x - 12)/(3x - 2)$
12. $(6x^2 + x - 13)/(2x + 3)$
13. $(3x^2 + 13x - 12)/(3x - 2)$
14. $(2x^2 - 7x - 1)/(2x + 1)$

B

15. $(x^2 - 4)/(x - 2)$
16. $(y^2 - 9)/(y + 3)$
17. $(m^2 - 7)/(m - 3)$
18. $(u^2 - 18)/(u + 4)$
19. $(8c + 4 + 5c^2)/(c + 2)$
20. $(4a^2 - 22 - 7a)/(a - 3)$
21. $(9x^2 - 8)/(3x - 2)$
22. $(8x^2 + 7)/(2x - 3)$

23. $(5y^2 - y + 2y^3 - 6)/(y + 2)$

24. $(x - 5x^2 + 10 + x^3)/(x + 2)$

25. $(x^3 - 1)/(x - 1)$

26. $(x^3 + 27)/(x + 3)$

27. $(x^4 - 16)/(x + 2)$

28. $(x^5 + 32)/(x - 2)$

29. $(3y - y^2 + 2y^3 - 1)/(y + 2)$

30. $(3 + x^3 - x)/(x - 3)$

C **31.** $(4x^4 - 10x - 9x^2 - 10)/(2x + 3)$

32. $(9x^4 - 2 - 6x - x^2)/(3x - 1)$

33. $(16x - 5x^3 - 4 + 6x^4 - 8x^2)/(2x - 4 + 3x^2)$

34. $(8x^2 - 7 - 13x + 24x^4)/(3x + 5 + 6x^2)$

Exercise 35 Chapter Review

A **1.** Add: $2x^2 + x - 3$, $2x - 3$, and $3x^2 + 2$

2. Subtract: $3x^2 - x - 2$ from $5x^2 - 2x + 5$

3. Multiply: $(3x - 2)(2x + 5)$

Factor in the integers, if possible.

4. $x^2 - 9x + 14$

5. $3x^2 - 10x + 8$

6. $4x^2y - 6xy^2$

7. $x^3 - 5x^2 + 6x$

8. $3u^2 - 12$

9. Eliminate common factors from numerator and denominator:

$$\frac{x^3 + 4x^2}{3x + 12}$$

10. Divide, using long division: $(6x^2 + 5x - 2)/(2x - 1)$

B **11.** Multiply: $(9x^2 - 4)(3x^2 + 7x - 6)$

12. Subtract $2x^2 - 5x - 6$ from the product $(2x - 1)(2x + 1)$.

13. Given the polynomial $3x^5 - 2x^3 + 7x^2 - x + 2$, (A) what is its degree? (B) what is the degree of the second term?

Factor in the integers, if possible.

14. $x^2 - 2xy - 5y^2$

15. $2x^2 - xy - 3y^2$

16. $6y^3 + 3y^2 - 45y$

17. $12x^3y + 27xy^3$

18. $x^2 - xy + 4x - 4y$

19. Eliminate common factors from numerator and denominator:

$$\frac{9x^2 - 24x + 16}{9x^2 - 16}$$

20. Divide, using long division. $(2 - 10x + 9x^3)/(3x - 2)$

C **21.** Simplify: $[(3x^2 - x + 1) - (x^2 - 4)] - [(2x - 5)(x + 3)]$

22. Simplify: $[-2xy(x^2 - 4y^2)] - [-2xy(x - 2y)(x + 2y)]$

23. Factor in the integers, if possible: $15x^2 + 28xy - 32y^2$

24. Find all integers p such that $x^2 + px + 8$ can be factored.

Factor as far as possible in the integers.

25. $36x^3y + 24x^2y^2 - 45xy^3$

26. $12u^4 - 12u^3v - 20u^3v^2$

27. $8x^3 + 1$

28. Eliminate common factors from numerator and denominator:

$$\frac{2x^3 - 6x^2 - 4x^2y + 12xy}{4x^4 - 16x^2y^2}$$

29. Divide, using long division: $(5x - 5 + 8x^4)/(x + 2x^2 - 1)$

CHAPTER 5
First-Degree Equations and Inequalities in One Variable

5.1 First-Degree Equations in One Variable

In the fairly large variety of applications we have already considered you might have noticed that all of them gave rise to essentially the same type of equation. In fact, if all of the terms in any one of these equations had been transferred to the left of the equal sign (leaving 0 on the right) and like terms combined, the equation would have been of the form

$$ax + b = 0$$

a *first-degree equation in one variable.*

We, of course, have already solved many first-degree equations in one variable so we will not go into the matter again here other than to say that any equation of the form $ax + b = 0$ with $a \neq 0$ always has a unique solution, namely $-b/a$. Showing that $-b/a$ is a solution is left as an exercise. To show that this solution is unique, we proceed as follows: assume p and q are both solutions of $ax + b = 0$, then

$$ap + b = aq + b \qquad \text{Why?}$$
$$ap = aq \qquad \text{Why?}$$
$$p = q \qquad \text{Why?}$$

In general, it is important to know under what conditions an equation has a solution and how many solutions are possible. We have now answered both of these questions for equations of the type $ax + b = 0$ with $a \neq 0$. Other types of equations will be studied later which have more than one solution; for example, -2 and $+2$ are both solutions of

$$x^2 - 4 = 0$$

Rather than considering all types of equations, hit or miss, it is far more efficient to group equations by type and study each type in detail. We have spent time on first-degree equations as a group. Higher-degree equations will be considered as we progress.

5.2 Applications

This section contains a wide variety of applications that lead to first-degree equations. The problems are grouped by subject area, and, in addition, many are cross referenced below by problem types for added convenience.

Rate-time problems: 1, 6, 12, 28, 29, 39, 50, 51.
Mixture problems: 7, 8, 10.
Percent problems: 3, 4, 5, 10, 26.
Ratio and proportion problems: 8, 9, 15, 16, 23, 24, 31, 33, 34, 35, 36, 37.

Since many applications of first-degree polynomial equations have already been presented, this section may be treated briefly or returned to if time is of concern. In any case, every reader should spend some time scanning the problems to gain an appreciation of the number and variety of applications he now has within his command. Any time that is spent in this section will be amply rewarded. Algebra will be of limited value to you if you do not gain experience in applying it to a variety of different types of significant problems.

Exercise 36

The problems in this exercise are grouped in several subject areas: business, chemistry, earth sciences, economics, geometry, domestic, life science, music, physics-engineering, psychology, and puzzles.

No attempt has been made to arrange the problems in order of difficulty or to match them in pairs as in preceding exercises.

BUSINESS

1. A research chemist charges \$20 per hr for his services and \$8 per hr for his assistant. On a given job a customer received a bill for \$1,040. If the chemist worked 10 hr more on the job than his assistant, how much time did each spend?

2. It costs a book publisher \$9,000 to prepare a book for publishing (art work, plates, reviews, etc.); printing costs are \$2 per book. If the book is sold to bookstores for \$5 a copy, how many copies must be sold for the publisher to break even?

3. A variety store sells a particular item costing \$4 for \$7. If this markup represents the store's pricing policy for all items, what percent markup (on cost) is used? HINT: cost + markup = retail.

4. A man borrowed a sum of money from a bank at 6 percent simple interest. At the end of $2\frac{1}{2}$ years he repaid the bank \$575. How much did he borrow from the bank? HINT: $A = P + Prt$.

5. A man has $\frac{1}{2}$ of his money invested at 6 percent, $\frac{1}{4}$ at 5 percent, and the rest at 3 percent. If his total annual income from all investments is \$4,000, how much money has he used for investing purposes?

6. In a soft-drink bottling plant one machine can fill and cap 20 bottles per min; another newer machine can do 30 per min. What will be the total time required to complete a 30,000-bottle order if the older machine is brought on the job 3 hr earlier than the newer machine, and then both machines are used together until the job is finished?

7. A grocer wishes to blend 84-cents-per-lb coffee with 24 lb of 60-cents-per-lb coffee so that the resulting mixture will sell for 68 cents per lb. How much 84-cents-per-pound coffee is needed?

CHEMISTRY

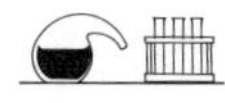

8. If there are 8 grams of sulfuric acid in 70 grams of solution, how many grams of sulfuric acid are there in 21 grams of the same solution?

9. A chemist wishing to produce hydrogen gas may pour hydrochloric acid (HCl) over zinc (Zn). If so, a chemical reaction takes place so as to produce two atoms of hydrogen for each atom of zinc used. Thus

$$2HCl + Zn \longrightarrow H_2 + ZnCl_2$$

To find how much hydrogen gas will be produced from, say, 100 grams of zinc, he will use the atomic weights of hydrogen (1.008) and zinc (65.38) and solve the proportion

$$\frac{x}{100} = \frac{(2)(1.008)}{65.38}$$

(A) Solve this proportion to two decimal places.
(B) Find the amount (to two decimal places) of zinc needed to produce 10 grams of hydrogen gas.

10. How many gallons of pure alcohol must be added to 3 gal of a 20 percent solution to get a 40 percent solution?

11. In the study of gases there is a simple law called Boyle's law that expresses a relationship between volume and pressure. It states that the product of the pressure and volume, as these quantities change and all other variables are held fixed, remains constant. Stated as a formula, $P_1V_1 = P_2V_2$. If 500 cc of air at 70-cm pressure were converted to 100-cm pressure, what volume would it have?

EARTH SCIENCE

12. An earthquake emits a primary wave and a secondary wave. Near the surface of the earth the primary wave travels at about 5 miles per sec, and the secondary wave at about 3 miles per sec. From the time lag between the two waves arriving at a given seismic station, it is possible to estimate the distance to the quake. (The "epicenter" can be located by getting distance bearings at three or more stations.) Suppose a station measured a time difference of 12 sec between the arrival of the two waves. How far would the earthquake be from the station?

13. As dry air moves upward, it expands and in so doing cools at the rate of about 5.5°F for each 1,000 ft in rise. This ascent is known as the "adiabatic process." If the ground temperature is 80°F, write an equation that relates temperature T with altitude h (in thousands of feet). How high is an airplane if the pilot observes that the temperature is 25°F?

14. Pressure in sea water increases by 1 atmosphere (15 lb per sq ft) for each 33 ft of depth; it is 15 lb per sq ft at the surface. Thus, $p = 15 + 15(d/33)$ where p is the pressure in pounds per square foot at a depth of d ft below the surface. How deep is a diver if he observes that the pressure is 165 lb per sq ft?

15. In a science class a scale model of the earth-sun part of the solar system is to be constructed. (A fairly large field, such as a football field or a baseball diamond is recommended.) The diameter of the sun is approximately 866,000 miles, the diameter of the earth is approximately 8,000 miles, and the distance between the sun and earth is approximately 93 million miles. If we use a 12-in.-diameter ball for the sun, what diameter sphere should we use for the earth, and how many feet should it be placed away from the ball? Express the answers to one decimal place.

ECONOMICS

16. *(A)* What would a net monthly salary have to be in 1940 to have the same purchasing power of a net monthly salary of $550 in 1965? HINT: Use an appropriate proportion. *(B)* Answer the same question for 1950 and 1960.

YEAR	COST-OF-LIVING INDEX
1940	48
1945	62
1950	82
1955	93
1960	103
1965	110

17. Gross national product (GNP) and net national product (NNP) are both measures of national income and are related as follows:

$$\text{GNP} = \text{NNP} + \text{depreciation}$$

A good estimate for depreciation is $\frac{1}{11}$ of GNP or $\frac{1}{10}$ of NNP.
(A) Write a formula for GNP in terms of NNP only.
(B) Use part *A* to replace the question marks in the table.

DATE	NNP (BILLIONS OF DOLLARS)	GNP (BILLIONS OF DOLLARS)
1929	96	?
1933	48	?

18. Henry Schultz, an economist, formulated a price-demand equation for sugar in the United States as follows: $q = 70.62 - 2.26p$, where p = wholesale price in cents of 1 lb of sugar, and q = per capita consumption in pounds of sugar in the United States in any year. At what price per pound would the per capita consumption per year be 25.42 lb?

GEOMETRY

19. The perimeter of a tennis court for singles is 210 ft. Find the dimensions of the court if its length is 3 ft less than 3 times its width.

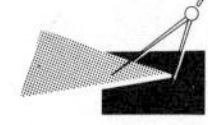

20. Find the dimensions of a rectangle with perimeter 72 in. if its length is 25 percent longer than its width.

21. What is the size of each angle of a triangle if the smallest is $\frac{1}{3}$ the size of the third, and the second is $\frac{2}{3}$ the size of the third?

22. A water reed sticks 1 ft out of the water. If it were pulled over to the side until the top just reached the surface, it would be at a point 3 ft from where it originally protruded. How deep is the water? HINT: Recall the Pythagorean theorem in Sec. 3.7.

DOMESTIC

23. If $1\frac{1}{2}$ cups of milk and 2 cups of flour are needed in a recipe for four people, how much of each is needed in a recipe for six people?

24. If on a trip your car goes 391 miles on 23 gal of gas, how many gallons of gas would be required for a 850-mile trip?

25. A father, in order to encourage his daughter to do better in algebra, agrees to pay her 25 cents for each problem she gets right on a test and to fine her 10 cents for each problem she misses. On a test with 20 problems she received \$3.60. How many problems did she get right?

26. A friend of yours paid \$72 for a pair of skis after receiving a discount of 20 percent. What was the list price for the skis?

27. A car rental company charges \$5 per day and 5 cents per mile. If a car was rented for two days, how far was it driven if the total rental bill was \$30?

28. A contractor has just finished constructing a swimming pool in your back yard, and you have turned on the large water valve to fill it. The large valve lets water into the pool at a rate of 60 gal per min. After 2 hr you get impatient and turn on the garden hose which lets water in at 15 gal per min. If the swimming pool holds 30,000 gal of water, what will be the total time required to fill it?

29. The cruising speed of an airplane is 150 mph (relative to ground). You wish to hire the plane for a 3-hr sightseeing trip. You instruct the pilot to fly north as far as he can and still return to the airport at the end of the allotted time.
(*A*) How far north should the pilot fly if there is a 30-mph wind blowing from the north?
(*B*) How far north should the pilot fly if there is no wind blowing?

LIFE SCIENCE

30. A fairly good approximation for the normal weight of a person over 60 in. (5 ft) tall is given by the formula $w = 5.5h - 220$, where h is height in inches and w is weight in pounds. How tall should a 121-lb person be?

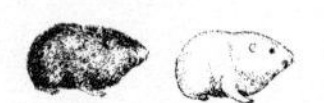

31. The naturalists in Yosemite National Park decided to estimate the number of bears in the most popular part of the park, Yosemite Valley. They used live traps (Fig. 1) to capture 50 bears; these bears were marked and released. A week later 50 more were captured, and it was observed that 10 of these were marked. Use this information to estimate the total number of bears in the valley.

Figure 1

32. In biology there is an approximate rule, called the bioclimatic rule for temperate climates, that states that in spring and early summer, periodic phenomena, such as blossoming for a given species, appearance of certain insects, and ripening of fruit, usually come about 4 days later for each 500 ft of altitude increase or 1° latitude increase from any given base. In terms of formulas we have

$$d = 4\left(\frac{h}{500}\right) \qquad \text{and} \qquad d = 4L$$

where d = change in days, h = change in altitude in feet, and L = change in latitude in degrees.

What change in altitude would delay pear trees from blossoming for 16 days? What change in latitude would accomplish the same thing?

33. Gregor Mendel (1822), a Bavarian monk and biologist whose name is known to almost everyone today, made discoveries which revolutionized the science of heredity. Out of many experiments in which he crossed peas of one characteristic with those of another, Mendel evolved his now famous laws of heredity. In one experiment he crossed dihybrid yellow round peas (which contained green and wrinkled as recessive genes) and obtained 560 peas of the following types: 319 yellow round, 101 yellow wrinkled, 108 green round, and 32 green wrinkled. From his laws of heredity he predicted the ratio 9:3:3:1. Using the ratio, calculate the theoretical expected number of each type of pea from this cross, and compare it with the experimental results.

MUSIC

34. Starting with a string tuned to a given note, one can move up and down the scale simply by decreasing or increasing its length (while maintaining the same tension) according to simple whole-number ratios (see Fig. 2). For example, $\frac{8}{9}$ of the C string gives the next higher note D, $\frac{2}{3}$ of the C string gives G, and $\frac{1}{2}$ of the C string gives C one octave higher. (The reciprocals of these fractions, $\frac{9}{8}$, $\frac{3}{2}$, and 2, respectively, are proportional to the fre-

	C	D	E	F	G	A	B	C	D	E	F	G	A	B	C
Relative string length	2	$\frac{16}{9}$	$\frac{8}{5}$	$\frac{3}{2}$	$\frac{4}{3}$	$\frac{6}{5}$	$\frac{16}{15}$	1	$\frac{8}{9}$	$\frac{4}{5}$	$\frac{3}{4}$	$\frac{2}{3}$	$\frac{3}{5}$	$\frac{8}{15}$	$\frac{1}{2}$
Scale ratios (proportional to frequencies)	$\frac{1}{2}$	$\frac{9}{16}$	$\frac{5}{8}$	$\frac{2}{3}$	$\frac{3}{4}$	$\frac{5}{6}$	$\frac{15}{16}$	1	$\frac{9}{8}$	$\frac{5}{4}$	$\frac{4}{3}$	$\frac{3}{2}$	$\frac{5}{3}$	$\frac{15}{8}$	2
Frequencies	132	149	165	176	198	220	248	264	297	330	352	396	440	495	528

Figure 2 Diatonic scale.

quencies of these notes.) Find the lengths of 7 strings (each less than 30 in.) that will produce the following seven chords when paired with a 30-in. string:

(*A*)	Octave	1:2	(*B*)	Fifth	2:3
(*C*)	Fourth	3:4	(*D*)	Major third	4:5
(*E*)	Minor third	5:6	(*F*)	Major sixth	3:5
(*G*)	Minor sixth	5:8			

35. The three minor chords are composed of notes whose frequencies are in the ratio 10:12:15. If the first note of the minor chord is A with a frequency of 220, what are the frequencies of the other two notes? Compute, then compare results with figure.

PHYSICS—ENGINEERING

36. An engineer knows that a $3\frac{1}{2}$-ft piece of steel rod weighs 25 lb. How much would a 14-ft piece of this rod weigh?

37. The magnifying power M of a telescope is the ratio of the focal length F of the objective lens to the focal length F of the eyepiece. If the focal length of the objective lens is 36 in. what must the focal length of the eyepiece be to produce a magnification of 108?

38. If a small solid object is thrown downward with an initial velocity of 50 fps, its velocity after time t is given approximately by

$$v = 50 + 32t$$

How many seconds are required for the object to attain a velocity of 306 fps?

39. In 1849, during a celebrated experiment, the Frenchman Fizeau made the first accurate approximation of the speed of light. By using a rotating disc with notches equally spaced on the circumference and a reflecting mirror 5 miles away (Fig. 3), he was able to measure the elapsed time for the light traveling to the mirror and back. Calculate his estimate for the speed of light (in miles per second) if his measurement for the elapsed time was $\frac{1}{20,000}$ sec?

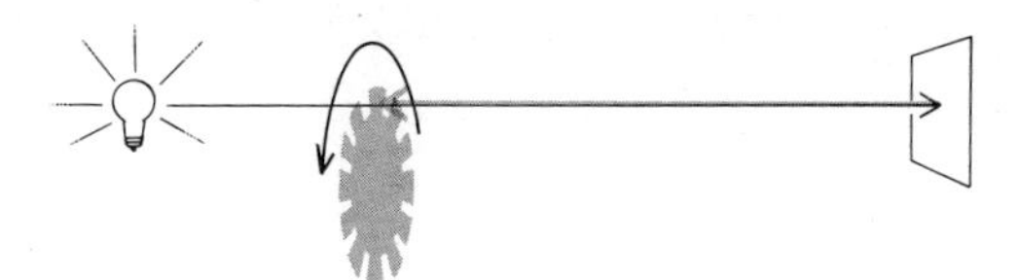

Figure 3

40. A type of physics problem with wide applications is the *lever problem.* For a lever, relative to a fulcrum, to be in static equilibrium (balanced) the sum of the downward forces times their respective distances on one side of the fulcrum must equal the sum of the downward forces times their respective distances on the other side of the fulcrum (Fig. 4).

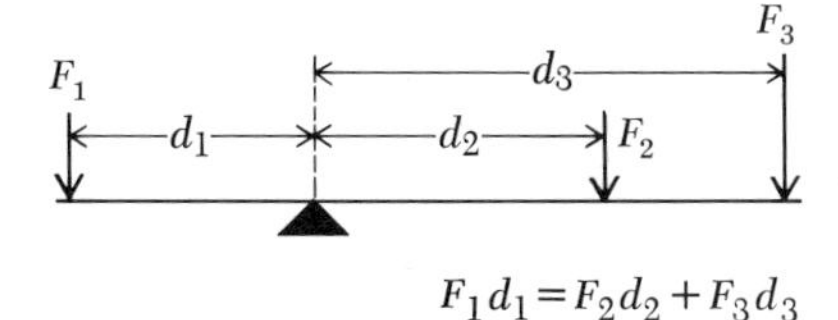

$$F_1 d_1 = F_2 d_2 + F_3 d_3$$

Figure 4

If a person has a 3-ft wrecking bar and places a fulcrum 3 in. from one end, how much can he lift if he applies a force of 50 lb to the long end?

41. Where would a fulcrum have to be placed to balance an 8-ft bar with a 6-lb weight on one end and a 7-lb weight on the other?

42. Two men decided to move a 1,920-lb rock by use of a 9-ft steel bar (Fig. 5). If they place the fulcrum 1 ft from the rock and one of the men applies a force of 150 lb on the other end, how much force will the second man have to apply 2 ft from that end to lift the rock?

Figure 5

43. If two pulleys are fastened together as in the diagram (Fig. 6), and the radius of the larger pulley is 10 in. and the smaller one 2 in., how heavy a weight can one lift by exerting an 80-lb pull on the free rope?

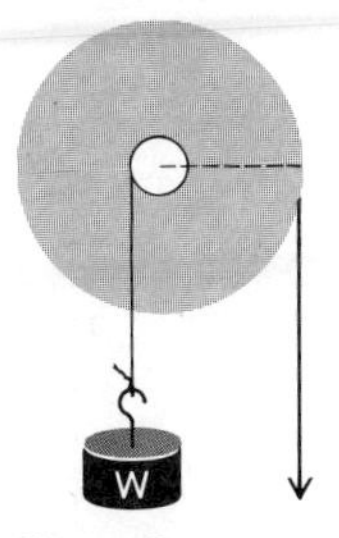

Figure 6

44. In engineering and physics it is often necessary to measure the work done by a machine or a force. Work done is defined to be the product of the force f and the distance d that the object is moved; that is,

$$w = fd$$

If you lift 50 lb a distance of 3 ft you will have done 150 ft-lb of work.
(A) How much work is done by a horse lifting 112 lb a distance of 196 ft?
(B) If in a vertical mine shaft an engine does 1,600,000 ft-lb of work in lifting 1 ton of ore, how far is the ore lifted? (1 ton = 2,000 lb.)

45. In a simple electric circuit, such as that found in a flashlight, the voltage provided by the batteries is related to the resistance and current in the circuit by Ohm's law,

$$E = IR$$

where E = electromotive force, volts
I = current, amperes
R = resistance, ohms

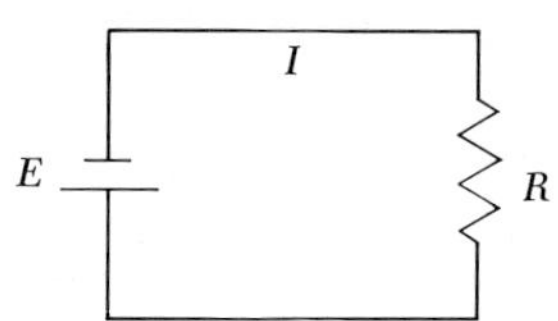

(A) If a two-cell battery puts out 3 volts and a current of 0.2 ampere flows through the circuit, what is the total resistance in the circuit?
(B) How much current will flow through a five-cell flashlight circuit (putting out 1.5 volts per cell), if the total resistance in the circuit is 25 ohms?

PSYCHOLOGY

46. Psychologists define IQ (Intelligence Quotient) as

$$\text{IQ} = \frac{\text{mental age}}{\text{chronological age}} \times 100$$

$$= \frac{\text{MA}}{\text{CA}} \times 100$$

If a person has an IQ of 120 and a mental age of 18, what is his chronological age (actual age)?

47. In 1948 Professor Brown, a psychologist, trained a group of rats (in an experiment on motivation) to run down a narrow passage in a cage to receive food in a goal box. He then put a harness on each rat and connected it to an overhead wire that was attached to a scale (Fig. 7). In this way he could place the rat at different distances from the food and measure the pull (in grams) of the rat toward the food. He found that a relation between motivation (pull) and position was given approximately by the equation

$$p = -\tfrac{1}{5}d + 70 \qquad 30 \le d \le 175$$

where pull p is measured in grams and distance d is measured in centimeters. If the pull registered was 40 grams, how far was the rat from the goal box?

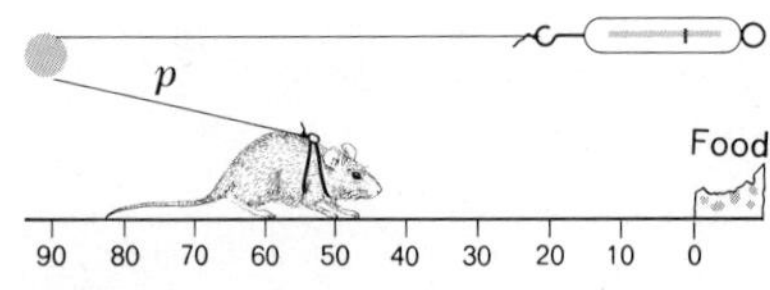

Figure 7

48. Professor Brown performed the same kind of experiment as described in the preceding problem except that he replaced the food in the goal box with a mild electric shock. With the same kind of apparatus, he was able to measure the avoidance strength relative to the distance from the object to be avoided. He found that the avoidance strength a (measured in grams) was related to the distance d that the rat was from the shock (measured in centimeters) approximately by the equation

$$a = -\tfrac{4}{3}d + 230 \qquad 30 \leq d \leq 175$$

If the same rat was trained as described in this and the last problem, at what distance (to one decimal place) from the goal box would the approach and avoidance strength be the same? (What do you think that the rat would do at this point?)

PUZZLES

49. A friend of yours came out of a post office having spent $1.32 on thirty 4-cent and 5-cent stamps. How many of each type did he buy?

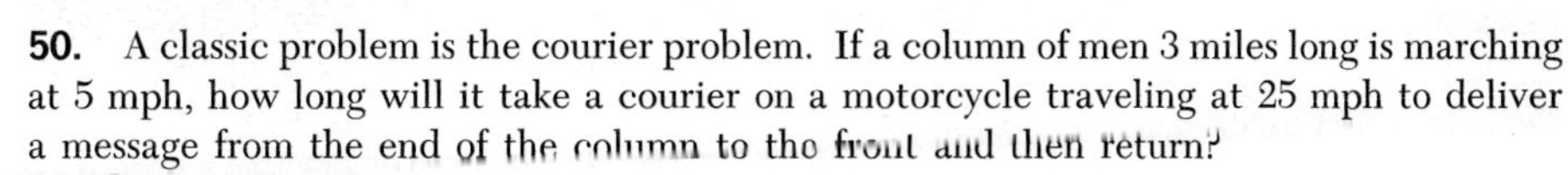

50. A classic problem is the courier problem. If a column of men 3 miles long is marching at 5 mph, how long will it take a courier on a motorcycle traveling at 25 mph to deliver a message from the end of the column to the front and then return?

51. After 12:00 noon exactly, what time will the hands of a clock be together again?

5.3 First-Degree Inequalities in One Variable

In the preceding chapters we worked with simple inequalities with obvious solutions; for example, $x < 5$, $-3 \leq t < 4$, and $m \geq -7$. In this section we will consider inequalities that do not have obvious solutions. Can you guess the real number solutions for

$$2x - 3 < 4x + 5$$

By the end of this section you will be able to solve this type of inequality almost as easily as you solved equations of the same type.

The familiar ideas and procedures which apply to equations also apply to inequalities, but with certain important modifications. As with equations, we are interested in performing operations on inequalities that will produce simpler *equivalent inequalities* (inequalities that have the same solution set), leading eventually to an inequality with an obvious solution. The operations in Theorem 1 produce equivalent inequalities.

THEOREM 1 If a, b, and c are real numbers and $a > b$, then

(A) $a + c > b + c$

(B) $a - c > b - c$

(C) $ca > cb$ if c is positive

(*D*) $ca < cb$ if c is negative

(*E*) $\dfrac{a}{c} > \dfrac{b}{c}$ if c is positive

(*F*) $\dfrac{a}{c} < \dfrac{b}{c}$ if c is negative

(*G*) $b < a$

The theorem also holds if $>$ is replaced with $\geq$ and $<$ is replaced with $\leq$.

To get a better understanding of the theorem, let us use it to solve a few inequalities. (Proofs of parts of the theorem will be found in the exercises.)

EXAMPLE 1 (*A*) Solve and graph: $x + 3 > -2$

SOLUTION

$$x + 3 > -2$$
$$x + 3 - 3 > -2 - 3$$
$$x > -5$$

(*B*) Solve and graph: $-3x > 12$

SOLUTION

$$-3x > 12$$
$$\frac{-3x}{-3} < \frac{12}{-3}$$
$$x < -4$$

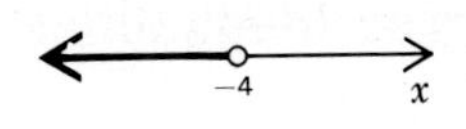

(*C*) Solve and graph: $2x - 3 \leq 4x + 5$

SOLUTION

$$2x - 3 \leq 4x + 5$$
$$2x - 3 + 3 \leq 4x + 5 + 3$$
$$2x \leq 4x + 8$$
$$2x - 4x \leq 4x + 8 - 4x$$
$$-2x \leq 8$$
$$\frac{-2x}{-2} \geq \frac{8}{-2}$$
$$x \geq -4$$

(*D*) Solve and graph:

$$\frac{3x - 2}{2} - 5 > 1 - \frac{x}{4}$$

SOLUTION

$$4\left(\frac{3x - 2}{2} - 5\right) > 4\left(1 - \frac{x}{4}\right)$$

$$2(3x - 2) - 20 > 4 - x$$
$$6x - 4 - 20 > 4 - x$$
$$7x > 28$$
$$x > 4$$

PROBLEM 1 Solve and graph:

(A) $x - 2 > -4$ (B) $-2x > 10$

(C) $3x + 2 \leq 5x - 6$ (D) $\dfrac{2x - 3}{3} - 2 > \dfrac{x}{6} - 1$

ANSWER (A) $x > -2$ (B) $x < -5$

(C) $x \geq 4$ (D) $x > 4$

EXAMPLE 2 Solve and graph: $-8 \leq 3x - 5 < 7$

SOLUTION We proceed as above, except we try to isolate x in the middle:

$$-8 \leq 3x - 5 < 7$$
$$-8 + 5 \leq 3x - 5 + 5 < 7 + 5$$
$$-3 \leq 3x < 12$$
$$\frac{-3}{3} \leq \frac{3x}{3} < \frac{12}{3}$$
$$-1 \leq x < 4$$

PROBLEM 2 Solve and graph: $-3 < 2x + 3 \leq 9$

ANSWER $-3 < x \leq 3$

Exercise 37

A *Indicate true (T) or false (F): If $6 > -2$, then*

1. $6 + 3 > -2 + 3$ **2.** $6 + 7 > -2 + 7$

3. $6 - 2 > -2 - 2$ **4.** $6 - 7 > -2 - 7$

5. $3(6) > 3(-2)$ **6.** $7(6) > 7(-2)$

7. $(-3)(6) < (-3)(-2)$ **8.** $(-7)(6) < (-7)(-2)$

9. $\frac{6}{2} > \frac{-2}{2}$

10. $\frac{6}{3} > \frac{-2}{3}$

11. $\frac{6}{-2} < \frac{-2}{-2}$

12. $\frac{6}{-3} < \frac{-2}{-2}$

Solve:

13. $x - 2 > 5$

14. $x - 4 < -1$

15. $x + 5 < -2$

16. $x + 3 > -4$

17. $2x > 8$

18. $3x < 6$

19. $-2x > 8$

20. $-3x < 6$

21. $\frac{x}{3} < -7$

22. $\frac{x}{5} > -2$

23. $\frac{x}{-3} < -7$

24. $\frac{x}{-5} > -2$

Solve and graph:

25. $3x + 7 < 13$

26. $2x - 3 > 5$

27. $-2x + 8 < 4$

28. $-4x - 7 > 5$

29. $7x - 8 \leq 4x + 7$

30. $6m + 2 \leq 4m + 6$

31. $9y + 3 \leq 4y - 7$

32. $4x + 8 \geq x - 1$

B **33.** $3y > 5y - 4$

34. $4x < 7x + 12$

35. $2 < x + 3 < 5$

36. $-3 \leq x - 5 \leq 8$

37. $-6 \leq 3x \leq 9$

38. $-24 \leq 6x \leq 6$

39. $-4 < 5x + 6 < 21$

40. $2 < 3x - 7 < 14$

41. $3 - x \geq 5(3 - x)$

42. $2(x - 3) + 5 < 5 - x$

43. $3 - (2 + x) > -9$

44. $2(1 - x) \geq 5x$

45. $m - \frac{1}{2} > \frac{8}{3}$

46. $\frac{x}{5} - 3 < \frac{3}{5} - x$

47. $-2 - \frac{x}{4} < \frac{1 + x}{3}$

48. $\frac{x - 3}{2} - 1 > \frac{x}{4}$

49. $-4 \leq \frac{9}{5}C + 32 \leq 68$

50. $-1 \leq \frac{2}{3}m + 5 \leq 11$

51. $-10 \leq \frac{5}{9}(F - 32) \leq 25$

52. $-5 \leq \frac{5}{9}(F - 32) \leq 10$

C **53.** Supply the reasons for the proof of Theorem 1A:

PROOF

	STATEMENTS	REASONS
1	$a > b$	1
2	$b < a$	2
3	$b + p = a$ for same $p > 0$	3
4	$(b + c) + p = a + c$	4
5	$b + c < a + c$	5
6	$a + c > b + c$	6

54. Prove Theorem 1*B*. (See the preceding problem.)

55. Supply the reason for the proof of Theorem 1*D*:

PROOF

	STATEMENTS	REASONS
1	$a > b$	1
2	$b < a$	2
3	$b + p = a$ for some $p > 0$	3
4	c is negative	4
5	$c(b + p) = ca$	5
6	$cb + cp = ca$	6
7	$ca - cp = cb$	7
8	$ca < cb$	8

56. Prove Theorem 1*F*. (See the preceding problem.)

57. Find the mistake: Assume that $a > b > 0$. Then

$$a > b$$

$$ab > b^2$$

$$ab - a^2 > b^2 - a^2$$

$$a(b - a) > (b + a)(b - a)$$

$$a > b + a$$

$$0 > b$$

5.4 Applications

This section contains a variety of applications from several different fields. Even though other methods can be used to solve some of the problems, you should use inequality methods.

EXAMPLE 3 What numbers satisfy the conditions, "4 more than twice a number is less than or equal to that number"?

SOLUTION Let $x =$ the number, then

$$2x + 4 \le x$$
$$x \le -4$$

PROBLEM 3 What numbers satisfy the condition, "6 less than 3 times a number is greater than or equal to 9"?

ANSWER $x \ge 5$

EXAMPLE 4 If the temperature for a 24-hr period in the Antarctica ranged between $-49°$F and $14°$F (that is, $-49 \le F \le 14$), what was the range in Celsius degrees? (Recall $F = \frac{9}{5}C + 32$)

SOLUTION Since $F = \frac{9}{5}C + 32$, we replace F in $-49 \le F \le 14$ with $\frac{9}{5}C + 32$ and solve the double inequality:

$$-49 \le \tfrac{9}{5}C + 32 \le 14$$

$$-49 - 32 \le \tfrac{9}{5}C + 32 - 32 \le 14 - 32$$

$$-81 \le \tfrac{9}{5}C \le -18$$

$$(\tfrac{5}{9})(-81) \le (\tfrac{5}{9})(\tfrac{9}{5}C) \le (\tfrac{5}{9})(-18)$$

$$-45 \le C \le -10$$

PROBLEM 4 Repeat Example 4 for $-31 \le F \le 5$.

ANSWER $-35 \le C \le -15$

Exercise 38

A *Inequality methods are to be used to solve the problems in this exercise.*

1. What numbers satisfy the condition, "5 more than twice the number is less than or equal to 7"?

2. What numbers satisfy the condition, "3 less than twice the number is greater than or equal to -6"?

3. What numbers satisfy the condition, "5 less than 3 times the number is less than or equal to 4 times the number"?

4. What numbers satisfy the condition, "If 15 is diminished by 3 times the number, the result is less than 6"?

5. A doctor told a woman, "Even if you lost 30 lb you would not weigh less than 125 lb." What can you say about the woman's weight?

6. If in driving the 420 miles between San Francisco and Los Angeles you know that your average speed is less than 50 mph, what can you say about the time it will take you to make the trip? (Use $d = rt$.)

7. If the perimeter of a rectangle with a length of 10 in. must be smaller than 30 in., how large may the width be?

8. If the area of a rectangle of length 10 in. must be greater than 65 sq in., how large may the width be?

B

9. How long would you have to leave \$100 invested at 6 percent simple interest for it to amount to more than \$133? (Use $A = P + Prt$.)

10. For a business to make a profit it is clear that revenue R must be greater than costs C; in short, a profit will result only if $R > C$. If a company manufactures records and its cost equation for a week is $C = 300 + 1.5x$, where x is the number of records manufactured in a week, and its revenue equation is $R = 2x$, where x is the number of records sold in a week, how many records must be sold for the company to realize a profit?

11. A family wishes to spend between \$20,000 and \$30,000 on a cabin. If a contractor estimates the building costs to be around \$20 per sq ft, then the number of square feet designed into the cabin should lie in what range?

12. A person's IQ is found by dividing his mental age, as indicated by standard tests, by his chronological age and then multiplying this ratio by 100. In terms of a formula,

$$\text{IQ} = \frac{\text{MA} \cdot 100}{\text{CA}}$$

If the IQ range of a group of 12-year olds is $70 \leq \text{IQ} \leq 120$, what is the mental age range of this group?

13. A photographic developer is to be kept between 68 and 77°F. What is the range of temperature in Celsius degrees? $(\text{F} = \frac{9}{5}\text{C} + 32)$

14. In a chemistry experiment the solution of hydrochloric acid is to be kept between 30 and 35°C. What would the range of temperature be in Fahrenheit degrees? $[\text{C} = \frac{5}{9}(\text{F} - 32)]$

15. To be eligible for a certain university, a student must have an average grade of not below 70 on 3 entrance examinations. If a student received a 55 and a 73 on the first 2 examinations, what must he receive on the third examination to be eligible to enter the university?

16. At the same university (referred to in the preceding problem), if a student attains an average grade above 90 on the 3 examinations, he will be eligible for certain honors courses. If no score above 100 is possible on any one of the tests, is it possible for the student in the preceding problem to receive a high enough score on the third test to make him eligible for an honors course?

C **17.** It is customary in supersonic studies to specify the velocity of an object relative to the velocity of sound. The ratio between these two velocities is called the Mach number, and it is given by the formula

$$M(\text{Mach number}) = \frac{V(\text{speed of object})}{S(\text{speed of sound})}$$

If a supersonic transport is designed to operate between Mach 1.7 and 2.4, what is the speed range of the transport in miles per hour? (Assume that the speed of sound is 740 mph.)

18. It is known that the temperature has a small but measurable effect upon the velocity of sound. For each degree Celsius rise in temperature in the air the velocity of sound increases about 2 fps. If at 0°C sound travels at 1,087 fps, its speed at other temperatures is given by

$$V = 1{,}087 + 2T$$

(*A*) What temperature range will correspond to a velocity range of $1{,}097 \leq V \leq 1{,}137$?
(*B*) What is the corresponding temperature range in Fahrenheit degrees $[\text{C} = \frac{5}{9}(\text{F} - 32)]$?

19. In a 110-volt electrical house circuit a 30-ampere fuse is used. In order to keep the fuse from "blowing," the total wattages in all of the appliances on that circuit must be kept below what figure? ($W = EI$, where $W =$ power in watts, $E =$ pressure in volts, and $I =$ current in amperes.) HINT: $I = W/E$ and $I \leq 30$.

20. If the power demands in an 110-volt electrical circuit in a home range between 220 and 2,750 watts, what is the range of current flowing through the circuit? (Use $W = EI$, where $W =$ power in watts, $E =$ pressure in volts, and $I =$ current in amperes.)

21. If in a given storm the time it takes thunder to be heard after lightning strikes varies between 5 and 10 sec, how does the distance (in miles) to the lightning vary? (Use $d = 1{,}088t$ and 1 mile $= 5{,}290$ ft.)

5.5 Cartesian Coordinate System

You already know how to graph equations and inequalities such as

$$2x + 3 = 5$$

$$2x + 3 \leq 5$$

on a number line, but how would you graph an equation involving two variables such as

$$2x + 3y = 6$$

A solution to $2x + 3y = 6$ is not just one number, but a pair of numbers listed in a certain order. For example

$$(3,0),\ (6,-2),\ (0,2)$$

where the first number in the ordered pair replaces x and the second y, are three of many solutions to $2x + 3y = 6$. Similarly

$$(0,0),\ (1,1),\ (0,2)$$

are three of many solutions to $2x + 3y \leq 6$.

Scientists and others are frequently involved with problems where two variables are needed. For example, distances that objects fall at different times, pressures of an enclosed gas at different temperatures, the approach drive of a rat at different distances from a goal box, and the pitch of musical notes and lengths of strings. Each of these examples involves two variable quantities.

It might be useful to have a picture (a graph) to visually show how two variables are related. How do we form such a graph? We can do this by means of a very widely used system called the cartesian or rectangular coordinate system.

To form a cartesian coordinate system we start with two sets of objects, the set of all points in a plane (a set of geometric objects) and the set of all ordered pairs of real numbers. In a plane select two number lines, one vertical and one horizontal, and let them cross at their respective origins (Fig. 8). Up and to the right are the usual choices for the positive directions. These two number lines are called the *vertical axis* and the *horizontal axis* or (together) the *coordinate axes*. The coordinate axes divide the plane into four parts called *quadrants*. The quadrants are numbered counterclockwise from I to IV.

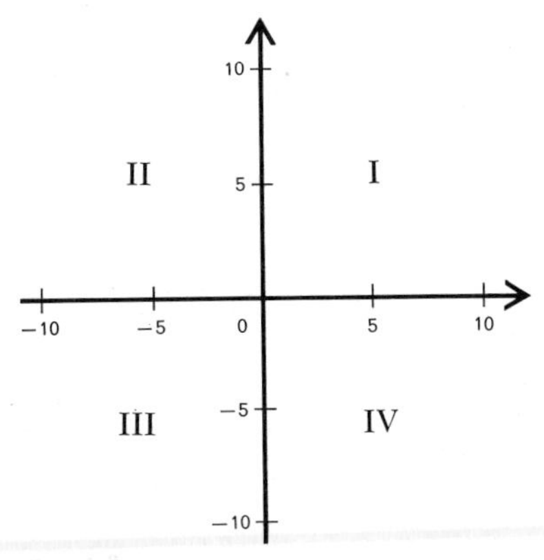

Figure 8

Pick a point P in the plane at random (see Fig. 9). Pass horizontal and vertical lines through the point. The vertical line will intersect the horizontal axis at a point with coordinate a, and the horizontal line will intersect the vertical axis at a point with coordinate b. The coordinate of each point of intersection, a and b, respectively, form the *coordinates*

$$(a,b)$$

of the point P in the plane.

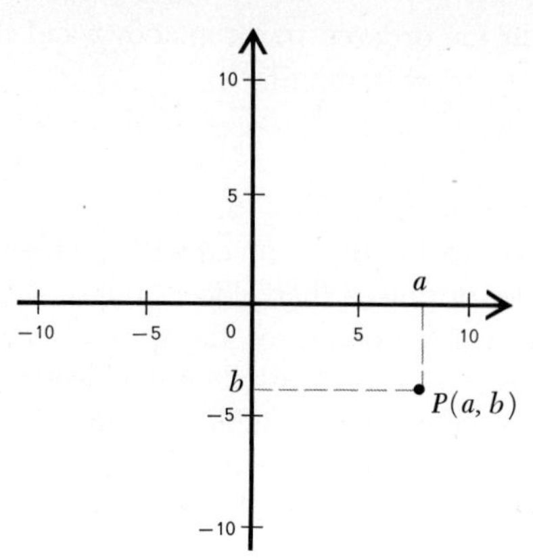

Figure 9

The first number a (called the *abscissa* of P) in the ordered pair (a,b) is the directed distance of the point from the vertical axis (measured on the horizontal scale); the second number b (called the *ordinate* of P) is the directed distance of the point from the horizontal axis (measured on the vertical scale).

We know from Sec. 3.7 that a and b exist for each point in the plane since every point on each axis has a real number associated with it. Hence, by the procedure described, every point in the plane can be labeled with a pair of real numbers. Conversely, by reversing the process, every pair of real numbers can be associated with a point in the extended plane. Thus we have the following important theorem.

THEOREM 2 There exists a one-to-one correspondence between the set of points in a plane and the set of all ordered pairs of real numbers.

The system that we have just defined to produce this correspondence is called a *cartesian coordinate system* (sometimes referred to as a rectangular coordinate system).

EXAMPLE 5 Find the coordinates of each of the points A, B, C, and D.

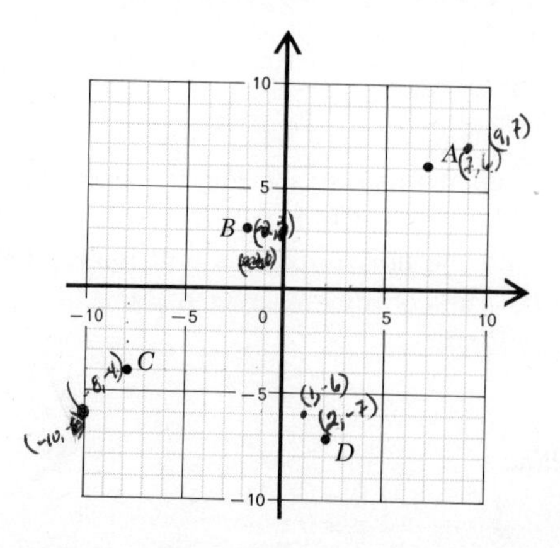

SOLUTION $A(7,6)$ $B(-2,3)$
$C(-8,-4)$ $D(2,-7)$

PROBLEM 5 Find the coordinates, using the figure in Example 5, for each of the following points:

(*A*) 2 units to the right and 1 unit up from A
(*B*) 2 units to the left and 2 units down from C
(*C*) 1 unit up and 1 unit to the left of D
(*D*) 2 units to the right of B.

ANSWER (*A*) (9,7) (*B*) (−10,−6) (*C*) (1,−6) (*D*) (0,3)

EXAMPLE 6 Graph (associate each ordered pair of numbers with a point in the cartesian coordinate system):

(2,7), (7,2), (−8,4), (4,−8), (−8,−4), (−4,−8)

SOLUTION

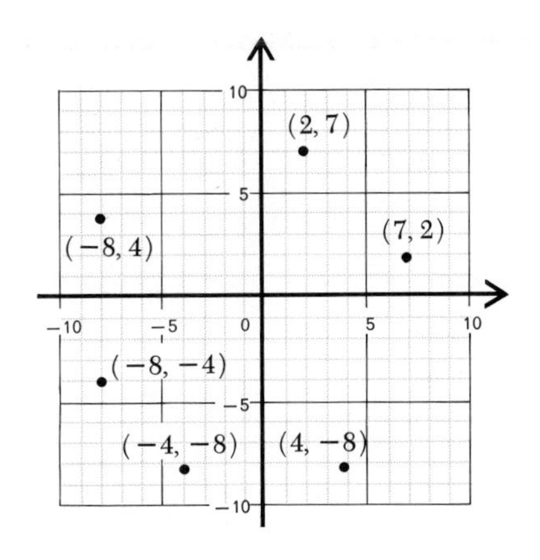

It is very important to note that the ordered pair (2,7) and the set {2, 7} are not the same thing; {2, 7} = {7, 2}, but (2,7) ≠ (7,2).

PROBLEM 6 Graph: (3,4), (−3,2), (−2,−2), (4,−2), (0,1), and (−4,0)

ANSWER

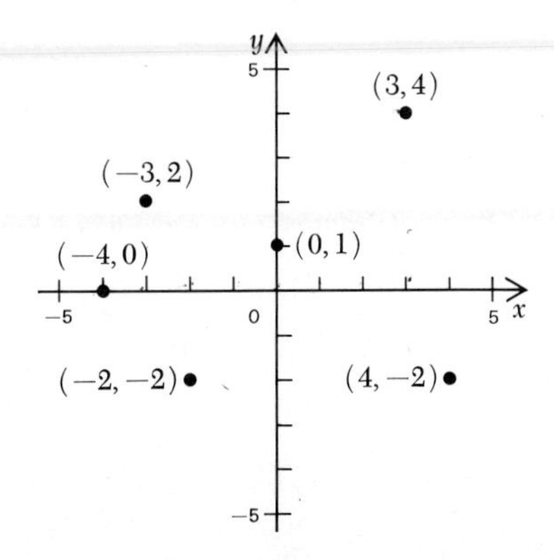

Exercise 39

A *Write down the coordinates to each labeled point.*

1.

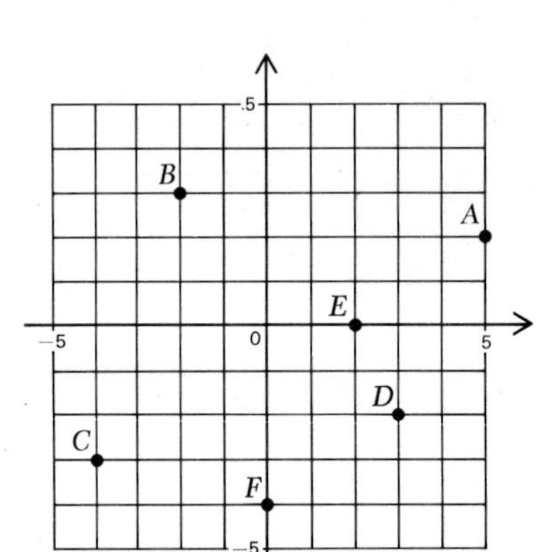

2.

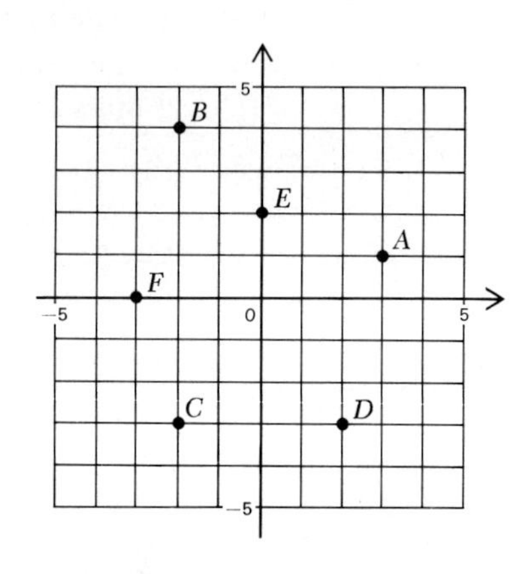

3.

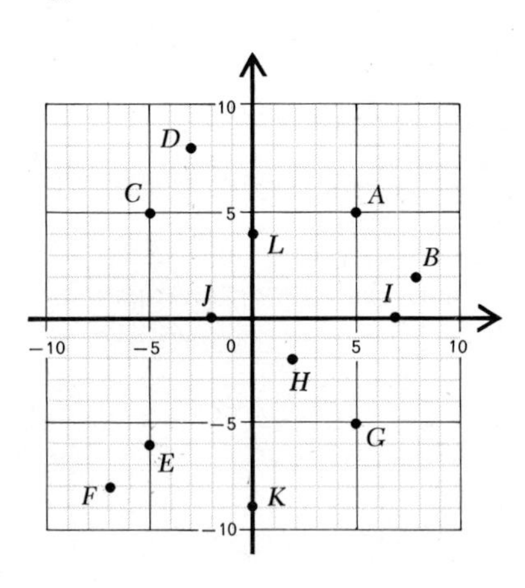

4.

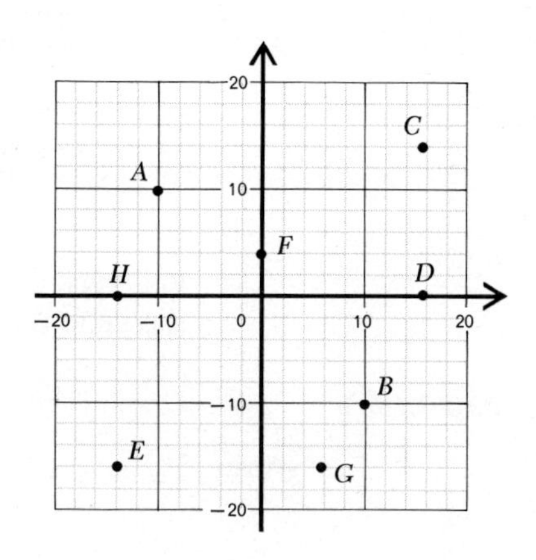

Graph each set of ordered pairs of numbers on the same coordinate system.

5. $(3,1)$, $(-2,3)$, $(-5,-1)$, $(2,-1)$, $(4,0)$, $(0,-5)$.

6. $(4,4)$, $(-4,1)$, $(-3,-3)$, $(5,-1)$, $(0,2)$, $(-2,0)$.

7. $(-9,8)$, $(8,-9)$, $(0,5)$, $(4,-8)$, $(-3,0)$, $(7,7)$, and $(-6,-6)$.

8. $(2,7)$, $(7,2)$, $(-6,3)$, $(-4, -7)$, $(2,3)$, $(0,-8)$, and $(9,0)$.

B *Write down the coordinates of each labeled point to the nearest quarter of a unit.*

9.

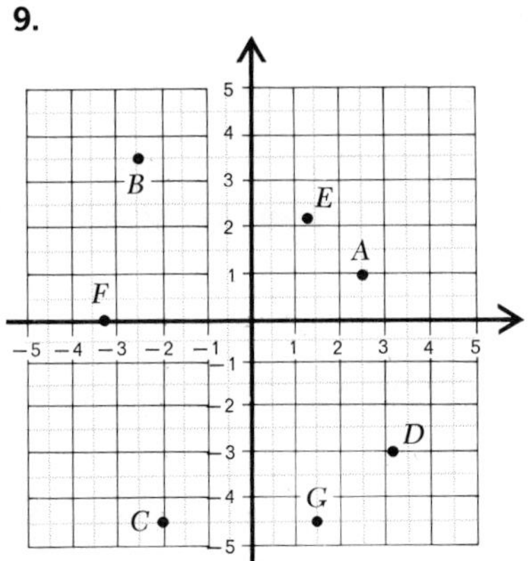

10.

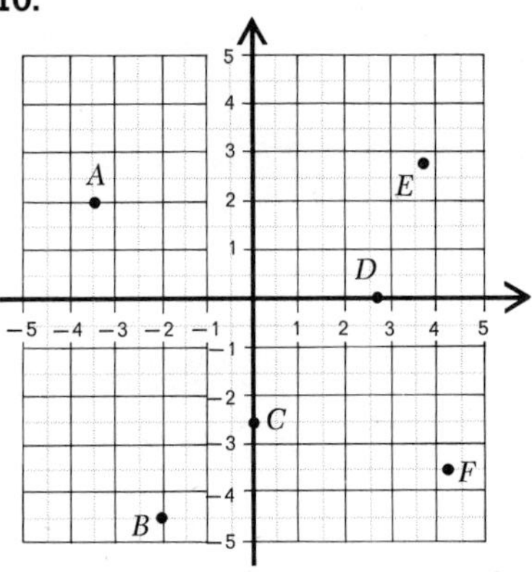

11. Graph the following ordered pairs of numbers on the same coordinate system: $A(1\frac{1}{2},3\frac{1}{2})$, $B(-3\frac{1}{4},0)$, $C(3,-2\frac{1}{2})$, $D(-4\frac{1}{2},1\frac{3}{4})$, and $E(-2\frac{1}{2},-4\frac{1}{4})$.

12. Graph the following ordered pairs of numbers on the same coordinate system: $A(3\frac{1}{2},2\frac{1}{2})$, $B(-4\frac{1}{2},3)$, $C(0,-3\frac{3}{4})$, $D(-2\frac{3}{4},-3\frac{3}{4})$, and $E(4\frac{1}{4},-3\frac{3}{4})$.

13. Without graphing, tell which quadrants contain the graph of each of the following ordered pairs:

(*A*) $(-20,-4)$ (*B*) $(-3,22\frac{3}{4})$
(*C*) $(4, 35{,}000)$ (*D*) $(\sqrt{2},-3)$

14. Without graphing, tell which quadrants contain the graph of each of the following ordered pairs (see Fig. 8):

(*A*) $(-23,403)$ (*B*) $(32\frac{1}{2},-430)$
(*C*) $(2{,}001, 25)$ (*D*) $(-0.008, -3.2)$

C *What is the abscissa of point A in*

15. Exercise 1 5 **16.** Exercise 2

17. Exercise 3 5 **18.** Exercise 4

What is the ordinate of point B in

19. Exercise 1 3 **20.** Exercise 2

21. Exercise 3 2 **22.** Exercise 4

5.6 First-Degree Equations and Straight Lines

The invention of the cartesian coordinate system represented a very important advance in mathematics. It was through the use of this system that René Descartes (1596–1650), a French philosopher-mathematician, was able to transform geometric problems requiring long tedious reasoning into algebraic problems which could be solved quite mechanically. This joining of algebra and geometry has now become known as *analytic geometry*.

Two fundamental problems of analytic geometry are the following:

1 Given an equation, find its graph.
2 Given a geometric figure, such as a straight line, circle, or ellipse, find its equation.

In this course we will be mainly interested in the first problem, with particular emphasis on equations whose graphs are straight lines.

Let us start by trying to find the solution set for

$$y = 2x - 4$$

Recall that a solution of an equation in two variables is an ordered pair of real numbers that satisfy the equation. If we agree that the first element in the ordered pair will replace x and the second y, then

$$(0, -4)$$

is a solution of $y = 2x - 4$, as can easily be checked. How do we find other solutions? The answer is easy: We simply assign to x in $y = 2x - 4$ any convenient value and solve for y. For example, if $x = 3$ then

$$\begin{aligned} y &= 2(3) - 4 \\ &= 2 \end{aligned}$$

Hence,

$$(3,2)$$

is another solution of $y = 2x - 4$. It is clear that by proceeding in this manner, we can get solutions to this equation without end. Thus, the solution set is infinite. Let us make up a table of some solutions and graph them in a cartesian coordinate system, identifying the horizontal axis with x and the vertical axis with y.

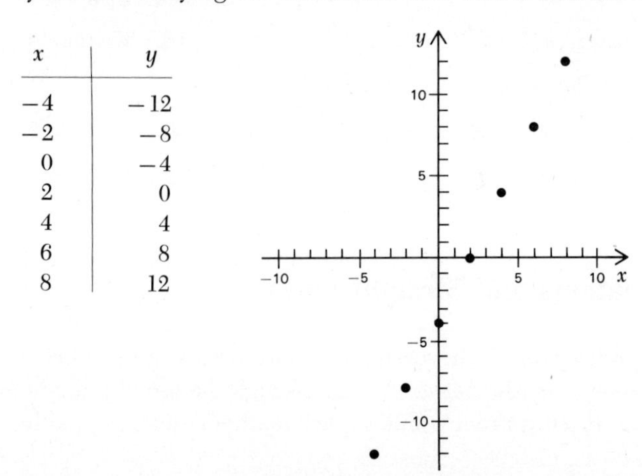

x	y
-4	-12
-2	-8
0	-4
2	0
4	4
6	8
8	12

It appears that the graph of the equation is a straight line. If we knew this for a fact, then graphing $y = 2x - 4$ would be easy. We would simply find two solutions of the equation, plot them, then plot as much of $y = 2x - 4$ as we like by drawing a line through the two points using a straightedge. It turns out that it is true that the graph of $y = 2x - 4$ is a straight line. In fact, we have the following general theorem, which we state without proof.

THEOREM 3 The graph of any equation of the form

$$Ax + By = C$$

where A, B, and C are constants (A and B both not 0) and x and y are variables, is a straight line. Every straight line in a cartesian coordinate system is the graph of an equation of this form.

It immediately follows that any equation of the form

$$y = mx + b$$

where m and b are constants, is also a straight line since it can be written in the form $-mx + y = b$, which is a special case of $Ax + By = C$.

Thus, to graph any equation of the form

$$Ax + By = C$$

or

$$y = mx + b$$

we plot any two points of the solution set and use a straightedge to draw a line through these two points. It is sometimes wise to find a third point as a check point.

It should be obvious that we cannot draw a straight line extending indefinitely in either direction. We will settle for the part of the line in which we are interested—usually the part close to the origin unless otherwise stated.

EXAMPLE 7 (A) The graph of $y = 2x - 4$ is

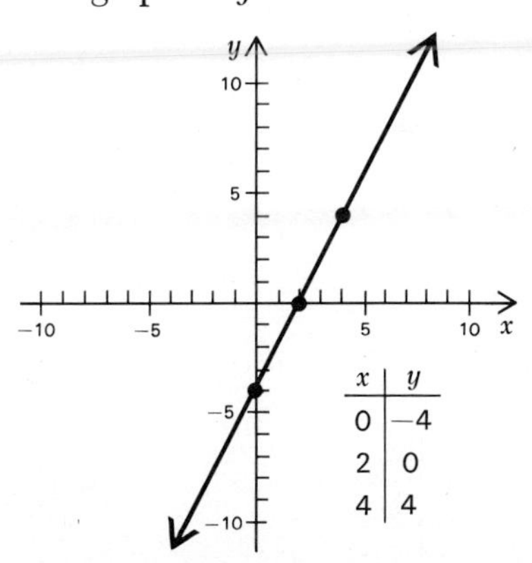

x	y
0	−4
2	0
4	4

(*B*) To graph $x + 3y = 6$, assign to either x or y any convenient value and solve for the other variable. If we let $x = 0$, a convenient value, then

$$\begin{aligned} 0 + 3y &= 6 \\ 3y &= 6 \\ y &= 2 \end{aligned}$$

Thus, (0,2) is a solution.

If we let $y = 0$, another convenient choice, then

$$\begin{aligned} x + 3(0) &= 6 \\ x + 0 &= 6 \\ x &= 6 \end{aligned}$$

Thus, (6,0) is a solution.

To find a check point, choose another value for x or y, say $x = -6$, then

$$\begin{aligned} -6 + 3y &= 6 \\ 3y &= 12 \\ y &= 4 \end{aligned}$$

Thus, $(-6,4)$ is also a solution.

Now plot these three points and draw a line through them. (If a straight line does not pass through all three points, then you have made a mistake and must go back and check your work.)

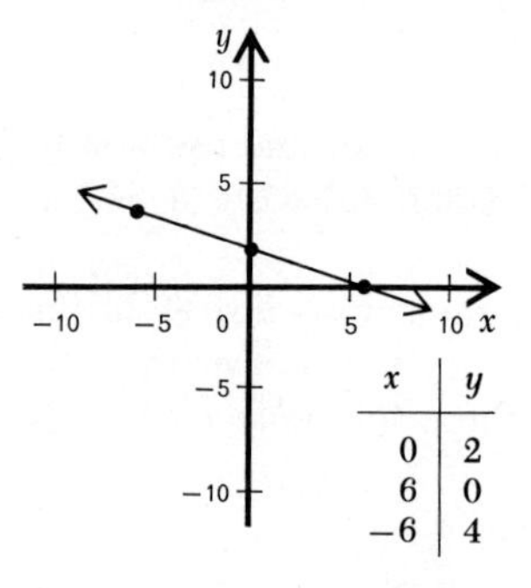

x	y
0	2
6	0
−6	4

PROBLEM 7 Graph: (*A*) $y = 2x - 6$ (*B*) $3x + y = 6$

ANSWER (*A*)

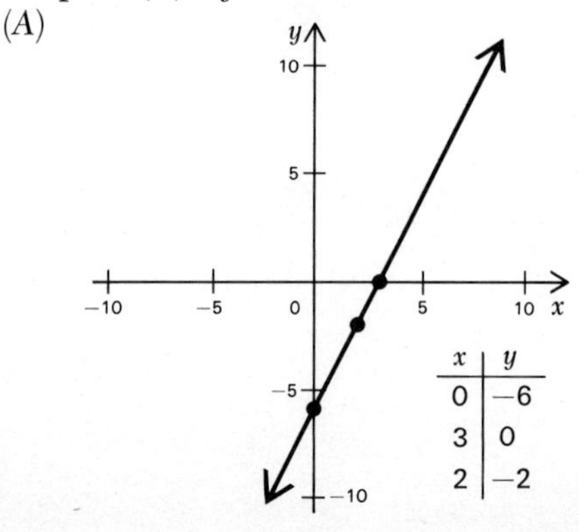

x	y
0	−6
3	0
2	−2

(*B*)

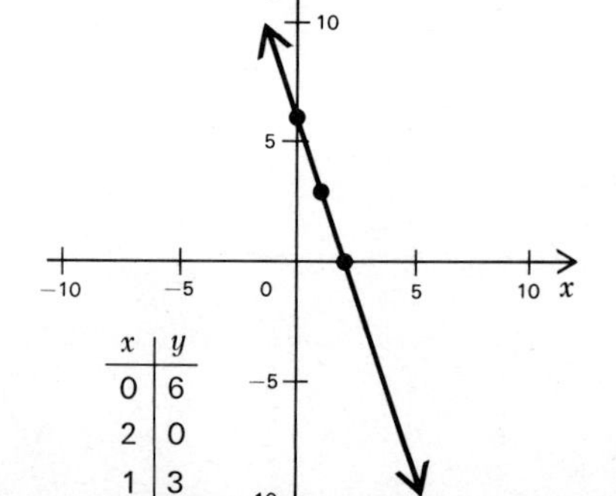

x	y
0	6
2	0
1	3

Vertical and horizontal lines in rectangular coordinate systems have particularly simple equations. For example, to graph

$$y = 4$$

we recognize that this is just a short way of writing

$$0x + y = 4$$

No matter what values are assigned to x, $0x = 0$; thus, as long as $y = 4$, x can assume any value. A moment's thought will reveal that the graph of $y = 4$ is a horizontal straight line crossing the y axis at 4.

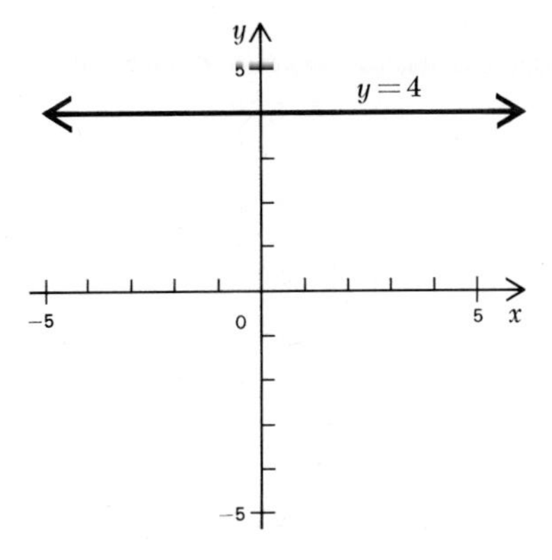

Similarly the graph of

$$x = 3$$

is a vertical line crossing the x axis at 3.

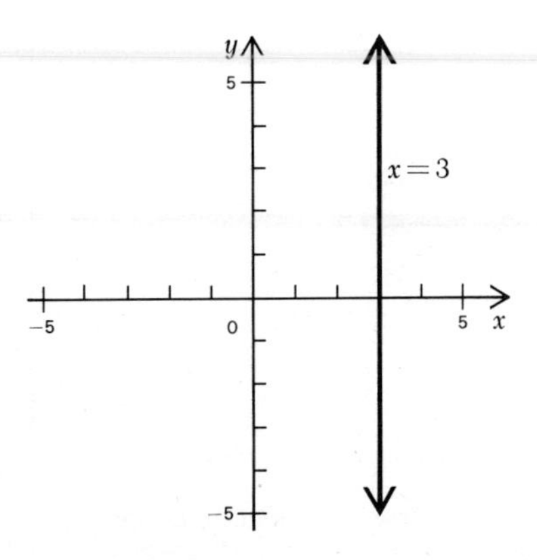

EXAMPLE 8 (*A*) The graph of $y = -2$ is

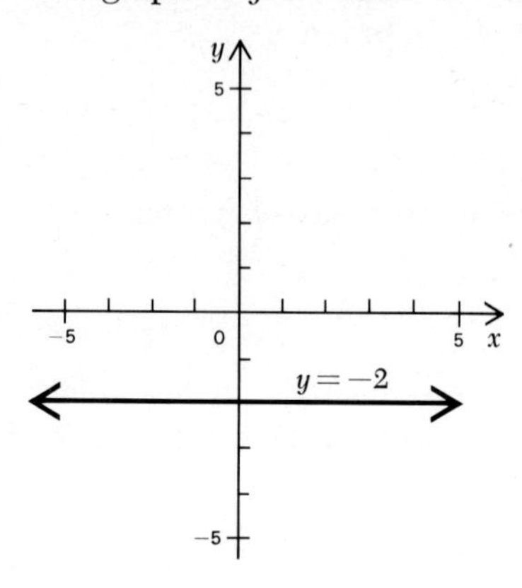

(*B*) The graph of $x = -4$ is

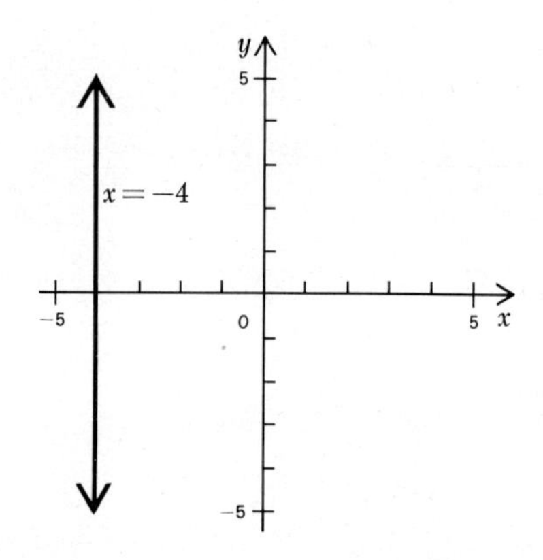

PROBLEM 8 Graph: (*A*) $y = 1$ and (*B*) $x = -2$

ANSWER (*A*)

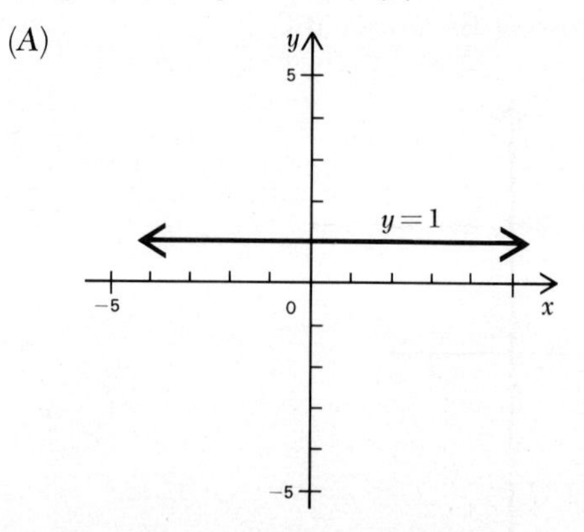

(*B*)

y
x = −2
x

It should now be clear why first-degree equations in two variables are often referred to as *linear equations*.

Exercise 40

A *Graph in a rectangular coordinate system.*

1. $y = x$ **2.** $y = 2x$ **3.** $y = x - 1$

4. $y = 2x - 3$ **5.** $y = \frac{x}{2}$ **6.** $y = \frac{x}{3}$

7. $y = \frac{x}{2} + 1$ **8.** $y = \frac{x}{3} + 2$ **9.** $x + y = 6$

10. $x + y = -4$ **11.** $x - y = 5$ **12.** $x - y = 3$

13. $2x + 3y = 12$ **14.** $3x + 4y = 12$ **15.** $3x - 5y = 15$

16. $8x - 3y = 24$ **17.** $x = 2$ **18.** $y = 3$

19. $y = -3$ **20.** $x = -4$

B **21.** $y = \frac{1}{4}x$ **22.** $y = \frac{1}{2}x$ **23.** $y = \frac{1}{4}x + 1$

24. $y = \frac{1}{2}x - 1$ **25.** $y = -2x + 6$ **26.** $y = -x + 2$

27. $y = -\frac{1}{2}x + 2$ **28.** $y = -\frac{1}{3}x - 1$ **29.** $2x + y = 7$

30. $3x + 2y = 10$ **31.** $7x - 4y = 21$ **32.** $5x - 6y = 15$

33. $x = 0$ **34.** $y = 0$

Write in the form $y = mx + b$ and graph.

35. $y - x - 2 = x + 1$ **36.** $x + 6 = 3x + 2 - y$

Write in the form $Ax + By = C$ and graph.

37. $6x - 3 + y = 2y + 4x + 5$ **38.** $y + 8 = 2 - x - y$

Use a different scale on the vertical axis to keep the size of the graph within reason.

39. $d = 60t,\ 0 \le t \le 10$

40. $I = 6t,\ 0 \le t \le 10$

41. $A = 100 + 10t,\ 0 \le t \le 10$

42. $v = 10 + 32t,\ 0 \le t \le 5$

43. Graph $x + y = 3$ and $x - 2y = 0$ on the same coordinate system. Determine by inspection the coordinates of the point where the two graphs cross. Show that the coordinates of the point of intersection satisfies both equations.

44. Repeat the preceding problem with the equations

$2x - 3y = -6$ and $x + 2y = 11$

C **45.** Graph $y = -\frac{1}{2}x + b$ for $b = -6, b = 0$, and $b = 6$ all on the same coordinate system.

46. Graph $y = mx - 2$ for $m = 2$, $m = \frac{1}{2}$, and $m = -2$, all on the same coordinate system.

47. Graph $y = |x|$. HINT: Graph $y = x$ for $x \geq 0$ and $y = -x$ for $x < 0$.

48. Graph $y = |2x|$ and $y = |\frac{1}{2}x|$ on the same coordinate system.

Choose horizontal and vertical scales to produce maximum clarity in graphs.

49. In biology there is an approximate rule, called the bioclimatic rule, for temperate climates that states that in spring and early summer periodic phenomena such as blossoming for a given species, appearance of certain insects, and ripening of fruit usually come about 4 days later for each 500 ft of altitude. Stated as a formula,

$$d = 4\left(\frac{h}{500}\right)$$

where d = change in days and h = change in altitude in feet. Graph the equation for $0 \leq h \leq 4{,}000$.

50. In 1948 Professor Brown, a psychologist, trained a group of rats (in an experiment on motivation) to run down a narrow passage in a cage to receive food in a goal box. He then put a harness on each rat and connected it to an overhead wire that was attached to a scale. In this way he could place the rat at different distances (in centimeters) from the food and measure the pull (in grams) of the rat toward the food. He found that a relation between motivation (pull) and position was given approximately by the equation $p = \frac{1}{5}d + 70$, $30 \leq d \leq 175$. Graph this equation for the indicated values of d.

51. In a simple electric circuit, such as found in a flashlight, if the resistance is 30 ohms, the current in the circuit I (in amperes) and the electromotive force E (in volts) are related by the equation $E = 30I$. Graph this equation for $0 \leq I \leq 1$.

5.7 Systems of Linear Equations

Many practical problems can be solved conveniently using two-equation–two-unknown methods. For example, if a 12-ft board is cut in two pieces so that one piece is 4 ft longer than the other piece, how long is each piece? We could solve this problem using one-equation–one-unknown methods studied earlier, but we can also proceed as follows using two variables.

Let x = the length of the longer piece
y = the length of the shorter piece

then

$$x + y = 12$$

$$x - y = 4$$

To solve this system is to find all the ordered pairs of real numbers that satisfy both equations at the same time.

There are several methods of solving this type of problem, each with its own merits. At this time we will only study two:

1. Solution by graphing
2. Solution by elimination

SOLVING SYSTEMS OF EQUATIONS BY GRAPHING

The method of solving a system of equations by graphing is perhaps the easiest of the two methods to understand. For this reason it will be considered first.

We proceed by graphing both equations on the same coordinate system. Then the coordinates of any points that the graphs have in common must be solutions to the system since they must satisfy both equations. (Why?)

EXAMPLE 9 Solve by graphing:

$$x + y = 12$$
$$x - y = 4$$

SOLUTION

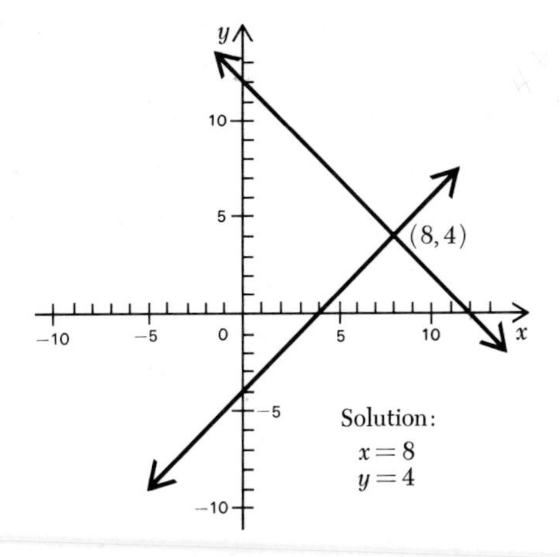

PROBLEM 9 Solve by graphing:

$$x + y = 10$$
$$x - y = 6$$

ANSWER $x = 8$ $\quad y = 2$

It is clear that Example 9 and Problem 9 each has exactly one solution since the two lines in each case intersect in exactly one point. Let us look at three typical cases that illustrate the three possible ways two lines can be related to each other in a rectangular coordinate system.

Solving the following three systems graphically

(A) $2x - 3y = 2$
$x + 2y = 8$

(B) $4x + 6y = 12$
$2x + 3y = -6$

(C) $2x - 3y = -6$
$-x + \frac{3}{2}y = 3$

we obtain

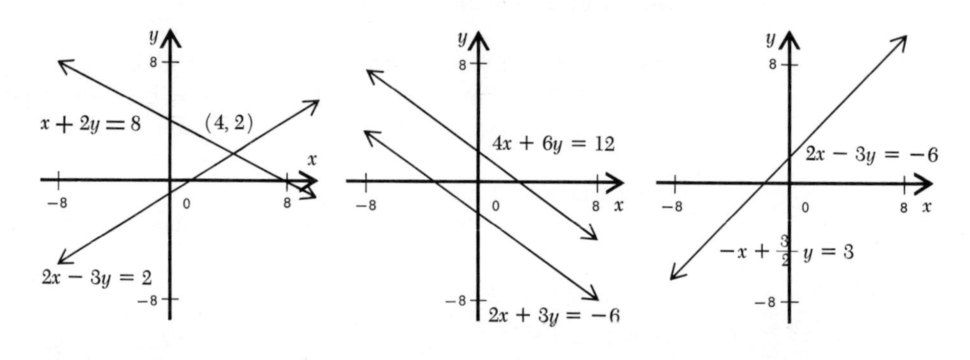

Lines intersect at one point only: *Exactly one solution*

$x = 4$, $y = 2$

Lines are parallel: *No solution*

Lines coincide: *Infinite number of solutions*

Now we know exactly what to expect when solving a system of two linear equations in two unknowns:

Exactly one pair of numbers as a solution.

No solutions.

An infinite number of solutions.

In most applications the first case prevails. If we find a pair of numbers that satisfy the system of equations and the graphs of the equations meet at only one point, then that pair of numbers is the only solution of the system, and we need not look further for others.

The graphical method of solving systems of equations yields considerable information as to what to expect in the way of solutions to a system of two linear equations in two unknowns. In addition, graphs frequently reveal relationships in problems that would otherwise be hidden. On the other hand, if one is interested in solutions with several-decimal-place accuracy, the graphical method is often not practical. The method of elimination, to be considered next, will take care of this deficiency.

Exercise 41

A *Solve by graphing and check.*

1. $x + y = 5$
$x - y = 1$

2. $x + y = 6$
$x - y = 2$

3. $x + y = 5$
$2x - y = 4$

4. $2x + y = 6$
$x - y = -3$

5. $x - 3y = 3$
$x - y = 7$

6. $x - 2y = -4$
$2x - y = 10$

B

7. $3x - y = 2$
$x + 2y = 10$

8. $x - 2y = 2$
$2x + y = 9$

9. $-2x + 3y = 12$
$2x - y = 4$

10. $3x - 2y = 12$
$7x + 2y = 8$

11. $-3x + y = 9$
$3x + 4y = -24$

12. $x + 5y = -10$
$5x + y = 24$

13. $x + 2y = 4$
$2x + 4y = -8$

14. $3x + 5y = 15$
$6x + 10y = -30$

15. $\frac{1}{2}x - y = -3$
$-x + 2y = 6$

16. $3x - 5y = 15$
$x - \frac{5}{3}y = 5$

C *It can be shown that if two linear equations are written in the form*

$$y = mx + b$$

and the coefficient of x is the same for both, then their graphs are parallel.

17. Show that the lines in Exercise 13 are parallel.

18. Show that the lines in Exercise 14 are parallel.

19. How can you use the idea above to show that the two lines in Exercise 15 coincide?

20. Show that the lines in Exercise 14 coincide.

SOLVING SYSTEMS OF EQUATIONS BY ELIMINATION

The elimination method we are now going to develop can produce answers to any decimal accuracy desired. This method has to do with the replacement of systems of equations with simpler equivalent systems (by performing appropriate operations) until we get a system whose solution is obvious. *Equivalent systems* are, as you would expect, systems with the same solution set. What operations on a system produce equivalent systems? The following theorem, stated but not proved, answers this question.

THEOREM 4 Equivalent systems result if

(*A*) Any algebraic expression in an equation is replaced with its equal.

(*B*) Both sides of the equations are multiplied by a nonzero constant.

(*C*) The two equations are combined by addition or subtraction and the result is paired with either of the two original equations.

Solving a system of equations by use of this theorem is best illustrated by examples.

EXAMPLE 10 Solve the system:

$$3x + 2y = 13$$
$$2x - y = 4$$

SOLUTION We use Theorem 4 to eliminate one of the variables, and thus obtain a system whose solution is obvious:

$$(A)\quad 3x + 2y = 13$$
$$(B)\quad 2x - y = 4$$

If we multiply equation (*B*) by 2 and add the result to equation (*A*), we can eliminate y.

$$(A)\quad 3x + 2y = 13$$
$$2(B)\quad 4x - 2y = 8$$
$$(A) + 2(B)\quad 7x = 21$$

Eliminate y by addition.

$$\boxed{x = 3}$$

$$2 \cdot 3 - y = 4$$
$$-y = -2$$
$$\boxed{y = 2}$$

Substitute $x = 3$ back into either equation (*A*) or equation (*B*), the simpler of the two, and solve for y.

CHECK

(*A*)

$$3x + 2y = 13$$
$$3 \cdot 3 + 2 \cdot 2 \stackrel{?}{=} 13$$
$$9 + 4 \stackrel{\checkmark}{=} 13$$

(*B*)

$$2x - y = 4$$
$$2 \cdot 3 - 2 \stackrel{?}{=} 4$$
$$6 - 2 \stackrel{\checkmark}{=} 4$$

PROBLEM 10 Solve the system:

$$2x + 3y = 7$$

$$3x - y = 5$$

ANSWER $x = 2,\ y = 1$

EXAMPLE 11 Solve the system:

$$2x + 3y = 1$$

$$5x - 2y = 12$$

SOLUTION

$$(A)\quad 2x + 3y = 1$$
$$(B)\quad 5x - 2y = 12$$

If we multiply equation (A) by 2 and equation (B) by 3 and add, we can eliminate y.

$$\begin{aligned} 4x + 6y &= 2 \\ 15x - 6y &= 36 \\ \hline 19x \qquad &= 38 \end{aligned}$$

$$\boxed{x = 2}$$

$$\begin{aligned} 2 \cdot 2 + 3y &= 1 \\ 3y &= -3 \end{aligned}$$

Substitute $x = 2$ back into either equation (A) or equation (B).

$$\boxed{y = -1}$$

CHECK

(A)

$$\begin{aligned} 2x + 3y &= 1 \\ 2 \cdot 2 + 3(-1) &\stackrel{?}{=} 1 \\ 4 - 3 &\stackrel{\checkmark}{=} 1 \end{aligned}$$

(B)

$$\begin{aligned} 5x - 2y &= 12 \\ 5 \cdot 2 - 2(-1) &\stackrel{?}{=} 12 \\ 10 + 2 &\stackrel{\checkmark}{=} 12 \end{aligned}$$

PROBLEM 11 Solve the system:

$$3x - 2y = 8$$

$$2x + 5y = -1$$

ANSWER $x = 2, \qquad y = -1$

EXAMPLE 12 Solve the system:

$$x + 3y = 2$$

$$2x + 6y = -3$$

SOLUTION

$$(A)\quad x + 3y = 2$$
$$(B)\quad 2x + 6y = -3$$

$$\begin{aligned} 2x + 6y &= 4 \\ 2x + 6y &= -3 \\ \hline 0 &= 7 \end{aligned}$$

A contradiction!

Hence, no solution.

Our assumption that there are values for x and y that satisfy equation (A) and equation (B) simultaneously must be false (otherwise, we have proved that $0 = 7$); thus, the system has no solutions. Systems of this type are said to be *inconsistent*—conditions have been placed on the unknowns x and y that are impossible to meet.

PROBLEM 12 Solve the system:

$$2x - y = 2$$
$$-4x + 2y = 1$$

ANSWER No solution

EXAMPLE 13 Solve the system:

$$-2x + y = -8$$
$$x - \tfrac{1}{2}y = 4$$

SOLUTION

$$\begin{aligned} (A)\quad -2x + y &= -8 \\ (B)\quad x - \tfrac{1}{2}y &= 4 \end{aligned}$$

$$\begin{aligned} -2x + y &= -8 \\ 2x - y &= 8 \\ \hline 0 &= 0 \end{aligned}$$

Both unknowns have been eliminated! Actually, if we had multiplied equation (B) by -2, we would have obtained equation (A). When one equation is a constant multiple of the other, the system is said to be *dependent,* and their graphs will coincide. There are infinitely many solutions to the system—any solution of one equation will be a solution to the other.

PROBLEM 13 Solve the system:

$$4x - 2y = 3$$
$$-2x + y = -\tfrac{3}{2}$$

ANSWER An infinite number of solutions—any solution of one equation is a solution of the other.

Exercise 42

A *Solve by elimination method and check.*

1. $x + y = 5$, $x - y = 1$

2. $x - y = 6$, $x + y = 10$

3. $x + 3y = 13$, $-x + y = 3$

4. $-x + y = 1$, $x - 2y = -5$

5. $2x + y = 0$, $3x + y = 2$

6. $x + 5y = 16$, $x - 2y = 2$

7. $2x + 3y = 1$, $3x - y = 7$

8. $3x - y = -3$, $5x + 3y = -19$

9. $3x - 4y = 1$, $-x + 3y = 3$

10. $-x + 5y = -3$
$2x - 3y = -1$

11. $2x + 4y = 6$
$-3x + y = 5$

12. $3x + y = -8$
$-5x + 3y = 4$

B

13. $11x + 2y = 1$
$9x - 3y = 24$

14. $3x - 11y = -7$
$4x + 3y = 26$

15. $3p + 8q = 4$
$15p + 10q = -10$

16. $5m - 3n = 7$
$7m + 12n = -1$

17. $4m + 6n = 2$
$6m - 9n = 15$

18. $5a - 4b = 1$
$3a - 6b = 6$

19. $3x + 5y = 15$
$6x + 10y = -5$

20. $x + 2y = 4$
$2x + 4y = -9$

21. $3x - 5y = 15$
$x - \frac{5}{3}y = 5$

22. $\frac{1}{2}x - y = -3$
$-x + 2y = 6$

Write each of the following in standard form

$$ax + by = c$$

$$dx + ey = f$$

and solve.

23. $y = 3x - 3$
$6x = 8 + 3y$

24. $3x = 2y$
$y = -7 - 2x$

25. $3m + 2n = 2m + 2$
$2m + 3n = 2n - 2$

26. $2x - 3y = 1 - 3x$
$4y = 7x - 2$

27. If 3 limes and 12 lemons cost 81 cents, and 2 limes and 5 lemons cost 42 cents, what is the cost of 1 lime and 1 lemon?

28. Find the capacity of each of two trucks if three trips of the larger and four trips of the smaller results in a total haul of 41 tons, and if four trips of the larger and three trips of the smaller results in a total haul of 43 tons.

C *Solve by elimination method.*

29. $0.3x - 0.6y = 0.18$
$0.5x + 0.2y = 0.54$

30. $0.8x - 0.3y = 0.79$
$0.2x - 0.5y = 0.07$

31. $\frac{x}{3} + \frac{y}{2} = 4$
$\frac{x}{3} - \frac{y}{2} = 0$

32. $\frac{x}{4} + \frac{y}{3} = 0$
$-\frac{x}{4} + \frac{y}{3} = -8$

33. $\frac{x}{2} + \frac{y}{3} = 1$
$\frac{2x}{3} + \frac{y}{2} = 2$

34. $\frac{a}{4} - \frac{2b}{3} = -2$
$\frac{a}{2} - b = -2$

35. Show that any ordered pair of numbers (a,b) that satisfies the system

$$Ax + By + C = 0$$

$$Dx + Ey + F = 0$$

will also satisfy the equation $m(Ax + By + C) + n(Dx + Ey + F) = 0$ for any real numbers m and n.

5.8 Applications

This section contains a wide variety of applications grouped by subject areas similar to those found in Sec. 5.2. All problems should be solved using a two-equation–two-unknown method discussed in the preceding section.

EXAMPLE 14 If you have 25 dimes and quarters in your pocket worth \$4, how many of each do you have?

SOLUTION Let $x =$ the number of dimes
$y =$ the number of quarters

then

$$\begin{aligned} x + \quad y &= 25 \\ 10x + 25y &= 400 \end{aligned}$$

Multiply the top equation by -10 and add:

$$\begin{aligned} -10x - 10y &= -250 \\ 10x + 25y &= 400 \\ \hline 15y &= 150 \\ y &= 10 \quad \text{quarters} \\ x + 10 &= 25 \\ x &= 15 \quad \text{dimes} \end{aligned}$$

CHECK: $10 + 15 = 25$ coins; $10(25) + 15(10) = 250 + 150 = 400$ cents or \$4

PROBLEM 14 If you have 25 nickels and quarters in your pocket worth \$2.25, how many of each do you have?

ANSWER 20 nickels, 5 quarters

EXAMPLE 15 The population of a town is 30,000, and it is decreasing at the rate of 550 per year. Another town has a population of 18,000 which is increasing at the rate of 1,450 per year. In how many years will both towns be the same size, and what will their population be at that time?

SOLUTION Let x be the population of a town after t years, then

$$x = 30{,}000 - 550t \qquad \text{first town}$$

$$x = 18{,}000 + 1{,}450t \qquad \text{second town}$$

Subtract to eliminate x (or substitute the right member of the second equation into the first member of the first equation):

$$\begin{aligned} 0 &= 12{,}000 - 2{,}000t \\ 2{,}000t &= 12{,}000 \\ t &= 6 \text{ years} \end{aligned}$$

(Both will be the same size.)

Use either equation to find the size of the towns after 6 years.

$$\begin{aligned} x &= 30{,}000 - 550(6) \\ &= 30{,}000 - 3{,}300 \\ &= 26{,}700 \text{ people} \end{aligned}$$

PROBLEM 15 Repeat Example 15 with the first town starting with a population of 150,000 and decreasing at 1,900 per year and the second town starting with a population of 100,000 and increasing at 3,100 per year.

ANSWER 10 years; 131,000 people

Exercise 43

The problems in this exercise are grouped in the following subject areas: business, chemistry, earth sciences, economics, geometry, domestic, life science, music, physics-engineering, psychology, and puzzles. No attempt has been made to arrange the problems in order of difficulty nor to match them in pairs as in preceding exercises.

BUSINESS

1. A packing carton contains 144 small packages, some weighing $\frac{1}{4}$ lb each and the others $\frac{1}{2}$ lb each. How many of each type are in the carton if the total contents of the carton weighs 51 lb?

2. Two architects contract to do the design and detail drawings for a building and agree to share the fee in the ratio of 3:5. If the total fee to the client is \$3,600, how much should each receive? HINT: Use the proportion $x/y = 3/5$ as one of the two equations.

3. A secretarial service charges \$4 per hr for a stenographer and \$2.50 per hr for a typist. On a particular job the bill from the service was \$44. If the typist worked 2 more hours than the stenographer, how much time did each spend on the job, and how much did each earn?

4. BREAKEVEN ANALYSIS. It costs a book publisher \$12,000 to prepare a book for publication (art work, plates, reviews, etc.); printing costs are \$3 per book.
(A) If the book is sold to bookstores for \$7 a copy, how many copies must be sold to break even, and what are the cost and revenue for this number. HINT: Solve the system

$$C = 12{,}000 + 3n$$

$$R = 7n$$

$$R = C$$

(B) Graph the first two equations on the same coordinate system for $0 \le n \le 20{,}000$. Interpret the regions between the lines to the left and to the right of the break-even point.

CHEMISTRY

5. A chemist has two concentrations of hydrochloric acid in stock, a 50 percent solution and an 80 percent solution. How much of each should he take to get 100 grams of a 68 percent solution?

6. A farmer placed an order with a chemical company for a fertilizer that would contain, among other things, 120 lb of nitrogen and 90 lb of phosphoric acid. The company had two mixtures on hand with the following compositions:

	NITROGEN	PHOSPHORIC ACID
Mixture *A*	20%	10%
Mixture *B*	6%	6%

How many pounds of each mixture should the chemist mix to fill the order?

EARTH SCIENCES

7. An earthquake emits a primary wave and a secondary wave. Near the surface of the earth the primary wave travels at about 5 miles per sec and the secondary wave at about 3 miles per sec. From the time lag between the two waves arriving at a given station, it is possible to estimate the distance to the quake. (The "epicenter" can be located by getting distance bearings at three or more stations.) Suppose a station measured a time difference of 16 sec between the arrival of the two waves. How long did each wave travel, and how far would the earthquake be from the station?

ECONOMICS

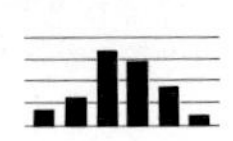

8. SUPPLY AND DEMAND. An important problem in economic studies has to do with supply and demand. The quantity of a product that people are willing to buy on a given day generally depends on its price; similarly, the quantity of a product that a supplier is willing to sell on a given day also depends on the price he is able to get for his product.

Let us assume that in a small town on a particular day the demand (in pounds) for hamburger is given by

$$d = 2{,}400 - 1{,}200p \qquad \$0.50 \le p \le \$1.75$$

and the supply by

$$s = -900 + 1{,}800p \qquad \$0.50 \le p \le \$1.75$$

Using these equations, we see that at \$1.50 per lb the people in the town would only be willing to buy 600 lb of hamburger on that day, whereas the suppliers would be willing to supply 1,800 lb. Hence, the supply would exceed the demand and force prices down. On the other hand, if the price were 75 cents per pound, the people would then be willing to buy 1,500 lb of hamburger on that day, but the supplier would only be willing to sell 450 lb. Thus the demand would exceed the supply, and prices would go up. At what price would hamburger stabilize for the day; that is, at what price would the supply actually equal the demand ($s = d$)?

(*A*) Solve graphically by graphing the supply and demand equations on the same coordinate system. (The point of intersection of the two graphs is called the *equilibrium point.*)
(*B*) Solve algebraically.
(*C*) Interpret the graph to the left and to the right of the equilibrium point.

GEOMETRY

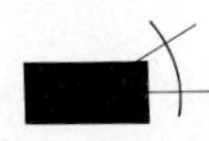

9. An 18-ft board is cut into 2 pieces so that one piece is 4 ft longer than the other piece. How long is each piece?

10. Find the dimensions of a rectangle with perimeter 72 in. if its length is 25 percent longer than its width.

11. If the sum of two angles in a right triangle is 90° and their difference is 14°, find the two angles.

DOMESTIC

12. A family wishes to invest its savings of $15,000. They decide to put part of it in the bank at 4 percent and the rest in a riskier real estate investment at 6 percent. How much should they invest in each if they want the annual return from each to be the same?

13. A school put on a musical comedy and sold 1,000 tickets for a total of $650. If tickets were sold to students for 50 cents and to adults for $1, how many of each type were sold?

14. Wishing to log some flying time, you have rented an airplane for 2 hr. You decide to fly due east until you have to turn around in order to be back at the airport at the end of the 2 hr. The cruising speed of the plane is 120 mph in still air.
(*A*) If there is a 30-mph wind blowing from the east, how long should you head east before you turn around, and how long will it take you to get back?
(*B*) How far from the airport were you when you turned back?
(*C*) Answer parts *A* and *B* with the assumption that no wind is blowing.

LIFE SCIENCE

15. The number of offspring that have brown-eyed parents who carry the recessive gene for blue eyes is 1,236. According to Mendel's laws of heredity, the expected ratio of offspring with brown eyes to those with blue eyes is 3:1. How many offspring with each eye color would you expect in the 1,236 sample? HINT: For one of the equations use the proportion $x/y = 3/1$.

16. A biologist, in a nutrition experiment, wants to prepare a special diet for his experimental animals. He requires a food mixture that contains, among other things, 20 oz of protein and 6 oz of fat. He is able to purchase food mixes of the following compositions:

	PROTEIN	FAT
Mix *A*	20%	2%
Mix *B*	10%	6%

How many ounces of each mix should he use to get the diet mix? Solve graphically and algebraically.

MUSIC

17. If a guitar string is divided in the ratio of 4:5, a major third will result. What will be the length of each part if a 36-in. string is used? HINT: Use the proportion $x/y = 4/5$ for one of the equations.

18. If a guitar string is divided in the ratio of 5:8, a minor sixth will result. How would you divide a 39-in. string to produce a minor sixth?

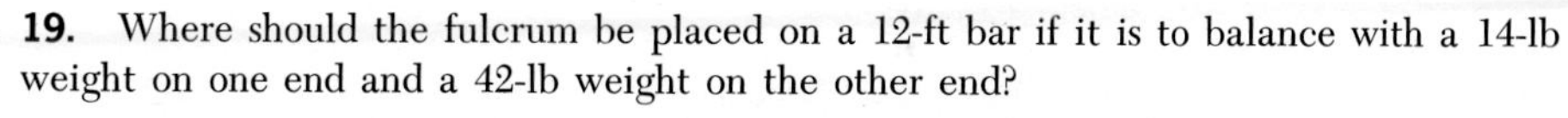

PHYSICS-ENGINEERING

19. Where should the fulcrum be placed on a 12-ft bar if it is to balance with a 14-lb weight on one end and a 42-lb weight on the other end?

20. In a Gemini-Agena rendezvous flight preparatory to placing man on the moon, the Agena passed over a tracking station in Carnoarvon, Australia, 6 min (0.1 hr) before the pursuing, astronaut-carrying Gemini. If the Agena was traveling at 17,000 mph and the Gemini at 18,700 mph, how long did it take the Gemini (after passing the tracking station) to rendezvous with the Agena, and how far from the tracking station, in the direction of motion, did this take place?

PSYCHOLOGY

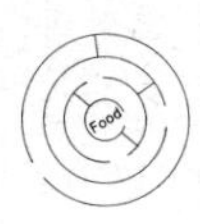

21. Professor Brown, a psychologist, trained a group of rats (in an experiment on motivation and avoidance) to run down a narrow passage in a cage to receive food in a goal box. He put a harness on each rat and connected the harness to an overhead wire that was attached to a scale. In this way he could place the rat at different distances from the food and measure the pull (in grams) of the rat toward the food. He found that a relation between motivation and distance was given approximately by the equation $p = -\frac{1}{5}d + 70$ with $30 \leq d \leq 175$, where p is the pull in grams and d is the distance from the goal box in centimeters.

Professor Brown then replaced the food with a mild electric shock, and with the same apparatus he was able to measure the avoidance strength relative to the distance from the object to be avoided. He found that the avoidance strength was given approximately by $a = -\frac{4}{3}d + 230$ with $30 \leq d \leq 175$, where a is the avoidance measured in grams and d is the distance from the goal box in centimeters.

If the rat were trained in both experiments, at what distance from the goal box should the approach and avoidance strength be the same? Solve algebraically and graphically. If the goal box is on the right, what would you predict that the rat will do if he is placed to the right of this point? To the left of this point?

PUZZLES

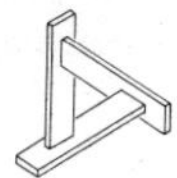

22. A friend of yours came out of a post office having spent \$1.32 on thirty 4-cent and 5-cent stamps. How many of each type did he buy?

23. A bank gave you \$1.50 in change consisting of only nickels and dimes. If there were 22 coins in all, how many of each type of coin did they give you?

24. If one flask and four mixing dishes balance 14 test tubes and two mixing dishes, and if two flasks balance two test tubes and 6 mixing dishes, how many test tubes will balance one flask, and how many test tubes will balance one mixing dish?

Exercise 44 Chapter Review

A

1. Which of the following equations has no solution: (*A*) $2x - 3 = 2(x + 4)$, (*B*) $3x - 3 = 2(x + 4)$?

2. Find the dimensions of a rectangle with perimenter 42 ft if the length is 3 ft less than twice the width.

3. Solve and graph: $4x + 9 \leq -3$

4. Solve and graph: $-14 \leq 3x - 2 < 7$

5. What numbers satisfy the condition, "5 less than 5 times the number is less than or equal to 10"?

6. Graph: $y = 2x - 3$

7. Graph: $2x + y = 6$

8. Solve graphically:

$$x - y = 5$$
$$x + y = 7$$

9. Solve by the elimination method:

$2x + 3y = 7$
$3x - y = 5$

10. Solve using two-equation–two-unknown methods: If you have 30 nickels and dimes in your pocket worth \$2.30, how many of each do you have?

B **11.** What is the maximum and minimum number of solutions for an equation of the type $ax + b = 0$, $a \neq 0$?

12. If you spent \$1.76 on nineteen 8-cent and 11-cent stamps, how many of each did you buy?

13. Solve and graph:

$$3x - 9 \leq 7x - 5$$

14. Solve and graph:

$$\frac{x-1}{3} - 1 \leq \frac{x}{2}$$

15. A chemical is to be kept between 59 and 86°F. What is the temperature range in Celsius degrees? (Use $F = \frac{9}{5}C + 32$.)

16. Graph: $y = \frac{1}{3}x - 2$

17. Graph: $3x - 2y = 9$

18. Solve graphically:

$2x - 3y = -3$
$3x + y = 12$

19. Solve by the elimination method:

$5m - 3n = 4$
$-2m + 4n = 10$

20. Part of \$6,000 is to be invested at 10 percent and the rest at 6 percent. How much should be invested at each rate if the total annual return from both investments is to be \$440? (Use 2-equation–2-unknown methods.)

C **21.** Show that $-b/a$, $a \neq 0$, is a solution to $ax + b = 0$.

22. How many gallons of pure acid must be mixed with 5 gal of 40 percent solution to get a 45 percent solution?

23. Indicate true (T) or false (F):

(A) If $x < y$ and $a > 0$, then $ax < ay$.
(B) If $x < y$ and $a < 0$, then $ax > ay$.

24. To develop a certain roll of photographic film, the temperature of the solution must be kept between 20 and 25°C (that is, $20 \leq C \leq 25$). Find the temperature range in Fahrenheit degrees [$C = \frac{5}{9}(F - 32)$].

25. Graph $y = x/2 + b$ for $b = -4$, 0, and 4, all on the same coordinate system.

26. Solve graphically:

$$2x - 6y = -3$$

$$-\frac{2}{3}x + 2y = 1$$

27. Solve by the elimination method:

$$x - 4y = 12$$

$$-\frac{x}{4} + x = 4$$

28. Solve using two-equation–two-unknown methods: A chemist has two concentrations of acid in stock, a 40 percent and a 70 percent solution. How much of each should he take to get 100 grams of a 49 percent solution?

CHAPTER 6 Algebraic Fractions

6.1 Rational Expressions

Fractional forms in which the numerator and denominator are polynomials are called *rational expressions*.

$$\frac{1}{x} \qquad \frac{1}{y-3} \qquad \frac{x-2}{2x^2-2x+5} \qquad \frac{x^2-3xy+y^2}{x^2-y^2}$$

are all rational expressions. (Recall that a constant is a polynomial of degree 0.)

In Chaps. 3 and 4 we worked with simple fractional forms. In this chapter we will consider more complicated forms, but we will use the basic ideas established earlier.

Since all of the polynomials that we will work with will represent real numbers for real number replacements of the variables, all of the properties of the real numbers listed at the end of Chap. 3 apply in what follows.

6.2 Multiplication and Division

We extend the earlier definition of multiplication of rational numbers to include the real numbers:

DEFINITION OF MULTIPLICATION OF REAL FRACTIONS

For a, b, c, and d any real numbers with $b, d \neq 0$, then

$$\frac{a}{b} \cdot \frac{c}{d} = \frac{a \cdot c}{b \cdot d}$$

This definition coupled with the fundamental principle of real fractions

$$\frac{ak}{bk} = \frac{a}{b} \qquad k \neq 0$$

provide the basic tools for multiplying and reducing rational expressions. The following examples should make the process clear.

EXAMPLE 1 (A)
$$\frac{3a^2b}{4c^2d} \cdot \frac{8c^2d^3}{9ab^2} = \frac{(3a^2b) \cdot (8c^2d^3)}{(4c^2d) \cdot (9ab^2)} = \frac{24a^2bc^2d^3}{36ab^2c^2d} = \frac{(2ad^2)(\cancel{12abc^2d})}{(3b)(\cancel{12abc^2d})} = \frac{2ad^2}{3b}$$

This process is easily shortened to the following when it is realized that, in effect, any factor in a numerator may "cancel" any like factor in a denominator. Thus,

$$\frac{\overset{1 \cdot a \cdot 1}{\cancel{3}\cancel{a^2}\cancel{b}}}{\underset{1 \cdot 1 \cdot 1}{\cancel{4}\cancel{c^2}\cancel{d}}} \cdot \frac{\overset{2 \cdot 1 \cdot d^2}{\cancel{8}\cancel{c^2}\cancel{d^3}}}{\underset{3 \cdot 1 \cdot b}{\cancel{9}\cancel{a}\cancel{b^2}}} = \frac{2ad^2}{3b}$$

(B)
$$(x^2 - 4) \cdot \frac{2x-3}{x+2} = \frac{\overset{1}{\cancel{(x+2)}}(x-2)}{1} \cdot \frac{2x-3}{\underset{1}{\cancel{x+2}}} = (x-2)(2x-3)$$

(C)
$$\frac{4a^2 - 9b^2}{4a^2 + 12ab + 9b^2} \cdot \frac{6a^2b}{8a^2b^2 - 12ab^3} = \frac{\overset{1}{\cancel{(2a-3b)}}\overset{1}{\cancel{(2a+3b)}}}{\underset{(2a+3b)}{\cancel{(2a+3b)^2}}} \cdot \frac{\overset{3a}{\cancel{6a^2b}}}{\underset{2b \quad\; 1}{\cancel{4ab^2}\cancel{(2a-3b)}}}$$

$$= \frac{3a}{2b(2a+3b)}$$

PROBLEM 1 Multiply and reduce to lowest terms:

(A) $\dfrac{4x^2y^3}{9w^2z} \cdot \dfrac{3wz^2}{2xy^4}$ (B) $\dfrac{x+5}{x^2-9} \cdot (x+3)$

(C) $\dfrac{x^2 - 9y^2}{x^2 - 6xy + 9y^2} \cdot \dfrac{6x^2y}{2x^2 + 6xy}$

ANSWER (A) $\dfrac{2xz}{3wy}$ (B) $\dfrac{x+5}{x-3}$ (C) $\dfrac{3xy}{x-3y}$

It follows from the definition of division (recall: $A \div B = Q$ if and only if $A = BQ$ and Q is unique) that for all nonzero real numbers b, c, and d, and any real number a,

$$\frac{a}{b} \div \frac{c}{d} = \frac{a}{b} \cdot \frac{d}{c}$$

To verify this statement, multiply the quotient by the divisor to obtain the dividend. Thus,

> To divide one rational expression by another, multiply by the reciprocal of the divisor.

EXAMPLE 2

(A) $\dfrac{6a^2b^3}{5cd} \div \dfrac{3a^2c}{10bd} = \dfrac{6a^2b^3}{5cd} \cdot \dfrac{10bd}{3a^2c} = \dfrac{4b^4}{c^2}$

(B) $(x + 4) \div \dfrac{2x^2 - 32}{6xy} = \dfrac{x+4}{1} \cdot \dfrac{6xy}{2(x-4)(x+4)} = \dfrac{3xy}{x-4}$

(C) $\dfrac{10x^3y}{3xy + 9y} \div \dfrac{4x^2 - 12x}{x^2 - 9} = \dfrac{10x^3y}{3y(x+3)} \cdot \dfrac{(x+3)(x-3)}{4x(x-3)} = \dfrac{5x^2}{6}$

PROBLEM 2 Divide and reduce to lowest terms:

(A) $\dfrac{8w^2z^2}{9x^2y} \div \dfrac{4wz}{6xy^2}$ (B) $\dfrac{2x^2 - 8}{4x} \div (x + 2)$

(C) $\dfrac{x^2 - 4x + 4}{4x^2y - 8xy} \div \dfrac{x^2 + x - 6}{6x^2 + 18x}$

ANSWER (A) $\dfrac{4wyz}{3xy}$ (B) $\dfrac{x-2}{2x}$ (C) $\dfrac{3}{2y}$

Exercise 45

A *Perform the indicated operations and simplify.*

1. $\dfrac{15}{16} \cdot \dfrac{24}{27}$ **2.** $\dfrac{6}{7} \cdot \dfrac{28}{9}$

3. $\frac{36}{8} \div \frac{9}{4}$

4. $\frac{4}{6} \div \frac{24}{8}$

5. $\frac{y^4}{3u^5} \cdot \frac{2u^3}{3y}$

6. $\frac{6x^3y}{7u} \cdot \frac{14u^3}{12xy}$

7. $\frac{uvw}{5xyz} \div \frac{5vy}{uwxz}$

8. $\frac{3c^2d}{a^3b^3} \div \frac{3a^3b^3}{cd}$

9. $\frac{x+3}{2x^2} \cdot \frac{4x}{x+3}$

10. $\frac{3x^2y}{x-y} \cdot \frac{x-y}{6xy}$

11. $\frac{a^2-a}{a-1} \cdot \frac{a+1}{a}$

12. $\frac{x+3}{x^3+3x^2} \cdot \frac{x^3}{x-3}$

13. $\frac{4x}{x-4} \div \frac{8x^2}{x^2-6x+8}$

14. $\frac{x-2}{4y} \div \frac{x^2-x-6}{12y^2}$

B **15.** $\frac{d^5}{3a} \div \left(\frac{d^2}{6a^2} \cdot \frac{a}{4d^3}\right)$

16. $\left(\frac{d^5}{3a} \div \frac{d^2}{6a^2}\right) \cdot \frac{a}{4d^3}$

17. $\frac{2x^2+4x}{12x^2y} \cdot \frac{6x}{x^2+6x+8}$

18. $\frac{6x^2}{4x^2y-12xy} \cdot \frac{x^2+x-12}{3x^2+12x}$

19. $\frac{2y^2+7y+3}{4y^2-1} \div (y+3)$

20. $(t^2-t-12) \div \frac{t^2-9}{t^2-3t}$

21. $\frac{x^2-6x+9}{x^2-x-6} \div \frac{x^2+2x-15}{x^2+2x}$

22. $\frac{m+n}{m^2-n^2} \div \frac{m^2-mn}{m^2-2mn+n^2}$

23. $-(x^2-4) \cdot \frac{3}{x+2}$

24. $-(x^2-3x) \cdot \frac{x-2}{x-3}$

C **25.** $\frac{2-m}{2m-m^2} \cdot \frac{m^2+4m+4}{m^2-4}$

26. $\frac{9-x^2}{x^2+5x+6} \cdot \frac{x+2}{x-3}$

27. $\left(\frac{x^2-xy}{xy+y^2} \div \frac{x^2-y^2}{x^2+2xy+y^2}\right) \div \frac{x^2-2xy+y^2}{x^2y+xy^2}$

28. $\frac{x^2-xy}{xy+y^2} \div \left(\frac{x^2-y^2}{x^2+2xy+y^2} \div \frac{x^2-2xy+y^2}{x^2y+xy^2}\right)$

29. $(x^2-1)/(x-1)$ and $x+1$ name the same real number for (all, all but one, no) replacements of x by real numbers.

30. $(x^2-x-6)/(x-3) = x+2$, except for what values of x?

31. Can you evaluate the following arithmetic problem in less than 3 min?

$$\frac{(108{,}641)^2 - (108{,}643)^2}{(108{,}642)(108{,}646) - (108{,}644)^2}$$

6.3 Addition and Subtraction

Addition and subtraction of rational expressions are based on the following properties of real fractions:

$$\frac{a}{b} + \frac{c}{b} = \frac{a+c}{b} \tag{1}$$

$$\frac{a}{b} - \frac{c}{b} = \frac{a-c}{b} \tag{2}$$

$$\frac{a}{b} = \frac{ak}{bk} \qquad k \neq 0 \tag{3}$$

Thus, if the denominators of two rational expressions are the same, we may either add or subtract the expressions by adding or subtracting the numerators and placing the result over the common denominator; if the denominators are not the same, we use property (3) to change the form of each fraction so they have a common denominator and then use either (1) or (2).

Even though any common denominator will do, the problem will generally become less involved if the least common denominator (LCD) is used. If the LCD is not obvious (often it is), then we proceed as we did when we studied rational numbers in Chap. 3, that is, factor each denominator completely, including numerical coefficients. The LCD should then contain each different factor in the denominators to the highest power it occurs in any one denominator.

EXAMPLE 3 (A)

$$\frac{4(x+4)}{4x(x-3)} - \frac{2(x-3)}{4x(x-3)} = \frac{4(x+4)-2(x-3)}{4x(x-3)} = \frac{4x+16-2x+6}{4x(x-3)}$$

$$= \frac{2x+22}{4x(x-3)} = \frac{2(x+11)}{4x(x-3)} = \frac{x+11}{2x(x-3)}$$

(B)

$$\frac{2}{2y} + \frac{1}{4y^2} - 1 = \frac{1(2y)}{(2y)(2y)} + \frac{1}{4y^2} - \frac{4y^2}{4y^2} = \frac{2y+1-4y^2}{4y^2} = \frac{1+2y-4y^2}{4y^2}$$

Note that the LCD is $4y^2$.

(C)

$$\frac{4}{x^2-4} - \frac{3}{x^2-x-2} = \frac{4}{(x-2)(x+2)} - \frac{3}{(x-2)(x+1)}$$

We see that the LCD is $(x-2)(x+2)(x+1)$. Then

$$\frac{4(x+1)}{(x-2)(x+2)(x+1)} - \frac{3(x+2)}{(x-2)(x+2)(x+1)} = \frac{4(x+1)-3(x+2)}{(x-2)(x+2)(x+1)}$$

$$= \frac{4x+4-3x-6}{(x-2)(x+2)(x+1)} = \frac{(x-2)}{(x-2)(x+2)(x+1)} = \frac{1}{(x+2)(x+1)}$$

PROBLEM 3 Combine into single fractions and simplify:

(A) $\dfrac{5(x-4)}{2(x+2)} - \dfrac{3(x-2)}{2(x+2)}$ (B) $\dfrac{1}{3y^2} - \dfrac{1}{6y} + 1$

(C) $\dfrac{3}{x^2-1} - \dfrac{2}{x^2+2x+1}$

ANSWER (A) $\dfrac{x-7}{x+2}$ (B) $\dfrac{2-y+6y^2}{6y^2}$ (C) $\dfrac{x+5}{(x-1)(x+1)^2}$

Exercise 46

A *Combine into single fractions and simplify.*

1. $\dfrac{5y}{4x} + \dfrac{3}{4x}$

2. $\dfrac{2x}{3y} + \dfrac{5}{3y}$

3. $\dfrac{3m}{2m^2} - \dfrac{1}{2m^2}$

4. $\dfrac{7x}{5x^2} - \dfrac{2}{5x^2}$

5. $\dfrac{2}{x} - \dfrac{1}{3}$

6. $\dfrac{3x}{y} + \dfrac{1}{4}$

7. $x + \dfrac{1}{x}$

8. $\dfrac{2}{x} + 1$

9. $\dfrac{1}{x} - \dfrac{y}{x^2} + \dfrac{y^2}{x^3}$

10. $\dfrac{u}{v^2} - \dfrac{1}{v} + \dfrac{u^2}{v^3}$

11. $\dfrac{3x-2}{2x} - \dfrac{x}{2x}$

12. $\dfrac{4t-1}{4mn} - \dfrac{3}{4mn}$

13. $\dfrac{5a}{a-1} - \dfrac{5}{a-1}$

14. $\dfrac{4x}{2x-1} - \dfrac{2}{2x-1}$

15. $\dfrac{y}{y^2-9} - \dfrac{3}{y^2-9}$

16. $\dfrac{2x}{4x^2-9} - \dfrac{3}{4x^2-9}$

17. $\dfrac{1}{x-2} + \dfrac{1}{x+3}$

18. $\dfrac{2}{x+1} + \dfrac{3}{x-2}$

19. $\dfrac{3}{2x} - \dfrac{2}{x+2}$

20. $\dfrac{2}{3y} - \dfrac{3}{y+3}$

B **21.** $\dfrac{3y+8}{4y^2} - \dfrac{2y-1}{y^3} - \dfrac{5}{8y}$

22. $\dfrac{4t-3}{18t^3} + \dfrac{3}{4t} - \dfrac{2t-1}{6t^2}$

23. $\dfrac{1}{2a^2} - \dfrac{2b-1}{2a^2}$

24. $\dfrac{5}{3k} - \dfrac{6x-4}{3k}$

25. $\dfrac{4}{2x - 3} - \dfrac{2x + 1}{(2x - 3)(x + 2)}$

26. $\dfrac{3}{x + 3} - \dfrac{3x + 1}{(x - 1)(x + 3)}$

27. $\dfrac{a}{a - 1} - \dfrac{2}{a^2 - 1}$

28. $\dfrac{m + 2}{m - 2} - \dfrac{m^2 + 4}{m^2 - 4}$

29. $\dfrac{3x - 1}{2x^2 + x - 3} - \dfrac{2}{x - 1}$

30. $\dfrac{1}{m^2 - n^2} + \dfrac{1}{m^2 + 2mn + n^2}$

31. $\dfrac{t + 1}{t - 1} - 1$

32. $2 + \dfrac{x + 1}{x - 3}$

33. $\dfrac{x^2 - 2x}{x + 2} + x - 3$

34. $x - 3 - \dfrac{x - 1}{x - 2}$

35. $\dfrac{1}{y + 2} + 3 - \dfrac{2}{y - 2}$

36. $5 + \dfrac{a}{a + 1} - \dfrac{a}{a - 1}$

37. $\dfrac{2}{x + 3} - \dfrac{1}{x - 3} + \dfrac{2x}{x^2 - 9}$

38. $\dfrac{2x}{x^2 - y^2} + \dfrac{1}{x + y} - \dfrac{1}{x - y}$

39. $\dfrac{2t}{3t^2 - 48} + \dfrac{t}{4t + t^2}$

40. $\dfrac{3s}{3s^2 - 12} + \dfrac{1}{2s^2 + 4s}$

C **41.** $\dfrac{3}{x - 1} + \dfrac{2}{1 - x}$

42. $\dfrac{5}{y - 3} - \dfrac{2}{3 - y}$

43. $\dfrac{1}{5x - 5} - \dfrac{1}{3x - 3} + \dfrac{1}{1 - x}$

44. $\dfrac{x + 7}{ax - bx} + \dfrac{y + 9}{by - ay}$

45. $\dfrac{1 + \dfrac{3}{x}}{x - \dfrac{9}{x}}$

46. $\dfrac{1 - \dfrac{y^2}{x^2}}{1 - \dfrac{y}{x}}$

47. $\dfrac{\dfrac{1}{x} + \dfrac{1}{y}}{\dfrac{y}{x} - \dfrac{x}{y}}$

48. $1 - \dfrac{1}{1 - \dfrac{1}{x}}$

6.4 Fractional Equations

We have already considered equations with rational coefficients, such as

$$\frac{x}{2} - 3 = \frac{2x + 5}{4}$$

(see Chap. 3), and found we could easily convert them to equivalent equations with integer coefficients by multiplying both sides by the LCD of all fractions in the equation—in the above case 4.

If an equation involves variables in one or more denominators, such as

$$\frac{3}{x} - \frac{1}{2} = \frac{4}{x}$$

we may proceed in essentially the same way as long as we stay away from any value that makes a denominator 0. In this case

$$x \neq 0$$

EXAMPLE 4 Solve and check:

$$\frac{3}{x} - \frac{1}{2} = \frac{4}{x}$$

SOLUTION

$$\frac{3}{x} - \frac{1}{2} = \frac{4}{x} \qquad x \neq 0$$

$$(2x)\frac{3}{x} - (2x)\frac{1}{2} = (2x)\frac{4}{x}$$

$$6 - x = 8$$
$$-x = 2$$
$$x = -2$$

CHECK

$$\frac{3}{-2} - \frac{1}{2} \stackrel{?}{=} \frac{4}{-2}$$

$$-\frac{4}{2} \stackrel{\checkmark}{=} -\frac{4}{2}$$

PROBLEM 4 Solve and check:

$$\frac{2}{3} - \frac{2}{x} = \frac{4}{x}$$

ANSWER $x = 9$

EXAMPLE 5 Solve and check:

$$\frac{3x}{x-2} - 4 = \frac{14-4x}{x-2}$$

SOLUTION

$$\frac{3x}{x-2} - 4 = \frac{14-4x}{x-2} \qquad x \neq 2$$

$$(x-2)\frac{3x}{x-2} - 4(x-2) = (x-2)\frac{14-4x}{x-2}$$

$$3x - 4x + 8 = 14 - 4x$$
$$3x = 6$$
$$x = 2$$

Since $x \neq 2$, the original equation has no solution.

PROBLEM 5 Solve and check:

$$\frac{2x}{x-1} - 3 = \frac{7-3x}{x-1}$$

ANSWER $x = 2$

Exercise 47

Solve:

A

1. $\frac{2}{x} - \frac{1}{3} = \frac{5}{x}$

2. $\frac{1}{2} - \frac{2}{x} = \frac{3}{x}$

3. $\frac{5}{6} - \frac{1}{y} = \frac{2}{3y}$

4. $\frac{1}{x} + \frac{2}{3} = \frac{1}{2}$

5. $\frac{2}{3x} + \frac{1}{2} = \frac{4}{x} + \frac{4}{3}$

6. $\frac{1}{m} - \frac{1}{9} = \frac{4}{9} - \frac{2}{3m}$

7. $\frac{1}{2t} + \frac{1}{8} = \frac{2}{t} - \frac{1}{4}$

8. $\frac{4}{3k} - 2 = \frac{k+4}{6k}$

B

9. $\frac{9}{L+1} - 1 = \frac{12}{L+1}$

10. $\frac{7}{y-2} - \frac{1}{2} = 3$

11. $\frac{3}{2x-1} + 4 = \frac{6x}{2x-1}$

12. $\frac{5x}{x+5} = 2 - \frac{25}{x+5}$

13. $\frac{3N}{N-2} - \frac{9}{4N} = 3$

14. $\frac{2E}{E-1} = 2 + \frac{5}{2E}$

15. $\frac{5}{x-3} = \frac{33-x}{x^2-6x+9}$

16. $\frac{D^2+2}{D^2-4} = \frac{D}{D-2}$

17. $\frac{n-5}{6n-6} = \frac{1}{9} - \frac{n-3}{4n-4}$

18. $\frac{1}{3} - \frac{s-2}{2s+4} = \frac{s+2}{3s+6}$

C

19. $\frac{2}{x-2} = 3 - \frac{5}{2-x}$

20. $\frac{3x}{x-4} - 2 = \frac{3}{4-x}$

21. $\frac{5t-22}{t^2-6t+9} - \frac{11}{t^2-3t} - \frac{5}{t} = 0$

22. $\frac{1}{c^2-c-2} - \frac{3}{c^2-2c-3} = \frac{1}{c^2-5c+6}$

6.5 Applications

This section includes applications involving fractional equations associated with optics, electricity, hydraulics, social science, distance-rate-time, work-rate-time, business, and puzzles. The problems are self contained and require no previous knowledge of the subjects concerned.

EXAMPLE 6

A speedboat takes 1.5 times longer to go 120 miles up a river than to return. If the boat cruises at 25 mph in still water, what is the rate of the current.

SOLUTION Let

$$x = \text{rate of current}$$
$$25 - x = \text{rate of boat upstream}$$
$$25 + x = \text{rate of boat downstream}$$

$$\text{time going upstream} = 1.5\ \text{time going downstream}$$

$$\frac{120}{25 - x} = (1.5)\frac{120}{25 + x} \qquad \text{Recall } t = d/r \text{ from } d = rt.$$

$$\frac{120}{25 - x} = \frac{180}{25 + x}$$

$$(25 + x)120 = (25 - x)180$$
$$3{,}000 + 120x = 4{,}500 - 180x$$
$$300x = 1{,}500$$
$$x = 5 \text{ mph (rate of current)}$$

CHECK

$$\text{Time upstream} = \frac{120}{20} = 6 \text{ hr}$$

$$\text{Time downstream} = \frac{120}{30} = 4 \text{ hr}$$

Therefore, time upstream is 1.5 times longer than time downstream.

PROBLEM 6

A fishing boat takes twice as long to go 24 miles up a river than to return. If the boat cruises at 9 mph in still water, what is the rate of the current?

ANSWER 3 mph

EXAMPLE 7

It takes 8 hr for one pipe to fill a swimming pool. Another pipe is installed to speed up the job, and together the pipes can now fill the pool in 3 hr. How long would it take the second pipe to fill the pool by itself?

SOLUTION Let

$x =$ the number of hours it takes the second pipe to fill the pool alone

$\frac{1}{8} =$ rate at which the first pipe fills the pool $\left(\frac{1}{8}\text{ of the pool per hour}\right)$

$\frac{1}{x} =$ rate at which the second pipe fills the pool $\left(\frac{1}{x}\text{ of the pool per hour}\right)$

$\frac{1}{8}(3) =$ part of pool filled by first pipe in 3 hr (rate · time)

$\frac{1}{x}(3) =$ part of pool filled by second pipe in 3 hr (rate · time)

$1 =$ one full pool (one whole job)

$$\frac{\text{part of pool filled by}}{\text{first pipe in 3 hr}} + \frac{\text{part of pool filled by}}{\text{second pipe in 3 hr}} = 1 \qquad \text{(one full pool)}$$

$$\frac{3}{8} \quad + \quad \frac{3}{x} \quad = 1$$

$$(8x)\frac{3}{8} + (8x)\frac{3}{x} = (8x)1$$

$$3x + 24 = 8x$$

$$x = 4\tfrac{4}{5}\text{ hr}$$

PROBLEM 7 Repeat Example 7 with the first pipe taking 6 hr alone and the two pipes together 2 hr.

ANSWER 3 hr

Exercise 48

OPTICS *In Problems 1 to 4 refer to Fig. 1 for appropriate formulas and interpretation.*

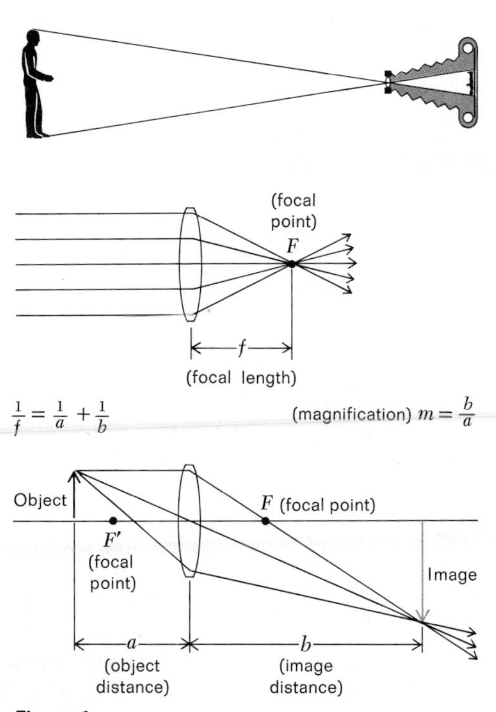

Figure 1

1. If the focal length f of a thin convex lens is 20 cm and an object is 30 cm from the lens, how far will the image be from the lens? How much will the object be magnified?

2. If the focal length f of a thin convex lens is 20 cm and an object is 80 cm from the lens, how far will the image be from the lens? How much will the object be magnified (or reduced)?

3. In an experiment to determine the focal length f of a thin convex lens the object and image distances were measured and found to be 15 and 60 cm, respectively. What is the focal length of the lens?

4. Solve the preceding problem with the object and image distances reversed.

HYDRAULICS *In Problems 5 and 6, refer to Fig. 2 for appropriate formulas and interpretation.*

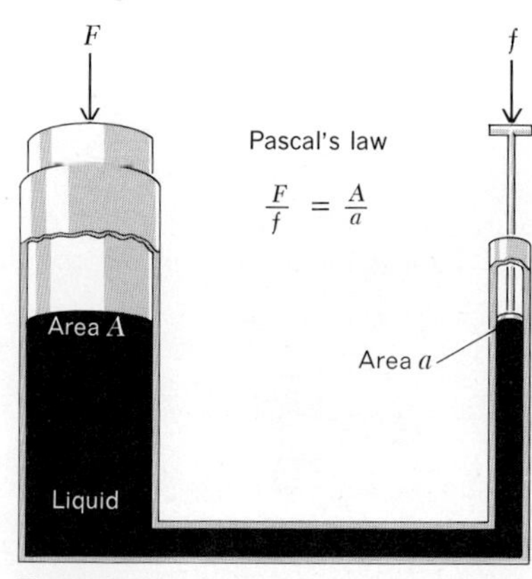

Figure 2

5. If the large cross-sectional area in a hydraulic lift is approximately 100 sq in. and a person wants to lift a weight of 5,000 lb with a 50-lb force, how large should the small cross-sectional area be?

6. If the large cross-sectional area of a hydraulic lift is 200 sq in. and the cross-sectional area of the small pipe is 3 sq in., how much force f will be required to lift 5,000 lb?

SOCIAL SCIENCE **7.** Anthropologists, in their study of race and human genetic groupings, use ratios called indices. One widely used index is the cephalic index, the ratio of the breadth of the head to its length expressed as a percent (looking down from above). Thus

$$C = \frac{100B}{L}$$

(long-headed, $C < 75$; intermediate, $75 \leq C \leq 80$; round-headed, $C > 80$.) If an Indian tribe in Baja, California had a cephalic index of 66 and the average breadth of their heads was 6.6 in., what was the average length of their heads?

8. Psychologists define intelligence quotient as the ratio of a person's mental age to his chronological age expressed as a percent: IQ = MA/CA $\times$ 100. If a student has a mental age of 12 and an IQ of 150, what is his chronological age?

DISTANCE-RATE-TIME

9. A jet airliner goes 900 miles with the wind in the same time it takes it to go 700 miles against the wind. If the wind velocity is 50 mph, what is the cruising speed of the airliner in still air?

10. A jet aircraft making a 600-mile trip travels at 200 mph for the first 300 miles and 600 mph for the last 300 miles. What was its average velocity for the total trip? The answer is not 400 mph! REMEMBER: average velocity is defined to be the ratio of the total distance to the total time.

11. A boat travels 40 miles upstream in the same time it takes it to travel 52 miles downstream. If the boat can cruise at 23 mph on still water, what is the speed of the current?

12. An explosion is set off on the surface of the water 11,000 ft from a ship. If the sound reaches the ship through the water 7.77 sec before it arrives through the air and if sound travels through water 4.5 times faster than through the air, how fast (to the nearest foot) does sound travel in air and in water?

WORK-RATE-TIME

13. At a family cabin, water is pumped and stored in a large water tank. Two pumps are used for this purpose. One can fill the tank by itself in 6 hr and the other can do the job in 9 hr. How long will it take both pumps operating together to fill the tank?

14. If one typist can do a job in 5 hr and another in 8 hr, how long will it take both together to do the job?

15. In an electronic computer center a card-sorter operator is given the job of alphabetizing a given quantity of IBM cards. He knows that an older sorter can do the job by itself in 3 hr. With the help of a newer machine the job is completed in 1 hr. How long would it take the new machine to do the job alone?

16. A small-town newspaper owns two printing presses, an older one and a new modern one. The new press operates 2.5 times faster than the older one. Together, they can get out the evening paper in 2 hr. How long would it take each press alone to prepare the evening paper?

BUSINESS

17. The simple interest formula is occasionally found in the form $P = A/(1 + rt)$ where r is the annual rate, P the principal, and A the amount due after t years. Find r if $A = 114$, $P = 100$, and $t = 4$.

18. In the preceding problem find t if $P = 100$, $A = 127$, and $r = 4\frac{1}{2}$ percent.

PUZZLES

19. The following famous Hindu puzzle is found in various forms in many different books (the version included here is from Mr. Maurice Kraitchik's "Mathematical Recreations," Dover, 1953): While three watchmen were guarding an orchard, a thief slipped in and stole some apples. On his way out he met the three watchmen one after the other, and to each in turn he gave a half of the apples he then had, and two besides. Thus he managed to escape with one apple. How many had he stolen originally?

6.6 Formulas and Equations with Several Variables

One of the immediate applications you will have for algebra in other courses is the changing of formulas or equations to alternate equivalent forms. The following examples are more or less typical.

EXAMPLE 8 Solve the formula $c = wrt/1{,}000$ for t. The formula gives the cost c of using an electrical appliance; w = power in watts, r = rate per kilowatt hour, t = time in hours.

SOLUTION

$$\frac{wrt}{1{,}000} = c \qquad \text{Recall: If } a = b, \text{ then } b = a.$$

$$\frac{1{,}000}{wr} \cdot \frac{wrt}{1{,}000} = \frac{1{,}000}{wr} \cdot c$$

$$t = \frac{1{,}000c}{wr}$$

PROBLEM 8 Solve the formula in Example 8 for w.

ANSWER

$$w = \frac{1{,}000c}{rt}$$

EXAMPLE 9 Solve the formula $A = P + Prt$ for r (simple interest formula).

SOLUTION

$$P + Prt = A$$

$$Prt = A - P$$

$$r = \frac{A - P}{Pt}$$

PROBLEM 9 Solve the formula $A = P + Prt$ for t.

ANSWER

$$t = \frac{A - P}{Pr}$$

EXAMPLE 10 Solve the formula $A = P + Prt$ for P.

SOLUTION

$$P + Prt = A$$

$$P(1 + rt) = A$$

$$P = \frac{A}{1 + rt}$$

PROBLEM 10 Solve $A = xy + xz$ for x.

ANSWER $x = \dfrac{A}{y + z}$

$xy + xz = A$
$x(y+z) = A$
$x = \dfrac{A}{y+z}$

Exercise 49

The following formulas and equations are widely used in science or mathematics.

A

1. Solve $A = P + I$ for I (SIMPLE INTEREST)
2. Solve $R = R_1 + R_2$ for R_2 (ELECTRICAL CIRCUITS—RESISTANCE IN SERIES)
3. Solve $d = rt$ for r (DISTANCE-RATE-TIME)
4. Solve $d = 1{,}100t$ for t (SOUND DISTANCE IN AIR)
5. Solve $I = Prt$ for t (SIMPLE INTEREST)
6. Solve $C = 2\pi r$ for r (CIRCUMFERENCE OF A CIRCLE)
7. Solve $C = \pi D$ for π (CIRCUMFERENCE OF A CIRCLE)
8. Solve $e = mc^2$ for m (MASS-ENERGY EQUATION)
9. Solve $ax + b = 0$ for x (FIRST-DEGREE POLYNOMIAL EQUATION IN ONE VARIABLE)
10. Solve $p = 2a + 2b$ for a (PERIMETER OF A RECTANGLE)
11. Solve $s = 2t - 5$ for t (SLOPE-INTERCEPT FORM FOR A LINE)
12. Solve $y = mx + b$ for m (SLOPE-INTERCEPT FORM FOR A LINE)

B

13. Solve $3x - 4y - 12 = 0$ for y (LINEAR EQUATION IN TWO VARIABLES)
14. Solve $Ax + By + C = 0$ for y (LINEAR EQUATION IN TWO VARIABLES)
15. Solve $I = \dfrac{E}{R}$ for E (ELECTRICAL CIRCUITS—OHM'S LAW)
16. Solve $m = \dfrac{b}{a}$ for a (OPTICS—MAGNIFICATION)
17. Solve $C = \dfrac{100B}{L}$ for L (ANTHROPOLOGY—CEPHALIC INDEX)
18. Solve $\text{IQ} = \dfrac{(100)(\text{MA})}{(\text{CA})}$ for (CA) (PSYCHOLOGY—INTELLIGENCE QUOTIENT)
19. Solve $F = G\dfrac{m_1m_2}{d^2}$ for m_1 (GRAVITATIONAL FORCE BETWEEN TWO MASSES)
20. Solve $F = G\dfrac{m_1m_2}{d^2}$ for G (GRAVITATIONAL FORCE BETWEEN TWO MASSES)

21. Solve $A = \frac{h}{2}(b_1 + b_2)$ for h (AREA OF A TRAPEZOID)

22. Solve $A = \frac{h}{2}(b_1 + b_2)$ for b_2 (AREA OF A TRAPEZOID)

23. Solve $C = \frac{5}{9}(F - 32)$ for F (CELSIUS-FAHRENHEIT)

24. Solve $F = \frac{9}{5}C + 32$ for C (FAHRENHEIT-CELSIUS)

C 25. Solve $\frac{1}{f} = \frac{1}{a} + \frac{1}{b}$ for f (OPTICS—FOCAL LENGTH)

26. Solve $\frac{1}{R} = \frac{1}{R_1} + \frac{1}{R_2}$ for R_1 (ELECTRICAL CIRCUITS)

Exercise 50 Chapter Review

A *Simplify:*

1. $\frac{4x^2y^3}{3a^2b^2} \div \frac{2xy^2}{3ab}$

2. $(d - 2)^2 + \frac{d^2 - 4}{d - 2}$

3. $\frac{2}{x} - \frac{1}{6x} + \frac{1}{3}$

4. $\frac{x + 1}{x + 2} - \frac{x + 2}{x + 3}$

5. $1 + \frac{2}{3x}$

6. $\frac{2}{3m} - \frac{1}{4m} = \frac{1}{12}$

7. $\frac{3x}{x - 5} - 8 = \frac{15}{x - 5}$

8. If $\frac{1}{2}$ is added to the reciprocal of a number, the sum is 2. Find the number.

9. Solve $W = I^2R$ for R (electrical circuits, power in watts).

10. Solve $A = \frac{bh}{2}$ for b (area of a triangle).

B *Simplify:*

11. $\frac{4x^2y}{3ab^2} \div 8\left(\frac{2a^2x^2}{b^2y} \cdot \frac{6a}{2y^2}\right)$

12. $\frac{x^3 - x}{x^2 - x} \div \frac{x^2 + 2x + 1}{x}$

13. $\frac{1}{10p^2} - \frac{3}{4pq} + \frac{2}{5q^2}$

14. $\frac{x}{4x + x^2} + \frac{2x}{3x^2 - 48}$

15. $1 - \frac{m - 1}{m + 1}$

16. $\frac{5}{2x + 3} - 5 = \frac{-5x}{2x + 3}$

Solve:

17. $\frac{3}{x} - \frac{2}{x + 1} = \frac{1}{2x}$

18. If an airplane can travel 300 miles against the wind in the same time it travels 400 miles with the wind and the speed of the wind is 25 mph, what is the cruising speed of the airplane in still air?

19. Given $I = E/R$ and $W = IE$, write a formula for W in terms of E and R.

20. Solve $S = \frac{n(a + L)}{2}$ for L (arithmetic progression).

C *Simplify:*

21. $\frac{y^2 - y - 6}{(y + 2)^2} \cdot \frac{2 + y}{3 - y}$

22. $\frac{2x + 4}{2x - y} + \frac{2x - y}{y - 2x}$

23. $\frac{\frac{x}{y} - \frac{y}{x}}{\frac{x}{y} + 1}$

24. $5 - \frac{2x}{3 - x} = \frac{6}{x - 3}$

25. One number is twice another. Find the two numbers if the reciprocal of the smaller minus the reciprocal of the larger is equal to the reciprocal of their product.

26. Solve $\frac{1}{f} = \frac{1}{f_1} + \frac{1}{f_2}$ for f_1 (OPTICS)

CHAPTER 7
Exponents and Radicals

7.1 Laws for Natural-Number Exponents

Earlier we defined a number raised to a natural number power. Recall

$$a^n = aa \cdots a \qquad n \text{ factors of } a$$

Then we introduced the first law of exponents: If m and n are positive integers and a is a real number, then

$$a^m a^n = a^{m+n}$$

By now you probably use this law almost unconsciously when multiplying polynomial forms.

When more complicated expressions involving exponents are encountered, other exponent laws combined with the first law provide an efficient tool for simplifying and manipulating these expressions. In this section we will introduce and discuss four additional exponent properties.

In each of the following expressions m and n are natural numbers and a and b are real numbers, excluding division by 0, of course.

3 factors | 4 factors | 3 + 4 factors

EXAMPLE 1

$$a^3a^4 = (a \cdot a \cdot a)(a \cdot a \cdot a \cdot a) = (a \cdot a \cdot a \cdot a \cdot a \cdot a \cdot a) = a^{3+4} = a^7$$

LAW 1

$$a^m a^n = a^{m+n}$$

PROBLEM 1

$$x^7x^9 = ?$$

ANSWER x^{16}

4 groups of 3 factors each

EXAMPLE 2

$$(a^3)^4 = a^3 \cdot a^3 \cdot a^3 \cdot a^3 = (a \cdot a \cdot a)(a \cdot a \cdot a)(a \cdot a \cdot a)(a \cdot a \cdot a)$$

$4 \cdot 3$ factors

$$= (a \cdot a \cdot a \cdot a \cdot a \cdot a \cdot a \cdot a \cdot a \cdot a \cdot a \cdot a) = a^{4 \cdot 3} = a^{12}$$

LAW 2

$$(a^n)^m = a^{mn}$$

PROBLEM 2

$$(x^2)^5 = ?$$

ANSWER x^{10}

4 factors of (ab) | 4 factors | 4 factors

EXAMPLE 3

$$(ab)^4 = (ab)(ab)(ab)(ab) = (a \cdot a \cdot a \cdot a)(b \cdot b \cdot b \cdot b) = a^4b^4$$

LAW 3

$$(ab)^m = a^m b^n$$

PROBLEM 3

$$(xy)^7 = ?$$

ANSWER x^7y^7

5 factors of a/b

EXAMPLE 4

$$\left(\frac{a}{b}\right)^5 = \left(\frac{a}{b} \cdot \frac{a}{b} \cdot \frac{a}{b} \cdot \frac{a}{b} \cdot \frac{a}{b}\right) = \frac{a \cdot a \cdot a \cdot a \cdot a}{b \cdot b \cdot b \cdot b \cdot b} = \frac{a^5}{b^5}$$

LAW 4

$$\left(\frac{a}{b}\right)^m = \frac{a^m}{b^m}$$

PROBLEM 4 $\left(\frac{x}{y}\right)^3 = ?$

ANSWER $\frac{x^3}{y^3}$

EXAMPLE 5

(A) $\frac{a^7}{a^3} = \frac{a \cdot a \cdot a \cdot a \cdot a \cdot a \cdot a}{a \cdot a \cdot a} = \frac{(a \cdot a \cdot a)(a \cdot a \cdot a \cdot a)}{(a \cdot a \cdot a)} = a^{7-3} = a^4$

(B) $\frac{a^3}{a^3} = \frac{a \cdot a \cdot a}{a \cdot a \cdot a} = 1$

(C) $\frac{a^4}{a^7} = \frac{a \cdot a \cdot a \cdot a}{a \cdot a \cdot a \cdot a \cdot a \cdot a \cdot a} = \frac{(a \cdot a \cdot a \cdot a)}{(a \cdot a \cdot a \cdot a)(a \cdot a \cdot a)} = \frac{1}{a^{7-4}} = \frac{1}{a^3}$

LAW 5

$$\frac{a^m}{a^n} = \begin{cases} a^{m-n} & \text{if } m > n \\ 1 & \text{if } m = n \\ \frac{1}{a^{n-m}} & \text{if } n > m \end{cases}$$

PROBLEM 5 (A) $x^8/x^3 = ?$ (B) $x^8/x^8 = ?$ (C) $x^3/x^8 = ?$

ANSWER (A) x^5 (B) 1 (C) $1/x^5$

The laws of exponents are theorems, and as such they require proofs. We have only given plausible arguments for each law; formal proofs of these laws require a property of the natural numbers, called the inductive property, which is beyond the scope of this course.

It is very important to observe and remember that the laws of exponents apply to products and quotients, and not to sums and differences. Many mistakes are made in algebra by people applying a law of exponents to the wrong algebraic form. The exponent laws are summarized below for convenient reference.

LAWS OF EXPONENTS

1 $a^m a^n = a^{m+n}$

2 $(a^n)^m = a^{mn}$

3 $(ab)^m = a^m b^m$

4 $\left(\frac{a}{b}\right)^m = \frac{a^m}{b^m}$

5 $\frac{a^m}{a^n} = \begin{cases} a^{m-n} & \text{if } m > n \\ 1 & \text{if } m = n \\ \frac{1}{a^{n-m}} & \text{if } n > m \end{cases}$

EXAMPLE 6

(A) $x^{12}x^{13} \quad = x^{12+13} \quad = x^{25}$

(B) $(t^7)^5 \quad = t^{5\cdot 7} \quad = t^{35}$

(C) $(xy)^5 = x^5y^5$

(D) $\left(\dfrac{u}{v}\right)^3 = \dfrac{u^3}{v^3}$

(E) $\dfrac{x^{12}}{x^4} \quad = x^{12-4} \quad = x^8$

(F) $\dfrac{t^4}{t^9} \quad = \dfrac{1}{t^{9-4}} \quad = \dfrac{1}{t^5}$

PROBLEM 6 Simplify:

(A) x^8x^6 (B) $(u^4)^5$ (C) $(xy)^9$

(D) $\left(\dfrac{x}{y}\right)^4$ (E) x^{10}/x^3 (F) x^3/x^{10}

ANSWER (A) x^{14} (B) u^{20} (C) x^9y^9 (D) $\dfrac{x^4}{y^4}$ (E) x^7 (F) $1/x^7$

EXAMPLE 7

(A) $(x^2y^3)^4 \quad = (x^2)^4(y^3)^4 \quad = x^8y^{12}$

(B) $\left(\dfrac{u^3}{v^4}\right)^3 \quad = \dfrac{(u^3)^3}{(v^4)^3} \quad = \dfrac{u^9}{v^{12}}$

(C) $\dfrac{2x^9y^{11}}{4x^{12}y^7} \quad = \dfrac{2}{4}\cdot\dfrac{x^9}{x^{12}}\cdot\dfrac{y^{11}}{y^7} = \dfrac{1}{2}\cdot\dfrac{1}{x^3}\cdot\dfrac{y^4}{1} \quad = \dfrac{y^4}{2x^3}$

PROBLEM 7 Simplify:

(A) $(u^3v^4)^2$ (B) $\left(\dfrac{u^3}{v^4}\right)^3$ (C) $\dfrac{9x^7y^2}{3x^5y^3}$

ANSWER (A) u^6v^8 (B) $\dfrac{u^9}{v^{12}}$ (C) $\dfrac{3x^2}{y}$

NOTE: As before, the "dotted boxes" are used to indicate steps that are usually carried out mentally.

Knowing the rules of the game of chess doesn't make one a good chess player; similarly, memorizing the laws of exponents doesn't necessarily make one good at using them. To acquire skill in their use, one must use these laws in a fairly large variety of problems. The following exercises should help you acquire this skill.

Exercise 51

A *Replace the question marks with appropriate symbols.*

1. $x^7x^5 = x^?$

2. $y^2y^7 = y^?$

3. $x^{10} = x^?x^6$

4. $y^8 = y^3y^?$

5. $(v^2)^3 = ?$

6. $(u^4)^3 = u^?$

7. $y^{12} = (y^6)^?$

8. $x^{10} = (x^?)^5$

9. $(xy)^5 = x^5y^?$

10. $(uv)^7 = ?$

11. $m^3n^3 = (mn)^?$

12. $p^4q^4 = (pq)^?$

13. $\left(\frac{x}{y}\right)^4 = \frac{x^?}{y^4}$

14. $\left(\frac{a}{b}\right)^8 = ?$

15. $\frac{x^7}{y^7} = \left(\frac{x}{y}\right)^?$

16. $\frac{m^3}{n^3} = \left(\frac{m}{n}\right)^?$

17. $\frac{x^7}{x^3} = x^?$

18. $\frac{n^{14}}{n^8} = n^?$

19. $x^3 = \frac{x^?}{x^4}$

20. $m^6 = \frac{m^8}{m^?}$

21. $\frac{a^5}{a^9} = \frac{a}{a^?}$

22. $\frac{x^4}{x^{11}} = \frac{1}{x^?}$

23. $\frac{1}{u^2} = \frac{u^?}{u^9}$

24. $\frac{1}{x^8} = \frac{x^4}{x^?}$

Simplify, using appropriate laws of exponents.

25. $(2x^3)(3x^7)$

26. $(5x^2)(2x^9)$

27. $\frac{4x^8}{2x^6}$

28. $\frac{9x^6}{3x^4}$

29. $\frac{4u^3}{2u^7}$

30. $\frac{6m^5}{8m^7}$

31. $(cd)^{12}$

32. $(xy)^{10}$

33. $\left(\frac{x}{y}\right)^6$

34. $\left(\frac{m}{n}\right)^5$

B **35.** $(2x^2)(3x^3)(x^4)$

36. $(4y^3)(3y)(y^6)$

37. $(2 \times 10^3)(3 \times 10^{12})$

38. $(5 \times 10^8)(7 \times 10^9)$

39. $(10^4)^5$

40. $(10^7)^2$

41. $(y^4)^5$ **42.** $(x^3)^2$ **43.** $(x^2y^3)^4$

44. $(m^2n^5)^3$ **45.** $\left(\frac{a^3}{b^2}\right)^4$ **46.** $\left(\frac{c^2}{d^5}\right)^3$

47. $\frac{2x^3y^8}{6x^7y^2}$ **48.** $\frac{9u^8v^6}{3u^4v^8}$ **49.** $(3a^3b^2)^3$

50. $(2s^2t^4)^4$ **51.** $2(x^2y)^4$ **52.** $6(xy^3)^5$

53. $\left(\frac{x^2y}{2w^2}\right)^3$ **54.** $\left(\frac{mn^3}{p^2q}\right)^4$

C **55.** $\frac{(2xy^3)^2}{(4x^2y)^3}$ **56.** $\frac{(4u^3v)^3}{(2uv^2)^6}$ **57.** $\frac{(-2x^2)^3}{(2^2x)^4}$

58. $\frac{(9x^3)^2}{(-3x)^2}$ **59.** $\frac{-2^2}{(-2)^2}$ **60.** $\frac{-x^2}{(-x)^2}$

61. $\frac{-2^4}{(-2a^2)^4}$ **62.** $\frac{(-x^2)^2}{(-x^3)^3}$

7.2 Integer Exponents

How should symbols such as

8^0 and 7^{-3}

be defined? In this section we will extend the meaning of exponent to include 0 and negative integers. Thus, typical scientific expressions such as

The diameter of a red corpuscle is approximately 8×10^{-5} cm.

The amount of water found in the air as vapor is about 9×10^{-6} times that found in seas.

The focal length of a thin lens is given by $f^{-1} = a^{-1} + b^{-1}$.

will then make sense.

In extending the concept of exponent beyond the natural numbers, we will require that any new exponent symbol be defined in such a way that all five laws of exponents for natural numbers continue to hold. Thus, we will need only one set of laws for all types of exponents rather than a new set for each new exponent.

We will start by defining the 0 exponent. If all the exponent laws must hold even if some of the exponents are 0, then a^0 $(a \neq 0)$ should be defined so that when the first law of exponents is applied,

$$a^0 \cdot a^2 = a^{0+2} = a^2$$

This suggests that a^0 should be defined as 1 for all nonzero real numbers a, since 1 is the only real number that gives a^2 when multiplied by a^2. If we let $a = 0$ and follow the same reasoning, we find that

$$0^0 \cdot 0^2 = 0^{0+2} = 0^2 = 0$$

and 0^0 could be any real number; hence it is not uniquely determined. For this reason 0^0 is not defined.

DEFINITION OF ZERO EXPONENT

For all real numbers $a \neq 0$.

$$a^0 = 1$$

0^0 is not defined

EXAMPLE 8

(A) $5^0 = 1$

(B) $325^0 = 1$

(C) $(\frac{1}{3})^0 = 1$

(D) $t^0 = 1\ (t \neq 0)$

(E) $(x^2y^3)^0 = 1\ (x \neq 0, y \neq 0)$

PROBLEM 8

Simplify:

(A) 12^0 (B) 999^0 (C) $(\frac{2}{7})^0$

(D) $x^0, x \neq 0$ (E) $(m^3n^3)^0, m,n \neq 0$

ANSWER All are equal to 1.

To get an idea of how a negative integer exponent should be defined, we can proceed as above. If the first law of exponents is to hold, then a^{-2} $(a \neq 0)$ must be defined so that

$$a^{-2} \cdot a^2 = a^{-2+2} = a^0 = 1$$

Thus a^{-2} must be the reciprocal of a^2; that is,

$$a^{-2} = \frac{1}{a^2}$$

This kind of reasoning leads us to the following general definition.

DEFINITION OF NEGATIVE INTEGER EXPONENT

If n is a positive integer and a is a nonzero real number, then

$$a^{-n} = \frac{1}{a^n}$$

Of course, it follows, using equality properties, that

$$a^n = \frac{1}{a^{-n}}$$

EXAMPLE 9 (A) $a^{-7} = \dfrac{1}{a^7}$

(B) $\dfrac{1}{x^{-8}} = x^8$

(C) $10^{-3} = \dfrac{1}{10^3}$ or $\dfrac{1}{1{,}000}$ or 0.001

(D) $\dfrac{x^{-3}}{y^{-5}} \boxed{= \dfrac{x^{-3}}{1} \cdot \dfrac{1}{y^{-5}} = \dfrac{1}{x^3} \cdot \dfrac{y^5}{1}} = \dfrac{y^5}{x^3}$

PROBLEM 9 Write using positive exponents or no exponents:

(A) x^{-5} (B) $\dfrac{1}{y^{-4}}$

(C) 10^{-2} (D) $\dfrac{m^{-2}}{n^{-3}}$

ANSWER (A) $\dfrac{1}{x^5}$ (B) y^4 (C) $\dfrac{1}{10^2}$ or $\dfrac{1}{100}$ or 0.01 (D) $\dfrac{n^3}{m^2}$

With the definition of negative exponent and 0 exponent behind us, we can now replace the fifth law of exponents with a simpler form that does not have any restrictions on the relative size of the exponents. Thus

$$\frac{a^m}{a^n} = a^{m-n} = \frac{1}{a^{n-m}}$$

EXAMPLE 10 (A) $\dfrac{2^5}{2^8} = 2^{5-8} = 2^{-3}$ or $\dfrac{2^5}{2^8} = \dfrac{1}{2^{8-5}} = \dfrac{1}{2^3}$

(B) $\dfrac{10^{-3}}{10^6} = 10^{-3-6} = 10^{-9}$ or $\dfrac{10^{-3}}{10^6} = \dfrac{1}{10^{6-(-3)}} = \dfrac{1}{10^{6+3}} = \dfrac{1}{10^9}$

PROBLEM 10 (A) Combine denominator with numerator: $\dfrac{3^4}{3^9}, \dfrac{x^{-2}}{x^3}$

(B) Combine numerator with denominator in part A.

ANSWER (A) $3^{-5}, x^{-5}$ (B) $\dfrac{1}{3^5}, \dfrac{1}{x^5}$

Table 1 provides a summary of all of our work on exponents to this point.

TABLE 1 INTEGER EXPONENTS AND THEIR LAWS (SUMMARY)

DEFINITION OF a^p p AN INTEGER AND a A REAL NUMBER	LAWS OF EXPONENTS n AND m INTEGERS, a AND b REAL NUMBERS
1 *If p is a positive integer, then* $a^p = a \cdot a \cdots a$ *p factors of a*	1 $a^m a^n = a^{m+n}$ 2 $(a^n)^m = a^{mn}$
EXAMPLE: $3^5 = 3 \cdot 3 \cdot 3 \cdot 3 \cdot 3$	3 $(ab)^m = a^m b^m$
2 *If* $p = 0$, *then* $a^p = 1 \quad a \neq 0$	4 $\left(\dfrac{a}{b}\right)^m = \dfrac{a^m}{b^m}$
EXAMPLE: $3^0 = 1$	5 $\dfrac{a^m}{a^n} = a^{m-n} = \dfrac{1}{a^{n-m}}$
3 *If p is a negative integer, then* $a^p = \dfrac{1}{a^{-p}} \quad a \neq 0$	
EXAMPLE: $3^{-4} = \dfrac{1}{3^{-(-4)}} = \dfrac{1}{3^4}$	

EXAMPLE 11 Simplify and express answers using positive exponents only.

(A) $$a^5a^{-2} = a^{5-2} = a^3$$

(B) $$(a^{-3}b^2)^{-2} = (a^{-3})^{-2}(b^2)^{-2} = a^6b^{-4} = \frac{a^6}{b^4}$$

(C) $$\left(\frac{a^{-5}}{a^{-2}}\right)^{-1} = \frac{(a^{-5})^{-1}}{(a^{-2})^{-1}} = \frac{a^5}{a^2} = a^3$$

(D) $$\frac{4x^{-3}y^{-5}}{6x^{-4}y^3} = \frac{2x^{-3-(-4)}}{3y^{3-(-5)}} = \frac{2x^{-3+4}}{3y^{3+5}} = \frac{2x}{3y^8}$$

or, changing to positive exponents first,

$$\frac{4x^{-3}y^{-5}}{6x^{-4}y^3} = \frac{2x^4}{3x^3y^3y^5} = \frac{2x}{3y^8}$$

(E) $$\frac{10^{-4} \cdot 10^2}{10^{-3} \cdot 10^5} = \frac{10^{-4+2}}{10^{-3+5}} = \frac{10^{-2}}{10^2} = \frac{1}{10^4} = \frac{1}{10{,}000} = 0.0001$$

(F) $$\left(\frac{m^{-3}m^3}{n^{-2}}\right)^{-2} = \left(\frac{m^{-3+3}}{n^{-2}}\right)^{-2} = \left(\frac{m^0}{n^{-2}}\right)^{-2} = \left(\frac{1}{n^{-2}}\right)^{-2} = \frac{1^{-2}}{(n^{-2})^{-2}} = \frac{1}{n^4}$$

PROBLEM 11 Simplify and express answers using positive exponents only:

(A) $x^{-2}x^6$ (B) $(x^3y^{-2})^{-2}$ (C) $\left(\frac{x^{-6}}{x^{-2}}\right)^{-1}$

(D) $\frac{8m^{-2}n^{-4}}{6m^{-5}n^2}$ (E) $\frac{10^{-3}\cdot 10^5}{10^{-2}\cdot 10^6}$

ANSWER (A) x^4 (B) y^4/x^6 (C) x^4 (D) $4m^3/3n^6$ (E) $1/10^2$ or 0.01

EXAMPLE 12 Simplify and express answers using positive exponents only.

(A) $$\frac{3^{-2}+2^{-1}}{11}=\frac{\frac{1}{3^2}+\frac{1}{2}}{11}=\frac{\frac{2}{18}+\frac{9}{18}}{11}=\frac{11}{18}\div 11=\frac{11}{18}\cdot\frac{1}{11}=\frac{1}{18}$$

(B) $$(a^{-1}-b^{-1})^2=\left(\frac{1}{a}-\frac{1}{b}\right)^2=\left(\frac{b-a}{ab}\right)^2=\frac{b^2-2ab+a^2}{a^2b^2}$$

PROBLEM 12 Simplify and express answers using positive exponents only:

(A) $\frac{2^{-2}+3^{-1}}{5}$ (B) $(x^{-1}+y^{-1})^2$

ANSWER (A) $\frac{7}{60}$ (B) $\frac{(x+y)^2}{x^2y^2}$ or $\frac{x^2+2xy+y^2}{x^2y^2}$

Exercise 52

A *Simplify and write answers using positive exponents only.*

1. 10^0
2. 23^0
3. x^0
4. y^0
5. 2^{-2}
6. 3^{-3}
7. x^{-4}
8. m^{-7}
9. $\frac{1}{3^{-2}}$
10. $\frac{1}{4^{-3}}$
11. $\frac{1}{x^{-3}}$
12. $\frac{1}{y^{-5}}$
13. $10^{-4}\cdot 10^6$
14. $10^7\cdot 10^{-5}$
15. x^6x^{-2}
16. $y^{-3}y^4$
17. $m^{-3}m^3$
18. u^5u^{-5}
19. $\frac{10^8}{10^{-3}}$
20. $\frac{10^3}{10^{-7}}$
21. $\frac{a^8}{a^{-4}}$
22. $\frac{x^9}{x^{-2}}$
23. $\frac{b^{-3}}{b^5}$
24. $\frac{z^{-2}}{z^3}$

25. $\frac{10^{-4}}{10^{2}}$ 26. $\frac{10^{-1}}{10^{6}}$ 27. $(2^{-3})^{-2}$

28. $(10^{-4})^{-3}$ 29. $(x^{-5})^{-2}$ 30. $(y^{-2})^{-4}$

31. $(x^{-3}y^{-2})^{-1}$ 32. $(u^{-5}v^{-3})^{-2}$ 33. $(x^{-2}y^{3})^{2}$

34. $(x^{2}y^{-3})^{2}$ 35. $(x^{2}y^{-3})^{-1}$ 36. $(x^{-2}y^{3})^{-1}$

B 37. $1{,}231^{0}$ 38. $(m^{2})^{0}$ 39. $\frac{10^{-2}}{10^{-4}}$

40. $\frac{10^{-3}}{10^{-5}}$ 41. $\frac{x^{-3}}{x^{-2}}$ 42. $\frac{y^{-2}}{y^{-3}}$

43. $\frac{10^{23}10^{-11}}{10^{-3}10^{-2}}$ 44. $\frac{10^{-13}10^{-4}}{10^{-21}10^{3}}$ 45. $\frac{8 \times 10^{-3}}{2 \times 10^{-5}}$

46. $\frac{18 \times 10^{12}}{6 \times 10^{-4}}$ 47. $\left(\frac{x^{2}}{x^{-1}}\right)^{2}$ 48. $\left(\frac{y}{y^{-2}}\right)^{3}$

49. $(2cd^{2})^{-3}$ 50. $\frac{1}{(3mn)^{-2}}$ 51. $(3x^{3}y^{-2})^{2}$

52. $(2mn^{-3})^{3}$ 53. $(x^{-3}y^{2})^{-2}$ 54. $(m^{4}n^{-5})^{-3}$

55. $(2^{-3}3^{2})^{-2}$ 56. $(2^{2}3^{-3})^{-1}$ 57. $(10^{2}3^{0})^{-2}$

58. $(10^{12}10^{-12})^{-1}$ 59. $\frac{9m^{-4}n^{3}}{12m^{-1}n^{-1}}$ 60. $\frac{8x^{-3}y^{-1}}{6x^{2}y^{-4}}$

61. $\frac{4x^{-2}y^{-3}}{2x^{-3}y^{-1}}$ 62. $\frac{2a^{6}b^{-2}}{16a^{-3}b^{2}}$ 63. $\left(\frac{n^{-3}}{n^{-2}}\right)^{-2}$

64. $\left(\frac{x^{-1}}{x^{-8}}\right)^{-1}$ 65. $\left(\frac{x^{4}y^{-1}}{x^{-2}y^{3}}\right)^{2}$ 66. $\left(\frac{m^{-2}n^{3}}{m^{4}n^{-1}}\right)^{2}$

67. $\left(\frac{2x^{-3}y^{2}}{4xy^{-1}}\right)^{-2}$ 68. $\left(\frac{6mn^{-2}}{3m^{-1}n^{2}}\right)^{-3}$

C 69. $(x + y)^{-2}$ 70. $(a^{2} - b^{2})^{-1}$ 71. $\frac{2^{-1} + 3^{-1}}{25}$

72. $\frac{x^{-1} + y^{-1}}{x + y}$ 73. $\frac{12}{2^{-2} + 3^{-1}}$ 74. $\frac{c - d}{c^{-1} - d^{-1}}$

75. $(2^{-2} + 3^{-2})^{-1}$ 76. $(x^{-1} + y^{-1})^{-1}$ 77. $(10^{-2} + 10^{-3})^{-1}$

78. $(x^{-1} - y^{-1})^{2}$

7.3 Power-of-Ten Notation and Applications

Work in science often involves the use of very, very large numbers:

The energy of a laser beam can go as high as 10,000,000,000,000 watts per sq cm.

Also involved is the use of very, very small numbers:

The mass of one water molecule is 0.00000000000000000000003 grams.

Writing and working with numbers of this type in standard decimal notation is generally awkward. Earlier in the course we used power-of-ten notation to represent very large numbers; now, with the introduction of negative exponents, we also can use power-of-ten notation to represent very small numbers (see Figs. 1 and 2). Together these power forms provide a valuable tool for the person working with large and small quantities.

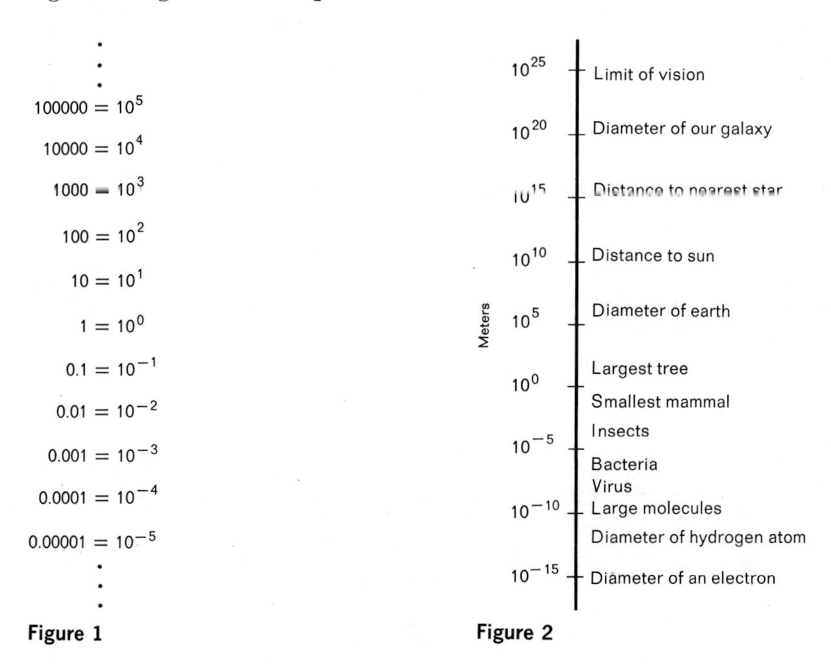

Figure 1 **Figure 2**

For computational purposes, or simply for convenience, it is often desirable to convert a decimal fraction to a power-of-ten form. In fact, any decimal fraction, however large or small, can be represented as the product of a number between 1 and 10 and a power of ten.

EXAMPLE 13

DECIMAL FRACTIONS AND POWER-OF-TEN NOTATION

$5 = 5 \times 10^0$	$0.7 = 7 \times 10^{-1}$
$35 = 3.5 \times 10$	$0.083 = 8.3 \times 10^{-2}$
$430 = 4.3 \times 10^2$	$0.0043 = 4.3 \times 10^{-3}$
$5{,}870 = 5.87 \times 10^3$	$0.000687 = 6.87 \times 10^{-4}$
$8{,}910{,}000 = 8.91 \times 10^6$	$0.00000036 = 3.6 \times 10^{-7}$

Can you discover a simple mechanical rule that relates the number of decimal places the decimal is moved with the power of 10 that is used?

PROBLEM 13 Write as a product of a number between 1 and 10 and a power of 10:

(*A*) 450 (*B*) 27,000

(*C*) 0.05 (*D*) 0.0000063

ANSWER (*A*) 4.5×10^2 (*B*) 2.7×10^4 (*C*) 5×10^{-2} (*D*) 6.3×10^{-6}

EXAMPLE 14 **EVALUATION OF A COMPLICATED ARITHMETIC PROBLEM**

$$\frac{(0.26)(720)}{(48{,}000{,}000)(0.0013)} = \frac{(2.6 \times 10^{-1})(7.2 \times 10^{2})}{(4.8 \times 10^{7})(1.3 \times 10^{-3})}$$

$$= \frac{(\overset{2}{\cancel{2.6}})(\overset{6}{\cancel{7.2}})}{(\underset{4}{\cancel{4.8}})(\underset{1}{\cancel{1.3}})} \cdot \frac{(10^{-1})(10^{2})}{(10^{7})(10^{-3})}$$

$$= 3 \times 10^{-3} \text{ or } 0.003$$

PROBLEM 14 Convert to power-of-ten form and evaluate: $\frac{(42{,}000)(0.009)}{(600)(0.000021)}$

ANSWER 3×10^4 or 30,000

Man is able to look back into time by looking out into space. Since light travels at a fast but finite rate, he is seeing heavenly bodies not as they exist now, but as they existed sometime in the past. If the distance between the sun and the Earth is approximately 9.3×10^7 miles and if light travels at the rate of approximately 1.86×10^5 miles per sec, we see the sun as it was how many minutes ago?

$$d = rt \qquad t = d/r$$

$$t = \frac{9.3 \times 10^7}{1.86 \times 10^5} = 5 \times 10^2 = 500 \text{ sec} \quad \text{or} \quad 500/60 = 8.3 \text{ min}$$

Hence, we always see the sun as it was 8.3 min ago.

Exercise 53

A *Write as the product of a number between 1 and 10 and a power of 10.*

1. 60 **2.** 80 **3.** 600

4. 800 **5.** 600,000 **6.** 80,000

7. 0.06 **8.** 0.008 **9.** 0.00006

10. 0.00000008 **11.** 35 **12.** 52

13. 0.72
14. 0.63
15. 270
16. 340
17. 0.032
18. 0.085
19. 5,200
20. 6,300
21. 0.00072
22. 0.0000068

Write as a decimal fraction.

23. 5×10^2
24. 8×10^2
25. 8×10^{-2}
26. 4×10^{-2}
27. 6×10^6
28. 3×10^5
29. 2×10^{-5}
30. 9×10^{-4}
31. 7.1×10^3
32. 5.6×10^4
33. 8.6×10^{-4}
34. 9.7×10^{-3}
35. 8.8×10^6
36. 4.3×10^5
37. 6.1×10^{-6}
38. 3.8×10^{-7}

B *Write as the product of a number between 1 and 10 and a power of 10.*

39. 42,700,000
40. 5,460,000,000
41. 0.0000723
42. 0.0000000729

43. The distance that light travels in one year is called a light-year. It is approximately 5,870,000,000,000 miles.

44. The energy of a laser beam can go as high as 10,000,000,000,000 watts.

45. The mass of one water molecule is 0.00000000000000000000003 grams.

46. The nucleus of an atom has a diameter of a little more than 1/100,000 that of the whole atom.

Write as a decimal fraction.

47. 3.46×10^9
48. 8.35×10^{10}
49. 6.23×10^{-7}
50. 6.14×10^{-12}

51. The distance from the earth to the sun is approximately 9.3×10^7 miles.

52. The diameter of the sun is approximately 8.65×10^5 miles.

53. The diameter of a red corpuscle is approximately 7.5×10^{-5} cm.

54. The probable mass of a hydrogen atom is 1.7×10^{-24} grams.

Simplify and express answer in power-of-ten form.

55. $(4 \times 10^5)(2 \times 10^{-3})$
56. $(3 \times 10^{-6})(3 \times 10^{10})$
57. $(4 \times 10^{-8})(2 \times 10^5)$
58. $(2 \times 10^3)(3 \times 10^{-7})$

59. $\dfrac{9 \times 10^8}{3 \times 10^5}$ **60.** $\dfrac{6 \times 10^{12}}{2 \times 10^7}$

61. $\dfrac{12 \times 10^3}{4 \times 10^{-4}}$ **62.** $\dfrac{15 \times 10^{-2}}{3 \times 10^{-6}}$

Convert each numeral to power-of-ten notation and simplify. Express answer in power-of-ten form and as a decimal fraction.

63. $\dfrac{(0.0006)(4000)}{0.00012}$ **64.** $\dfrac{(90{,}000)(0.000002)}{0.006}$

65. $\dfrac{(0.000039)(140)}{(130{,}000)(0.00021)}$ **66.** $\dfrac{(60{,}000)(0.000003)}{(0.0004)(1{,}500{,}000)}$

C **67.** In 1929 Vernadsky, a biologist, estimated that all of the free oxygen of the earth is 1.5×10^{21} grams and that it is produced by life alone. If one gram is approximately 2.2×10^{-3} lb, what is the amount of free oxygen in pounds?

68. If the mass of the earth is 6×10^{27} grams and each gram is 1.1×10^{-6} tons, find the mass of the earth in tons.

69. Some of the designers of high-speed computers are currently thinking of single-addition times of 10^{-7} sec (100 nanosec). How many additions would such a computer be able to perform in 1 sec? In 1 min?

70. If electricity travels in a computer circuit at the speed of light (1.86×10^5 miles per sec), how far will it travel in the time it takes the computer in the preceding problem to complete a single addition? (Size of circuits is becoming a critical problem in computer design.) Give the answer in miles and in feet.

7.4 Square Roots and Radicals

Going from exponents to radicals in the same chapter may seem unnatural; however, exponents and radicals have a lot more in common than one might first expect. We will comment briefly on this relationship before the end of the chapter. In more advanced courses the relationship is developed in detail.

In this and the following two sections we will take a careful look at the square root radical

$$\sqrt{}$$

and some of its properties. To start, we define a square root of a number:

x is a square root of y if $x^2 = y$

EXAMPLE 15 (*A*) 2 is a square root of 4 since $2^2 = 4$

(*B*) -2 is also a square root of 4 since $(-2)^2 = 4$

PROBLEM 15 Find two square roots of 9.

ANSWER 3 and -3

How many square roots of a real number are there? The following theorem, which we state without proof, answers this question.

THEOREM 1

(*A*) Every positive real number has exactly two real square roots, each the negative of the other.

(*B*) Square roots of negative numbers do not exist in the real numbers. (Why?)

(*C*) The square root of 0 is 0.

If a is a positive number, then we let

$$\sqrt{a}$$

represent the positive square root of a, and

$$-\sqrt{a}$$

the negative square root of a.

EXAMPLE 16

(*A*) $\sqrt{4} = 2$

(*B*) $-\sqrt{4} = -2$

(*C*) $\sqrt{0} = 0$

PROBLEM 16 Evaluate:

(*A*) $\sqrt{9}$ (*B*) $-\sqrt{9}$

ANSWER (*A*) 3 (*B*) -3

It can be shown that if a is a positive integer that is not the square of an integer, then

$$-\sqrt{a} \quad \text{and} \quad \sqrt{a}$$

name irrational numbers. Thus

$$-\sqrt{7} \quad \text{and} \quad \sqrt{7}$$

name the negative and positive irrational numbers whose squares are 7; that is, the positive and negative square roots of 7.

SQUARE ROOT PROPERTIES

Note that $\sqrt{4}\sqrt{36} = 2 \cdot 6 = 12$ and $\sqrt{4 \cdot 36} = \sqrt{144} = 12$, therefore,

$$\sqrt{4}\sqrt{36} = \sqrt{4 \cdot 36}$$

and that

$$\frac{\sqrt{36}}{\sqrt{4}} = \frac{6}{2} = 3 \qquad \text{and} \qquad \sqrt{\frac{36}{4}} = \sqrt{9} = 3$$

therefore,

$$\frac{\sqrt{36}}{\sqrt{4}} = \sqrt{\frac{36}{4}}$$

In general, the following theorem holds:

THEOREM 2 If a and b are nonnegative real numbers, then

$$\sqrt{a}\sqrt{b} = \sqrt{ab} \qquad \text{and} \qquad \frac{\sqrt{a}}{\sqrt{b}} = \sqrt{\frac{a}{b}} \quad b \neq 0$$

To see that the first part holds, let $N = \sqrt{a}$ and $M = \sqrt{b}$, then $N^2 = a$ and $M^2 = b$. Hence,

$$\sqrt{a}\sqrt{b} = NM = \sqrt{(NM)^2} = \sqrt{N^2M^2} = \sqrt{ab}$$

Note how properties of exponents are used. The proof of the quotient part is left as an exercise.

EXAMPLE 17

(*A*) $\sqrt{5}\sqrt{10} = \sqrt{5 \cdot 10} = \sqrt{50} = \sqrt{25 \cdot 2} = \sqrt{25}\sqrt{2} = 5\sqrt{2}$

(*B*) $\dfrac{\sqrt{32}}{\sqrt{8}} = \sqrt{\dfrac{32}{8}} = \sqrt{4} = 2$

(*C*) $\sqrt{\dfrac{7}{4}} = \dfrac{\sqrt{7}}{\sqrt{4}} = \dfrac{\sqrt{7}}{2}$ or $\dfrac{1}{2}\sqrt{7}$

PROBLEM 17 Simplify as in Example 17:

(*A*) $\sqrt{3}\sqrt{6}$ (*B*) $\dfrac{\sqrt{18}}{\sqrt{2}}$ (*C*) $\sqrt{\dfrac{11}{9}}$

ANSWER (*A*) $3\sqrt{2}$ (*B*) 3 (*C*) $\dfrac{\sqrt{11}}{3}$ or $\dfrac{1}{3}\sqrt{11}$

SIMPLEST RADICAL FORM

The foregoing definitions and theorems allow us to change algebraic expressions containing radicals to a variety of equivalent forms. One form that is often useful is called the "simplest radical form."

DEFINITION OF THE SIMPLEST RADICAL FORM

An algebraic expression that contains square root radicals is said to be in *simplest radical form* if all three of the following conditions are satisfied:

(*A*) No radicand (the expression within the radical sign) when expressed in a completely factored form contains a factor of power greater than 1.

(*B*) No radical appears in a denominator.

(*C*) No fraction appears within a radical.

It should be understood that forms other than the simplest radical form may be more useful on occasion. The situation dictates the choice.

EXAMPLE 18 Change to simplest radical form:

(*A*) $\sqrt{72} = \sqrt{6^2 \cdot 2} = \sqrt{6^2}\sqrt{2} = 6\sqrt{2}$

(*B*) $\sqrt{8x^3} = \sqrt{(4x^2)(2x)} = \sqrt{4x^2}\sqrt{2x} = 2x\sqrt{2x}$

(*C*) $\dfrac{3x}{\sqrt{3}} = \dfrac{3x}{\sqrt{3}} \cdot \dfrac{\sqrt{3}}{\sqrt{3}} = \dfrac{3x\sqrt{3}}{3} = x\sqrt{3}$

(*D*) $\sqrt{\dfrac{x}{2}} = \sqrt{\dfrac{x}{2} \cdot \dfrac{2}{2}} = \sqrt{\dfrac{2x}{4}} = \dfrac{\sqrt{2x}}{\sqrt{4}} = \dfrac{\sqrt{2x}}{2}$ or $\dfrac{1}{2}\sqrt{2x}$

PROBLEM 18 Change to simplest radical form:

(*A*) $\sqrt{32}$ (*B*) $\sqrt{18y^3}$

(*C*) $\dfrac{2x}{\sqrt{2}}$ (*D*) $\sqrt{\dfrac{y}{3}}$

ANSWER (*A*) $4\sqrt{2}$ (*B*) $3y\sqrt{2y}$ (*C*) $x\sqrt{2}$ (*D*) $\dfrac{\sqrt{3y}}{3}$ or $\dfrac{1}{3}\sqrt{y}$

Exercise 54

A *Simplify and express each answer in simplest radical form. All variables represent positive real numbers unless stated to the contrary.*

1. $\sqrt{16}$ **2.** $\sqrt{25}$ **3.** $-\sqrt{81}$

4. $-\sqrt{49}$ **5.** $\sqrt{x^2}$ **6.** $\sqrt{y^2}$

7. $\sqrt{9m^2}$ **8.** $\sqrt{4u^2}$ **9.** $\sqrt{8}$

10. $\sqrt{18}$ **11.** $\sqrt{x^3}$ **12.** $\sqrt{m^3}$

13. $\sqrt{18y^3}$ **14.** $\sqrt{8x^3}$ **15.** $\sqrt{\dfrac{1}{4}}$

16. $\sqrt{\dfrac{1}{9}}$ **17.** $-\sqrt{\dfrac{4}{9}}$ **18.** $-\sqrt{\dfrac{9}{16}}$

19. $\frac{1}{\sqrt{x^2}}$ **20.** $\frac{1}{\sqrt{y^2}}$ **21.** $\frac{1}{\sqrt{3}}$

22. $\frac{1}{\sqrt{5}}$ **23.** $\sqrt{\frac{1}{3}}$ **24.** $\sqrt{\frac{1}{5}}$

25. $\frac{1}{\sqrt{x}}$ **26.** $\frac{1}{\sqrt{y}}$ **27.** $\sqrt{\frac{1}{x}}$

28. $\sqrt{\frac{1}{y}}$ **29.** $\sqrt{25x^2y^4}$ **30.** $\sqrt{49x^4y^2}$

B **31.** $\sqrt{4x^5y^3}$ **32.** $\sqrt{9x^3y^5}$ **33.** $\sqrt{8x^7y^6}$

34. $\sqrt{18x^8y^5}$ **35.** $\frac{1}{\sqrt{3y}}$ **36.** $\frac{1}{\sqrt{2x}}$

37. $\frac{4xy}{\sqrt{2y}}$ **38.** $\frac{6x^2}{\sqrt{3x}}$ **39.** $\frac{2x^2y}{\sqrt{3xy}}$

40. $\frac{3a}{\sqrt{2ab}}$ **41.** $\sqrt{\frac{2}{3}}$ **42.** $\sqrt{\frac{3}{5}}$

43. $\sqrt{\frac{3m}{2n}}$ **44.** $\sqrt{\frac{6x}{7y}}$ **45.** $\sqrt{\frac{4a^3}{3b}}$

46. $\sqrt{\frac{9m^5}{2n}}$

Approximate with decimal fractions using the square root table in the Appendix.

EXAMPLE

(A) $\sqrt{35}\sqrt{40} = \sqrt{35 \cdot 40} = \sqrt{(2^2 \cdot 5^2)(2 \cdot 7)} = 10\sqrt{14} = (10)(3.742) = 37.42$

(B) $\sqrt{\frac{7}{5}} = \frac{\sqrt{7}}{\sqrt{5}} = \frac{\sqrt{7}\sqrt{5}}{\sqrt{5}\sqrt{5}} = \frac{\sqrt{35}}{5} = \frac{5.916}{5} = 1.183$

47. $\sqrt{6}\sqrt{3}$ **48.** $\sqrt{2}\sqrt{6}$ **49.** $\sqrt{\frac{1}{5}}$

50. $\sqrt{\frac{1}{3}}$ **51.** $\frac{\sqrt{33}}{\sqrt{2}}$ **52.** $\frac{\sqrt{23}}{\sqrt{5}}$

C *Express in simplest radical form.*

53. $\frac{\sqrt{2x}\sqrt{5}}{\sqrt{20x}}$ **54.** $\frac{\sqrt{6}\sqrt{8x}}{\sqrt{3x}}$ **55.** $\sqrt{a^2 + b^2}$

56. $\sqrt{m^2 + n^2}$ **57.** $\sqrt{x^4 - 2x^2}$ **58.** $\sqrt{m^3 + 4m^2}$

59. Explain why the square root of a negative number cannot be a real number.

60. Prove the second part of Theorem 2.

61. Is $\sqrt{x^2} = x$ true for $x = 4$? For $x = -4$?

62. Is $\sqrt{x^2} = |x|$ true for $x = 4$? For $x = -4$?

63. If we define the symbol $5^{\frac{1}{2}}$ in such a way that the laws of exponents continue to hold [in particular $(5^{\frac{1}{2}})^2 = 5^{2(\frac{1}{2})} = 5$ or $5^{\frac{1}{2}} \cdot 5^{\frac{1}{2}} = 5^{\frac{1}{2}+\frac{1}{2}} = 5$], how should it be defined?

64. If $x^2 = y^2$, does it necessarily follow that $x = y$? HINT: Can you find a pair of numbers that make the first equation true, but the second equation false?

65. Find the fallacy in the following "proof" that all real numbers are equal: If m and n are any real numbers, then

$$\begin{aligned}(m - n)^2 &= (n - m)^2 \\ m - n &= n - m \\ 2m &= 2n \\ m &= n\end{aligned}$$

7.5 Sums and Differences of Radicals

Algebraic expressions can often be simplified by adding or subtracting terms that contain the same radicals. We proceed in essentially the same way that we do when we combine like terms. You will recall that the distributive law played a central role in this process.

EXAMPLE 19

(A) $3\sqrt{2} + 5\sqrt{2} \quad = (3 + 5)\sqrt{2} \quad = 8\sqrt{2}$

(B) $2\sqrt{m} - 7\sqrt{m} \quad = (2 - 7)\sqrt{m} \quad = -5\sqrt{m}$

(C) $3\sqrt{x} - 2\sqrt{5} + 4\sqrt{x} - 7\sqrt{5} \quad = 3\sqrt{x} + 4\sqrt{x} - 2\sqrt{5} - 7\sqrt{5}$
$= 7\sqrt{x} - 9\sqrt{5}$

PROBLEM 19 Simplify:

(A) $2\sqrt{3} + 4\sqrt{3}$ (B) $3\sqrt{x} - 5\sqrt{x}$

(C) $2\sqrt{y} - 3\sqrt{7} + 4\sqrt{y} - 2\sqrt{7}$

ANSWER (A) $6\sqrt{3}$ (B) $-2\sqrt{x}$ (C) $6\sqrt{y} - 5\sqrt{7}$

Occasionally terms containing radicals can be combined after they have been expressed in simplest radical form.

EXAMPLE 20 (*A*)

$$\begin{aligned} 4\sqrt{8} - 2\sqrt{18} &= 4 \cdot \sqrt{4} \cdot \sqrt{2} - 2 \cdot \sqrt{9} \cdot \sqrt{2} \\ &= 4 \cdot 2 \cdot \sqrt{2} - 2 \cdot 3 \cdot \sqrt{2} \\ &= 8\sqrt{2} - 6\sqrt{2} \\ &= 2\sqrt{2} \end{aligned}$$

(*B*)

$$\begin{aligned} 2\sqrt{12} - \sqrt{\frac{1}{3}} &= 2 \cdot \sqrt{4} \cdot \sqrt{3} - \frac{1 \cdot \sqrt{3}}{\sqrt{3} \cdot \sqrt{3}} \\ &= 4\sqrt{3} - \frac{\sqrt{3}}{3} \\ &= \left(4 - \frac{1}{3}\right)\sqrt{3} \\ &= \frac{11}{3}\sqrt{3} \quad \text{or} \quad \frac{11\sqrt{3}}{3} \end{aligned}$$

PROBLEM 20 Express in simplest radical form and simplify:

(*A*) $5\sqrt{3} - 2\sqrt{12}$ (*B*) $3\sqrt{8} - \sqrt{\frac{1}{2}}$

ANSWER (*A*) $\sqrt{3}$ (*B*) $\frac{11\sqrt{2}}{2}$

Exercise 55

Simplify by combining as many terms as possible. All variables represent positive real numbers. Use exact radical forms only.

A

1. $5\sqrt{2} + 3\sqrt{2}$

2. $7\sqrt{3} + 2\sqrt{3}$

3. $6\sqrt{x} - 3\sqrt{x}$

4. $12\sqrt{m} - 3\sqrt{m}$

5. $4\sqrt{7} - 3\sqrt{5}$

6. $2\sqrt{3} + 5\sqrt{2}$

7. $\sqrt{y} - 4\sqrt{y}$

8. $2\sqrt{a} - 7\sqrt{a}$

9. $3\sqrt{5} - \sqrt{5} + 2\sqrt{5}$

10. $4\sqrt{7} - 6\sqrt{7} + \sqrt{7}$

11. $2\sqrt{x} - \sqrt{x} + 3\sqrt{x}$

12. $\sqrt{n} - 4\sqrt{n} - 2\sqrt{n}$

13. $3\sqrt{2} - 2\sqrt{3} - \sqrt{2}$

14. $\sqrt{5} - 2\sqrt{3} + 3\sqrt{5}$

15. $2\sqrt{x} - \sqrt{y} + 3\sqrt{y}$

16. $\sqrt{m} - \sqrt{n} - 2\sqrt{n}$

B

17. $\sqrt{8} - \sqrt{2}$

18. $\sqrt{18} + \sqrt{2}$

19. $\sqrt{27} - 3\sqrt{12}$

20. $\sqrt{8} - 2\sqrt{32}$

21. $\sqrt{8} + 2\sqrt{27}$

22. $2\sqrt{12} + 3\sqrt{18}$

23. $\sqrt{4x} - \sqrt{9x}$

24. $\sqrt{8mn} + 2\sqrt{18mn}$

25. $\sqrt{24} - \sqrt{12} + 3\sqrt{3}$

26. $\sqrt{8} - \sqrt{20} + 4\sqrt{2}$

C **27.** $\sqrt{\frac{2}{3}} - \sqrt{\frac{3}{2}}$

28. $\sqrt{\frac{1}{8}} + \sqrt{8}$

29. $\sqrt{\dfrac{xy}{2}} + \sqrt{8xy}$

30. $\sqrt{\dfrac{3uv}{2}} - \sqrt{24uv}$

31. $\sqrt{12} - \sqrt{\frac{1}{2}}$

32. $\sqrt{\frac{3}{5}} + 2\sqrt{20}$

33. $\sqrt{\dfrac{1}{2}} + \dfrac{\sqrt{2}}{2} + \sqrt{8}$

34. $\dfrac{\sqrt{3}}{3} + 2\sqrt{\dfrac{1}{3}} + \sqrt{12}$

7.6 Products and Quotients Involving Radicals

We will conclude this chapter by considering several special types of products and quotients that involve radicals. The distributive law plays a central role in our approach to these problems. In the examples that follow all variables represent positive real numbers.

SPECIAL PRODUCTS

EXAMPLE 21 Multiply and simplify:

(A) $\sqrt{2}(\sqrt{2} - 3) = \sqrt{2}\sqrt{2} - 3\sqrt{2} = 2 - 3\sqrt{2}$

(B) $\sqrt{x}(\sqrt{x} - 3) = \sqrt{x}\sqrt{x} - 3\sqrt{x} = x - 3\sqrt{x}$

(C)
$$\begin{aligned}(\sqrt{2} - 3)(\sqrt{2} + 5) &= \sqrt{2}\sqrt{2} - 3\sqrt{2} + 5\sqrt{2} - 15 \\ &= 2 + 2\sqrt{2} - 15 \\ &= 2\sqrt{2} - 13\end{aligned}$$

(D)
$$\begin{aligned}(\sqrt{x} - 3)(\sqrt{x} + 5) &= \sqrt{x}\sqrt{x} - 3\sqrt{x} + 5\sqrt{x} - 15 \\ &= x + 2\sqrt{x} - 15\end{aligned}$$

PROBLEM 21 Multiply and simplify:

(A) $\sqrt{3}(2 - \sqrt{3})$ (B) $\sqrt{y}(2 + \sqrt{y})$

(C) $(\sqrt{3} - 1)(\sqrt{3} + 4)$ (D) $(\sqrt{y} + 2)(\sqrt{y} - 5)$

ANSWER (A) $2\sqrt{3} - 3$ (B) $2\sqrt{y} + y$ (C) $3\sqrt{3} - 1$ (D) $y - 3\sqrt{y} - 10$

EXAMPLE 22 Show that $(2 - \sqrt{3})$ is a solution of the equation $x^2 - 4x + 1 = 0$.

SOLUTION

$$\begin{aligned} x^2 - 4x + 1 &= 0 \\ (2 - \sqrt{3})^2 - 4(2 - \sqrt{3}) + 1 &\stackrel{?}{=} 0 \\ 4 - 4\sqrt{3} + 3 - 8 + 4\sqrt{3} + 1 &\stackrel{?}{=} 0 \\ 0 &\stackrel{\checkmark}{=} 0 \end{aligned}$$

PROBLEM 22 Show that $(2 + \sqrt{3})$ is a solution of $x^2 - 4x + 1 = 0$.

ANSWER $(2 + \sqrt{3})^2 - 4(2 + \sqrt{3}) + 1 = 4 + 4\sqrt{3} + 3 - 8 - 4\sqrt{3} + 1 = 0$

SPECIAL QUOTIENTS—RATIONALIZING DENOMINATORS

Recall that to express $\sqrt{2}/\sqrt{3}$ in simplest radical form, we multiplied the numerator and denominator by $\sqrt{3}$ to clear the denominator of the radical:

$$\frac{\sqrt{2}}{\sqrt{3}} = \frac{\sqrt{2} \cdot \sqrt{3}}{\sqrt{3} \cdot \sqrt{3}} = \frac{\sqrt{6}}{3}$$

The denominator is thus converted to a rational number. The process of converting irrational denominators to rational forms is called *rationalizing the denominator.*

How can we rationalize the binomial denominator in

$$\frac{1}{\sqrt{3} - \sqrt{2}}$$

Multiplying the numerator and denominator by $\sqrt{3}$ or $\sqrt{2}$ does not help. Try it! Recalling the special product

$$(a - b)(a + b) = a^2 - b^2$$

this suggests that we multiply the numerator and denominator by the denominator, only with the middle sign changed. Thus,

$$\frac{1}{\sqrt{3} - \sqrt{2}} = \frac{1(\sqrt{3} + \sqrt{2})}{(\sqrt{3} - \sqrt{2})(\sqrt{3} + \sqrt{2})} = \frac{\sqrt{3} + \sqrt{2}}{3 - 2} = \sqrt{3} + \sqrt{2}$$

EXAMPLE 23 Rationalize denominators and simplify.

(A)

$$\frac{\sqrt{2}}{\sqrt{6} - 2} = \frac{\sqrt{2}(\sqrt{6} + 2)}{(\sqrt{6} - 2)(\sqrt{6} + 2)} = \frac{\sqrt{12} + 2\sqrt{2}}{6 - 4}$$

$$= \frac{2\sqrt{3} + 2\sqrt{2}}{2} = \frac{2(\sqrt{3} + \sqrt{2})}{2} = \sqrt{3} + \sqrt{2}$$

(B)

$$\frac{\sqrt{x} - \sqrt{y}}{\sqrt{x} + \sqrt{y}} = \frac{(\sqrt{x} - \sqrt{y})(\sqrt{x} - \sqrt{y})}{(\sqrt{x} + \sqrt{y})(\sqrt{x} - \sqrt{y})} = \frac{x - 2\sqrt{xy} + y}{x - y}$$

PROBLEM 23 Rationalize denominators and simplify:

(A) $\dfrac{\sqrt{2}}{\sqrt{2}+3}$ (B) $\dfrac{\sqrt{x}+\sqrt{y}}{\sqrt{x}-\sqrt{y}}$

ANSWER (A) $\dfrac{2-3\sqrt{2}}{-7}$, (B) $\dfrac{x+2\sqrt{xy}+y}{x-y}$

Exercise 56

A *Multiply and simplify where possible.*

1. $4(\sqrt{5}+2)$ **2.** $3(\sqrt{3}-4)$

3. $2(5-\sqrt{2})$ **4.** $5(3-\sqrt{5})$

5. $\sqrt{2}(\sqrt{2}+3)$ **6.** $\sqrt{3}(\sqrt{3}+2)$

7. $\sqrt{5}(\sqrt{5}-4)$ **8.** $\sqrt{7}(\sqrt{7}-2)$

9. $\sqrt{3}(2-\sqrt{3})$ **10.** $\sqrt{2}(3-\sqrt{2})$

11. $\sqrt{x}(\sqrt{x}-3)$ **12.** $\sqrt{y}(\sqrt{y}-8)$

13. $\sqrt{m}(3-\sqrt{m})$ **14.** $\sqrt{n}(4-\sqrt{n})$

B **15.** $\sqrt{6}(\sqrt{2}-1)$ **16.** $\sqrt{3}(5+\sqrt{6})$

17. $\sqrt{5}(\sqrt{10}+\sqrt{5})$ **18.** $\sqrt{20}(\sqrt{5}-1)$

19. $(\sqrt{2}-1)(\sqrt{2}+3)$ **20.** $(2-\sqrt{3})(3+\sqrt{3})$

21. $(\sqrt{x}+2)(\sqrt{x}-3)$ **22.** $(\sqrt{m}-3)(\sqrt{m}-4)$

23. $(\sqrt{5}+2)^2$ **24.** $(\sqrt{3}-3)^2$

25. $(2\sqrt{2}-5)(3\sqrt{2}+2)$ **26.** $(4\sqrt{3}-1)(3\sqrt{3}-2)$

27. $(3\sqrt{x}-2)(2\sqrt{x}-3)$ **28.** $(4\sqrt{y}-2)(3\sqrt{y}+1)$

29. Show that $2+\sqrt{3}$ is a solution to $x^2-4x+1=0$.

30. Show that $2-\sqrt{3}$ is a solution to $x^2-4x+1=0$.

Reduce by removing common factors from numerator and denominator.

31. $\dfrac{8+4\sqrt{2}}{12}$ **32.** $\dfrac{6-2\sqrt{3}}{6}$

33. $\dfrac{-3-6\sqrt{5}}{9}$ **34.** $\dfrac{-4+2\sqrt{7}}{4}$

35. $\dfrac{6-\sqrt{18}}{3}$ **36.** $\dfrac{10+\sqrt{8}}{2}$

C *Rationalize denominators and simplify.*

37. $\dfrac{1}{\sqrt{11}+3}$

38. $\dfrac{1}{\sqrt{5}+2}$

39. $\dfrac{2}{\sqrt{5}+1}$

40. $\dfrac{4}{\sqrt{6}-2}$

41. $\dfrac{\sqrt{y}}{\sqrt{y}+3}$

42. $\dfrac{\sqrt{x}}{\sqrt{x}-2}$

43. $\dfrac{\sqrt{3}+2}{\sqrt{3}-2}$

44. $\dfrac{\sqrt{2}-1}{\sqrt{2}+2}$

45. $\dfrac{\sqrt{x}+2}{\sqrt{x}-3}$

46. $\dfrac{\sqrt{a}-3}{\sqrt{a}-2}$

Exercise 57 Chapter Review

A *All variables represent positive real numbers.*

1. Evaluate: (*A*) 2^4, (*B*) 3^{-2}

2. Evaluate (*A*) $\left(\frac{1}{3}\right)^0$, (*B*) $\dfrac{1}{3^{-2}}$

Simplify and write answers using positive exponents only.

3. $\left(\dfrac{2x^2}{3y^3}\right)^2$

4. $(x^2y^{-3})^{-1}$

5. Multiply $(3 \times 10^4)(2 \times 10^{-6})$ and express answer in (*A*) power-of-ten form and (*B*) nonpower-of-ten form.

Simplify and express in simplest radical form.

6. $-\sqrt{25}$

7. $\sqrt{4x^2y^4}$

8. $\sqrt{\dfrac{25}{y^2}}$

9. $4\sqrt{x} - 7\sqrt{x}$

10. $\sqrt{5}(\sqrt{5}+2)$

B **11.** Evaluate: (*A*) $(3^{-2})^{-1}$, (*B*) $10^{-21}10^{19}$

12. Evaluate: (*A*) $\dfrac{3^{-2}}{3}$, (*B*) $(25^{-5})(25^5)$

Simplify and write answers using positive exponents only.

13. $\dfrac{1}{(2x^2y^{-3})^{-2}}$

14. $\dfrac{3m^4n^{-7}}{6m^2n^{-2}}$

15. Change $\frac{(480{,}000)(0.005)}{1{,}200{,}000}$ to power-of-ten notation and evaluate. Express answer in (*A*) power-of-ten form (*B*) nonpower-of-ten form.

Simplify and express in simplest radical form.

16. $\sqrt{36x^4y^7}$

17. $\frac{1}{\sqrt{2y}}$

18. $\sqrt{\frac{3x}{2y}}$

19. $\sqrt{\frac{2}{3}} + \sqrt{\frac{3}{2}}$

20. $(\sqrt{3} - 1)(\sqrt{3} + 2)$

C *Simplify and write answers using positive exponents only.*

21. $\left(\frac{9m^3n^{-3}}{3m^{-2}n^2}\right)^{-2}$

22. $(x^{-1} + y^{-1})^{-1}$

23. The volume of mercury increases linearly with temperature over a fairly wide temperature range (this is why mercury is often used in thermometers). If 1 cc of mercury at 0°C is heated to a temperature of T degrees Celsius, its volume is given by the formula

$$V = 1 + (1.8 \times 10^{-4})T$$

Express the volume of this sample at (2×10^2)°C as a decimal fraction.

Simplify and express in simplest radical form.

24. $\frac{\sqrt{8m^3n^4}}{\sqrt{12m^2}}$

25. $\sqrt{4x^4 + 16x^2}$

26. $\frac{\sqrt{x} - 2}{\sqrt{2} + 2}$

27. If a is a square root of b, then does $a^2 = b$ or does $b^2 = a$?

28. Describe the set $\{x \mid \sqrt{x^2} = |x|, x \text{ a real number}\}$.

CHAPTER 8
Quadratic Equations

8.1 Introductory Remarks

The equation

$$\tfrac{1}{2}x - \tfrac{1}{2}(x + 3) = 2 - x$$

is a first-degree equation in one variable since it can be transformed into the equivalent equation

$$7x - 21 = 0$$

which is a special case of

$$ax + b = 0 \qquad a \neq 0$$

We have solved many equations of this type and found that they always have a single solution. From a mathematical point of view this pretty well takes care of first-degree equations in one variable.

In this chapter we will consider the next class of polynomial equations called quadratic equations. A *quadratic equation* in one variable is any equation that can be written in the form

$$ax^2 + bx + c = 0 \qquad a \neq 0$$

where x is a variable and a, b, and c are constants. We will refer to this form as the *standard form* for the quadratic equation. The equations

$$2x^2 - 3x + 5 = 0$$

$$15 = 180t - 16t^2$$

are both quadratic equations since they are either in the standard form or can be transformed into this form.

Problems that give rise to quadratic equations are many and varied. For example, to find the dimensions of a rectangle with an area of 78 sq in. and a length twice its width, we are led to the equation

$$(2x)(x) = 78$$

or

$$2x^2 - 78 = 0$$

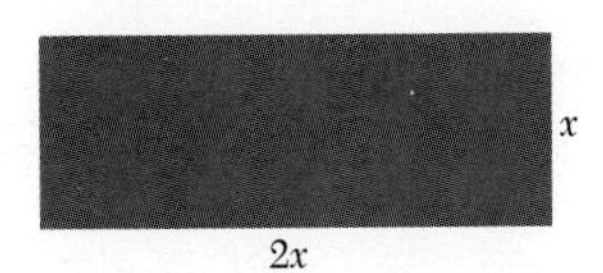

If an arrow is shot vertically in the air (from the ground) with an initial velocity of 176 fps, its distance y above the ground t sec after it is released (neglecting air resistance) is given by $y = 176t - 16t^2$. To find the times when y is 0, we are led to the equation

$$176t - 16t^2 = 0$$

or

$$16t^2 - 176t = 0$$

To find the times when the arrow is 16 ft off the ground, we are led to the equation

$$176t - 16t^2 = 16$$

or

$$16t^2 - 176t + 16 = 0$$

We actually have at hand, particularly since the last chapter on exponents and radicals, all of the tools we need to solve these equations—it is a matter of putting this material together in the right way. Putting this material together in the right way is the subject matter for this chapter.

8.2 Solution by Factoring

If the coefficients a, b, and c in the quadratic equation

$$ax^2 + bx + c = 0$$

are such that $ax^2 + bx + c$ can be written as the product of two first-degree factors with integer coefficients, then the quadratic equation can be quickly and easily solved. The method of solution by factoring rests on the following property of the real numbers: *If a and b are real numbers, then*

$ab = 0$ *if and only if* $a = 0$ *or* $b = 0$ *(or both)*

EXAMPLE 1 Solve $x^2 + 2x - 15 = 0$ by factoring.

SOLUTION

$$x^2 + 2x - 15 = 0$$
$$(x - 3)(x + 5) = 0$$
$$x - 3 = 0 \quad \text{or} \quad x + 5 = 0$$
$$x = 3 \quad \text{or} \quad x = -5$$

$(x - 3)(x + 5) = 0$ if and only if $(x - 3) = 0$ or $(x + 5) = 0$

CHECK

$x = 3$: $3^2 + 2(3) - 15 = 9 + 6 - 15 = 0$

$x = -5$: $(-5)^2 + 2(-5) - 15 = 25 - 10 - 15 = 0$

PROBLEM 1 Solve $x^2 - 2x - 8 = 0$ by factoring.

ANSWER $x = 4, -2$

EXAMPLE 2 Solve $2x^2 = 3x$.

SOLUTION

$$2x^2 = 3x$$

Why shouldn't both sides be divided by x?

$$2x^2 - 3x = 0$$
$$x(2x - 3) = 0$$
$$x = 0 \quad \text{or} \quad 2x - 3 = 0$$
$$x = 0 \quad \text{or} \quad x = \tfrac{3}{2}$$

$x(2x - 3) = 0$ if and only if $x = 0$ or $2x - 3 = 0$

CHECK

$x = 0$: $2(0)^2 \stackrel{?}{=} 3(0)$, $\quad 0 \stackrel{\checkmark}{=} 0$

$x = \frac{3}{2}$: $2(\frac{3}{2})^2 \stackrel{?}{=} 3(\frac{3}{2})$, $\quad \frac{9}{2} \stackrel{\checkmark}{=} \frac{9}{2}$

PROBLEM 2 Solve $3t^2 = 2t$.

ANSWER $t = 0, \frac{2}{3}$

EXAMPLE 3 Solve $2x^2 - 8x + 3 = 0$ by factoring, if possible.

SOLUTION $2x^2 - 8x + 3$ cannot be factored in the integers; hence another method must be used.

PROBLEM 3 Solve $x^2 - 3x - 3 = 0$ by factoring, if possible.

ANSWER Cannot be factored in the integers

Exercise 58

A *Solve:*

1. $(x - 3)(x - 4) = 0$
2. $(x - 9)(x - 4) = 0$
3. $(x + 6)(x - 5) = 0$
4. $(x - 0)(x + 3) = 0$
5. $(x + 4)(3x - 2) = 0$
6. $(2x - 1)(x + 2) = 0$
7. $(4t + 3)(5t - 2) = 0$
8. $(2m + 3)(3m - 2) = 0$
9. $u(4u - 1) = 0$
10. $z(3z + 5) = 0$

Solve by factoring.

11. $x^2 - 6x + 5 = 0$
12. $x^2 - 5x + 6 = 0$
13. $x^2 - 4x + 3 = 0$
14. $x^2 - 8x + 15 = 0$
15. $x^2 - 4x - 12 = 0$
16. $x^2 + 4x - 5 = 0$
17. $x^2 - 3x = 0$
18. $x^2 + 5x = 0$
19. $4t^2 - 8t = 0$
20. $3m^2 + 12m = 0$
21. $x^2 - 25 = 0$
22. $x^2 - 36 = 0$

B *Solve each equation by factoring. If an equation cannot be solved by factoring, state this as your answer.*

NOTE: *(A) Clear the equation of fractions (if they are present) by multiplying through by the least common multiple of all of the demoninators. (B) Write the equation in standard quadratic form. (C) If all numerical coefficients contain a common factor, divide it out. (D) Test for factorability. (E) If factorable, solve.*

23. $2x^2 = 3 - 5x$
24. $3x^2 = x + 2$
25. $3x(x - 2) = 2(x - 2)$
26. $2x(x - 1) = 3(x + 1)$
27. $4n^2 = 16n + 128$
28. $3m^2 + 12m = 36$
29. $3z^2 - 10z = 8$
30. $2y^2 + 15y = 8$
31. $3 = t^2 + 7t$
32. $y^2 = 5y - 2$

33. $\frac{u}{4}(u + 1) = 3$

34. $\frac{x^2}{2} = x + 4$

35. $y = \frac{9}{y}$

36. $\frac{t}{2} = \frac{2}{t}$

37. The width of a rectangle is 8 in. less than its length. If its area is 33 sq in., find its dimensions.

38. Find the base and height of a triangle with area 2 sq ft if its base is 3 ft longer than its height ($A = \frac{1}{2}bh$).

C *Solve:*

39. $y = \frac{15}{y - 2}$

40. $2x - 3 = \frac{2}{x}$

41. $2 + \frac{2}{x^2} = \frac{5}{x}$

42. $1 - \frac{3}{x} = \frac{10}{x^2}$

43. The sum of a number and its reciprocal is $\frac{13}{6}$. Find the number.

44. The difference between a number and its reciprocal is $\frac{7}{12}$. Find the number.

45. A flag has a cross of uniform width centered on a red background (Fig. 1). Find the width of the cross so that it takes up exactly half of the total area of a 4- by 3-ft flag.

Figure 1

8.3 Solution by Square Root

The method of square root is very fast when it applies, and it leads to a general method for solving all quadratic equations, as will be seen in the next section. A few examples should make the process clear.

EXAMPLE 4

$x^2 - 9 = 0$

$x^2 = 9$ What number squared is 9?

$x = \pm\sqrt{9}$ Short for $\sqrt{9}$ and $-\sqrt{9}$.

$x = \pm 3$

PROBLEM 4 Solve $x^2 = 16$ by the square root method.

ANSWER $x = \pm 4$

EXAMPLE 5

$$\begin{aligned} x^2 - 7 &= 0 \\ x^2 &= 7 \\ x &= \pm\sqrt{7} \end{aligned}$$

PROBLEM 5 Solve $x^2 - 8 = 0$.

ANSWER $\pm\sqrt{8}$ or $\pm 2\sqrt{2}$

EXAMPLE 6

$$\begin{aligned} 2x^2 - 3 &= 0 \\ 2x^2 &= 3 \\ x^2 &= \tfrac{3}{2} \\ x &= \pm\sqrt{\tfrac{3}{2}} \quad \text{or} \quad \pm\frac{\sqrt{6}}{2} \end{aligned}$$

PROBLEM 6 Solve $3x^2 - 2 = 0$.

ANSWER $\pm\sqrt{\dfrac{2}{3}}$ or $\pm\dfrac{\sqrt{6}}{3}$

EXAMPLE 7

$$\begin{aligned} (x - 2)^2 &= 16 \\ x - 2 &= \pm 4 \\ x &= 2 \pm 4 \\ x &= 6, -2 \end{aligned}$$

PROBLEM 7 Solve $(x + 3)^2 = 25$.

ANSWER $-8, 2$

EXAMPLE 8

$$\begin{aligned} (x + \tfrac{1}{2})^2 &= \tfrac{5}{4} \\ x + \tfrac{1}{2} &= \pm\sqrt{\tfrac{5}{4}} \\ x &= -\frac{1}{2} \pm \frac{\sqrt{5}}{2} \\ x &= \frac{-1 \pm \sqrt{5}}{2} \end{aligned}$$

PROBLEM 8 Solve $(x - \frac{1}{3})^2 = \frac{7}{9}$.

ANSWER $\dfrac{1 \pm \sqrt{7}}{3}$

EXAMPLE 9

$$x^2 = -9$$

No solution in the real numbers, since no real number squared is negative.

PROBLEM 9 Solve $x^2 + 4 = 0$

ANSWER No real solutions

Exercise 59

A *Solve by the square root method.*

1. $x^2 = 16$

2. $x^2 = 49$

3. $m^2 - 64 = 0$

4. $n^2 - 25 = 0$

5. $x^2 = 3$

6. $y^2 = 2$

7. $u^2 - 5 = 0$

8. $x^2 - 11 = 0$

9. $a^2 = 18$

10. $y^2 = 8$

11. $x^2 - 12 = 0$

12. $n^2 - 27 = 0$

13. $x^2 = \frac{4}{9}$

14. $y^2 = \frac{9}{16}$

15. $9x^2 = 4$

16. $16y^2 - 9$

17. $9x^2 - 4 = 0$

18. $16y^2 - 9 = 0$

B **19.** $25x^2 - 4 = 0$

20. $9x^2 - 1 = 0$

21. $4t^2 - 3 = 0$

22. $9x^2 - 7 = 0$

23. $2x^2 - 5 = 0$

24. $3m^2 - 7 = 0$

25. $3m^2 - 1 = 0$

26. $5n^2 - 1 = 0$

27. $(y - 2)^2 = 9$

28. $(x - 3)^2 = 4$

29. $(x + 2)^2 = 25$

30. $(y + 3)^2 = 16$

31. $(y - 2)^2 = 3$

32. $(y - 3)^2 = 5$

33. $(x - \frac{1}{2})^2 = \frac{9}{4}$

34. $(x - \frac{1}{3})^2 = \frac{4}{9}$

35. $(x - 3)^2 = -4$

36. $(t + 1)^2 = -9$

C **37.** $(x - \frac{3}{2})^2 = \frac{3}{2}$

38. $(y + \frac{5}{2})^2 = \frac{5}{2}$

39. Solve for b: $a^2 + b^2 = c^2$

40. Solve for v: $k = \frac{1}{2}mv^2$

41. The pressure p in pounds per sq ft from a wind blowing at v mph is $p = 0.003v^2$. If a pressure gauge on a bridge registers a wind pressure of 14.7 lb per sq ft, what is the velocity of the wind?

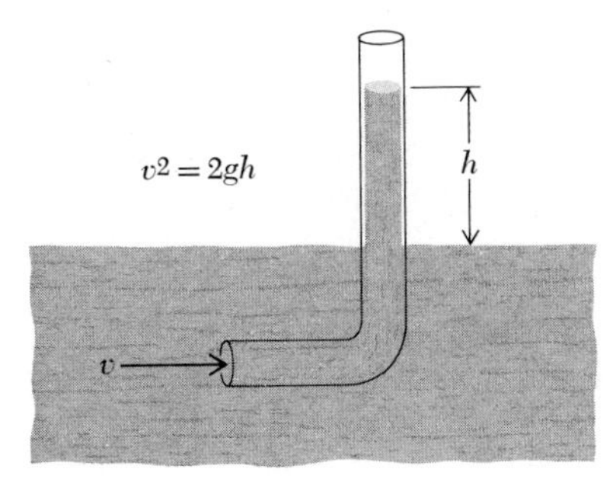

Figure 2

42. One method of measuring the velocity of water in a stream or river is to use an L-shaped tube as indicated in Fig. 2. Torricelli's law in physics tells us that the height (in feet) that the water is pushed up into the tube above the surface is related to the water's velocity (in feet per second) by the formula $v^2 = 2gh$, where g is approximately 32 ft per sec per sec. (NOTE: The device can also be used as a simple speedometer for a boat.) How fast is a stream flowing if $h = 0.5$ ft? Find the answer to two decimal places.

8.4 Solution by Completing the Square

The factoring and square root methods discussed in the last two sections are fast and easy to use when they apply. Unfortunately many quadratic equations will not yield to either method. For example, the very simple-looking polynomial in

$$x^2 + 6x - 2 = 0$$

cannot be factored in the integers. It requires a new method if it can be solved at all.

In this section we will discuss a method, called "solution by completing the square," that will work for all quadratic equations. In the next section we will use this method to develop a general formula that will be used in the future whenever the methods of the two preceding sections fail.

The method of completing the square is based on the process of transforming the standard quadratic equation,

$$ax^2 + bx + c = 0$$

into the form

$$(x + A)^2 = B$$

where A and B are constants. This last equation can easily be solved (assuming $B \geq 0$) by the method discussed in the last section. Thus

$$\begin{aligned} (x + A)^2 &= B \\ x + A &= \pm\sqrt{B} \\ x &= -A \pm \sqrt{B} \end{aligned}$$

Before considering how the first part is accomplished, let's pause for a moment and consider a related problem: What number must be added to $x^2 + 6x$ so that the result is the square of a linear expression? There is an easy mechanical rule for finding this number based on the squares of the following binomials:

$$(x + m)^2 = x^2 + 2mx + m^2$$

$$(x - m)^2 = x^2 - 2mx + m^2$$

In either case, we see that the third term on the right is the square of one-half of the coefficient of x in the second term on the right. This observation leads directly to the rule:

> To *complete the square* of a quadratic of the form
>
> $$x^2 + bx$$
>
> add the square of one-half of the coefficient of x, that is
>
> $$\left(\frac{b}{2}\right)^2$$

EXAMPLE 10 (*A*) To complete the square of $x^2 + 6x$, add $(\frac{6}{2})^2$, that is, 9; thus

$$x^2 + 6x + 9 = (x + 3)^2$$

(*B*) To complete the square of $x^2 - 3x$, add $(-\frac{3}{2})^2$; that is $\frac{9}{4}$; thus

$$x^2 - 3x + \tfrac{9}{4} = (x - \tfrac{3}{2})^2$$

PROBLEM 10 (*A*) Complete the square of $x^2 + 10x$ and factor.

(*B*) Complete the square of $x^2 + 5x$ and factor.

ANSWER (*A*) $x^2 + 10x + 25 = (x + 5)^2$ (*B*) $x^2 + 5x + \frac{25}{4} = (x + \frac{5}{2})^2$

It is important to note that the rule stated above applies only to quadratic forms where the coefficient of the second-degree term is 1.

SOLUTION OF QUADRATIC EQUATIONS BY COMPLETING THE SQUARE

Solving quadratic equations by the method of completing the square is best illustrated by examples. In this course we are only going to be interested in real solutions.

EXAMPLE 11 Solve $x^2 + 6x - 2 = 0$ by the method of completing the square.

SOLUTION

$x^2 + 6x - 2 = 0$ — Add 2 to both sides of the equation to remove -2 from the left side.

$x^2 + 6x = 2$ — To complete the square of the left side, add the square of one-half of the coefficient of x to each side of the equation.

$x^2 + 6x + 9 = 2 + 9$

$(x + 3)^2 = 11$ — Factor the left side.

$x + 3 = \pm\sqrt{11}$ — Proceed as in the last section.

$x = -3 \pm \sqrt{11}$

PROBLEM 11 Solve $x^2 - 8x + 10 = 0$ by method of completing the square.

ANSWER $x = 4 \pm \sqrt{6}$

EXAMPLE 12 Solve $2x^2 - 4x - 3 = 0$ by the method of completing the square.

SOLUTION

Note that the coefficient of x^2 is not 1. Divide through by the leading coefficient and proceed as in the last example.

$$2x^2 - 4x - 3 = 0$$
$$x^2 - 2x - \tfrac{3}{2} = 0$$
$$x^2 - 2x = \tfrac{3}{2}$$
$$x^2 - 2x + 1 = \tfrac{3}{2} + 1$$
$$(x - 1)^2 = \tfrac{5}{2}$$
$$x - 1 = \pm\sqrt{\tfrac{5}{2}}$$
$$x = 1 \pm \frac{\sqrt{10}}{2}$$
$$x = \frac{2 \pm \sqrt{10}}{2}$$

PROBLEM 12 Solve $2x^2 + 8x + 3 = 0$ by method of completing the square.

ANSWER $x = -2 \pm \sqrt{\dfrac{5}{2}}$ or $\dfrac{-4 \pm \sqrt{10}}{2}$

Exercise 60

A *Complete the square and factor.*

1. $x^2 + 4x$

2. $x^2 + 8x$

3. $x^2 - 6x$

4. $x^2 - 10x$

5. $x^2 + 12x$

6. $x^2 + 2x$

Solve by method of completing the square.

7. $x^2 + 4x + 2 = 0$

8. $x^2 + 8x + 3 = 0$

9. $x^2 - 6x - 3 = 0$

10. $x^2 - 10x - 3 = 0$

B *Complete the square and factor.*

11. $x^2 + 3x$

12. $x^2 + x$

13. $u^2 - 5u$

14. $m^2 - 7m$

Solve by method of completing the square.

15. $x^2 + x - 1 = 0$

16. $x^2 + 3x - 1 = 0$

17. $u^2 - 5u + 2 = 0$

18. $n^2 - 3n - 1 = 0$

19. $m^2 - 4m + 8 = 0$

20. $x^2 - 2x + 3 = 0$

21. $2y^2 - 4y + 1 = 0$

22. $2x^2 - 6x + 3 = 0$

23. $2u^2 + 3u - 1 = 0$

24. $3x^2 + x - 1 = 0$

C **25.** Solve for x: $x^2 + mx + n = 0$

26. Solve for x: $ax^2 + bx + c = 0,\ a \neq 0$

8.5 The Quadratic Formula

The method of completing the square can be used to solve any quadratic equation, but the process is often tedious. If you had a very large number of quadratic equations to solve by completing the square, before you finished you would probably ask yourself if the process could not be made more efficient. Why not take the general equation

$$ax^2 + bx + c = 0 \qquad a \neq 0$$

and solve it once and for all for x in terms of the coefficients a, b, and c by the method of completing the square, and thus, obtain a formula that could be memorized and used whenever a, b, and c are known?

We start by making the leading coefficient 1. How? Multiply both sides of the equation by $1/a$. Thus

$$x^2 + \frac{b}{a}x + \frac{c}{a} = 0$$

Adding $-c/a$ to each side, we get

$$x^2 + \frac{b}{a}x = -\frac{c}{a}$$

Add the square of one-half of the coefficient of x, which is $(b/2a)^2$, to each side to complete the square of the left side. Thus,

$$x^2 + \frac{b}{a}x + \frac{b^2}{4a^2} = \frac{b^2}{4a^2} - \frac{c}{a}$$

Factor the left side and combine the right side into a single term, leaving

$$\left(x + \frac{b}{2a}\right)^2 = \frac{b^2 - 4ac}{4a^2}$$

If $b^2 - 4ac \geq 0$, then by the definition of square root of positive real numbers,

$$x + \frac{b}{2a} = \pm\sqrt{\frac{b^2 - 4ac}{4a^2}}$$

$$x = -\frac{b}{2a} \pm \frac{\sqrt{b^2 - 4ac}}{2a}$$

$$\boxed{x = \frac{-b \pm \sqrt{b^2 - 4ac}}{2a} \qquad a \neq 0}$$

This last equation is called the *quadratic formula.* It should be memorized and used to solve quadratic equations when simpler methods fail. The following examples illustrate the use of the formula. Note that if $b^2 - 4ac$ is negative there are no real solutions.

EXAMPLE 13 Solve $2x + \frac{3}{2} = x^2$ by use of the quadratic formula.

SOLUTION

$$2x + \tfrac{3}{2} = x^2$$

$$4x + 3 = 2x^2$$ Clear the equation of fractions.

$$2x^2 - 4x - 3 = 0$$ Write in standard form.

$$x = \frac{-b \pm \sqrt{b^2 - 4ac}}{2a} \qquad a = 2,\ b = -4,\ c = -3$$ Write down the quadratic formula, and identify a, b, and c.

$$x = \frac{-(-4) \pm \sqrt{(-4)^2 - 4(2)(-3)}}{2(2)}$$ Substitute into formula and simplify.

$$x = \frac{4 \pm \sqrt{40}}{4} = \frac{4 \pm 2\sqrt{10}}{4}$$

$$x = \frac{2 \pm \sqrt{10}}{2}$$

PROBLEM 13 Solve $3x^2 = 2x + 2$ by use of the quadratic formula.

ANSWER $x = \dfrac{1 \pm \sqrt{7}}{3}$

Exercise 61

A *Specify the constants a, b, and c for each quadratic equation when written in the standard form $ax^2 + bx + c = 0$.*

1. $x^2 + 4x + 2 = 0$

2. $x^2 + 8x + 3 = 0$

3. $x^2 - 3x - 2 = 0$

4. $x^2 - 6x - 8 = 0$

5. $3x^2 - 2x + 1 = 0$

6. $2x^2 - 5x + 3 = 0$

7. $2u^2 = 1 - 3u$

8. $m = 1 - 3m^2$

9. $2x^2 - 5x = 0$

10. $3y^2 - 5 = 0$

Solve by use of the quadratic formula.

11. $x^2 + 4x + 2 = 0$

12. $x^2 + 8x + 3 = 0$

13. $y^2 - 6y - 3 = 0$

14. $y^2 - 10y - 3 = 0$

B

15. $3t + t^2 = 1$

16. $x^2 = 1 - x$

17. $2x^2 - 6x + 3 = 0$

18. $2x^2 - 4x + 1 = 0$

19. $3m^2 = 1 - m$

20. $3u + 2u^2 = 1$

21. $x^2 = 2x - 3$

22. $x^2 + 8 = 4x$

23. $2x = 3 + \dfrac{3}{x}$

24. $x + \dfrac{2}{x} = 6$

C

25. $m^2 = \dfrac{8m - 1}{5}$

26. $x^2 = 3x + \dfrac{1}{2}$

27. $3u^2 = \sqrt{3}u + 2$

28. $t^2 - \sqrt{5}t - 11 = 0$

True (T) or false (F)?

29. If $b^2 - 4ac < 0$, the quadratic equation has no real solution.

30. If $b^2 - 4ac > 0$, the quadratic equation has two real soutions.

31. If $b^2 - 4ac = 0$, the quadratic equation has one real solution.

32. If $b^2 - 4ac = 0$, the quadratic equation has two real solutions.

8.6 Which Method?

In normal practice the quadratic formula is used whenever the square root method or the factoring method do not produce results. These latter methods are generally faster when they apply, and should be used.

Note that any equation of the form

$$ax^2 + c = 0$$

can always be solved (if solutions exist in the real numbers) by the square root method. And any equation of the form

$$ax^2 + bx = 0$$

can always be solved by factoring since $ax^2 + bx = x(ax + b)$.

It is important to realize, however, that the quadratic equation can always be used and will produce the same results as any of the other methods. For example, let us solve

$$2x^2 + 7x - 15 = 0$$

Suppose you observe that the polynomial factors. Thus,

$$(2x - 3)(x + 5) = 0$$

$$2x - 3 = 0 \qquad \text{or} \qquad x + 5 = 0$$

$$x = \tfrac{3}{2} \qquad \text{or} \qquad x = -5$$

Suppose you had used the quadratic formula instead?

$$x = \frac{-b \pm \sqrt{b^2 - 4ac}}{2a} \qquad \begin{aligned} a &= 2 \\ b &= 7 \\ c &= -15 \end{aligned}$$

$$x = \frac{-(7) \pm \sqrt{7^2 - 4(2)(-15)}}{2(2)}$$

$$x = \frac{-7 \pm \sqrt{169}}{4} = \frac{-7 \pm 13}{4}$$

$$x = \tfrac{3}{2}, -5$$

The formula produces the same result (as it should), but with a little more work.

In the exercises for this section the problems will be mixed up, and it will be up to you to use the most efficient method—formula, factoring, or square root—for each particular problem.

Exercise 62

Use the most efficient method to find all real solutions to each equation.

A

1. $x^2 - x - 6 = 0$

2. $x^2 + 2x - 8 = 0$

3. $x^2 + 7x = 0$

4. $x^2 = 3x$

5. $2x^2 = 32$

6. $3x^2 - 27 = 0$

7. $x^2 + 2x - 2 = 0$

8. $m^2 - 3m - 1 = 0$

9. $2x^2 = 4x$

10. $2y^2 + 3y = 0$

11. $x^2 - 2x = 1$

12. $x^2 - 2 = 2x$

B

13. $u^2 = 3u - \frac{3}{2}$

14. $t^2 = \frac{3}{2}(t + 1)$

15. $M = M^2$

16. $t(t - 3) = 0$

17. $6y = \frac{1 - y}{y}$

18. $2x + 1 = \frac{6}{x}$

19. $I^2 - 50 = 0$

20. $72 = u^2$

21. $(B - 2)^2 = 3$

22. $(u + 3)^2 = 5$

C

23. $x^2 + 4 = 0$

24. $(x - 2)^2 = -9$

25. $\frac{24}{n} = 12n - 28$

26. $3x = \frac{84 - 9x}{x}$

27. $\frac{24}{10 + x} + 1 = \frac{24}{10 - x}$

28. $\frac{1.2}{x - 1} + \frac{1.2}{x} = 1$

Solve for the indicated letter in terms of the other letters.

29. $d = \frac{1}{2}gt^2$ for t (positive)

30. $a^2 + b^2 = c^2$ for a (positive)

31. $A = P(1 + r)^2$ for r (positive)

32. $P = EI - RI^2$ for I (positive)

8.7 Applications

We conclude this chapter with a number of applications from several different areas: communications, economics-business, geometry, number problems, police science, physics-engineering, and rate-time.

Since quadratic equations often have two solutions, it is important to check both of the solutions in the original problem to see if one or the other must be rejected.

EXAMPLE 14

An artist with a painting measuring 6 by 8 in. wishes to frame it with a frame of uniform width that has a total area equal to the area of the painting. How wide should he make the frame? Give the answer in simplest radical form and as a decimal fraction to two decimal places.

SOLUTION

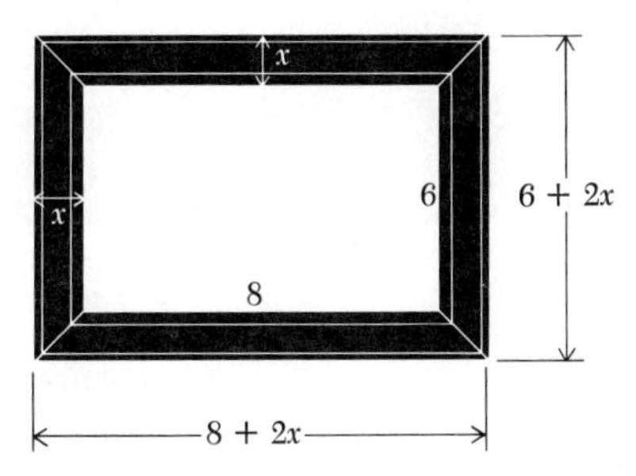

$$\text{Total area of picture and frame} = \text{Twice the area of the picture}$$

$$(6 + 2x)(8 + 2x) = 2(6 \cdot 8)$$
$$48 + 28x + 4x^2 = 96$$
$$x^2 + 7x - 12 = 0$$

$$x = \frac{-b \pm \sqrt{b^2 - 4ac}}{2a} \qquad a = 1,\ b = 7,\ c = 12$$

$$x = \frac{-7 \pm \sqrt{7^2 - 4(1)(-12)}}{2(1)}$$

$$x = \frac{-7 \pm \sqrt{97}}{2}$$

The negative answer must be rejected since it has no meaning relative to the original problem; hence

$$x = \frac{-7 + \sqrt{97}}{2} = 1.43 \text{ in.}$$

PROBLEM 14

If the length and width of a 4- by 2-in. rectangle are each increased by the same amount, the area of the new rectangle will be twice the old. What are the dimensions to two decimal places of the new rectangle?

ANSWER 5.12 by 3.12 in.

Exercise 63

COMMUNICATIONS

1. The number of telephone connections possible through a switchboard to which n telephones are connected is given by the formula $c = n(n - 1)/2$. How many telephones could be handled by a switchboard that had the capacity of 190 connections?

ECONOMICS AND BUSINESS

2. If P dollars is invested at r percent compounded annually, at the end of two years it will grow to $A = P(1 + r)^2$. At what interest rate will \$100 grow to \$144 in two years? NOTE: $A = 144$ and $P = 100$.

3. Cost equations for manufacturing companies are often quadratic in nature. (At very high or very low outputs the costs are more per unit because of inefficiency of plant operation at these extremes.) If the cost equation for manufacturing transistor radios is $C = x^2 - 10x + 31$, where C is the cost of manufacturing x units per week (both in thousands), find (*A*) the output for a \$15,000 weekly cost and (*B*) the output for a \$6,000 weekly cost.

4. The manufacturing company in the preceding problem sells its transistor radios for \$3 each. Thus its revenue equation is $R = 3x$, where R is revenue and x is the number of units sold per week (both in thousands). Find the break-even points for the company, that is, the output at which revenue equals cost.

5. In a certain city the demand equation for popular records is $q_d = 3{,}000/p$, where q_d would be the quantity of records demanded on a given day if the selling price were p dollars per record. (Notice as the price goes up, the number of records the people are willing to buy goes down, and vice versa.) On the other hand, the supply equation is $q_s = 1{,}000p - 500$, where q_s is the quantity of records a supplier is willing to supply at p dollars per record. (Notice as the price goes up, the number of records a supplier is willing to sell goes up, and vice versa.) At what price will supply equal demand; that is, at which price will $q_d = q_s$? In economic theory the price at which supply equals demand is called the *equilibrium point*, the point at which the price ceases to change.

GEOMETRY *The following theorem may be used where needed:*

PYTHAGOREAN THEOREM. *A triangle is a right triangle if and only if the square of the longest side is equal to the sum of the squares of the two shorter sides.*

$$c^2 = a^2 + b^2$$

c b a

6. Approximately how far would a person be able to see from the top of a mountain 2 miles high (Fig. 3)? Use the square root table to estimate the answer to the nearest mile.

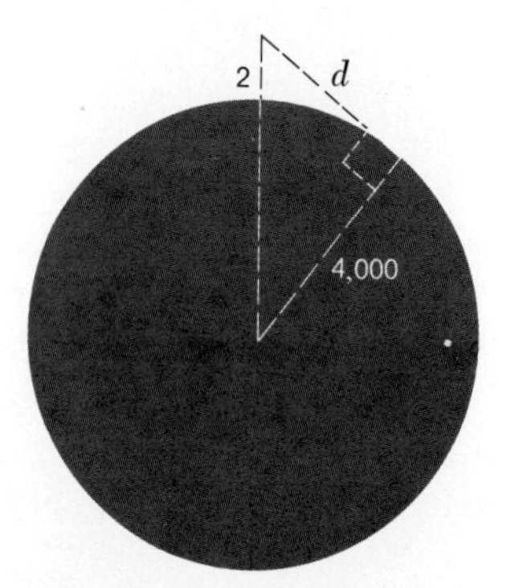

Figure 3

7. Find the length of each side of a right triangle if the second longest side is 1 in. longer than the shortest side and the longest side is 2 in. longer than the shortest side.

This right triangle has been well known for over 2,000 years. It has many interesting properties and has been put to many interesting uses. For example, it can be shown that any triangle whose sides are in the same ratio must be a right triangle. The ancients used this fact to lay out right angles when surveying land and constructing buildings. An appropriately knotted piece of rope is all of the equipment that is needed.

Also, by stretching a guitar string around three pulleys nailed to a board and spaced so that distances between them are in the ratio of the sides of the triangle under consideration, the three sides of the triangle thus formed will sound the same chord as the top three strings of a guitar if the longest side is tuned to G.

8. Find r in Fig. 4. Express the answer in simplest radical form.

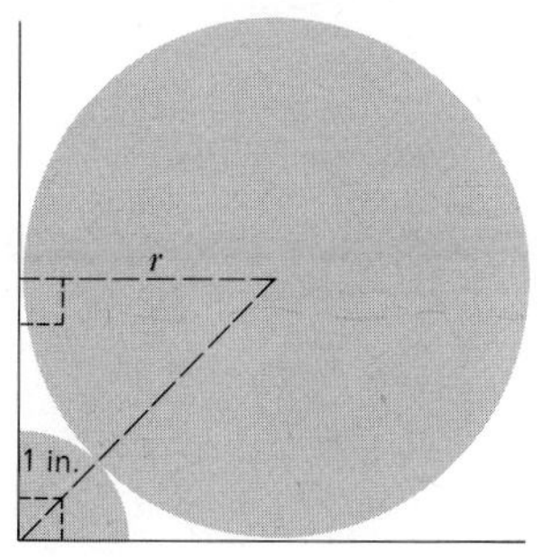

Figure 4

9. A golden rectangle is defined as one that has the property that when a square of the side equal to the short side of the rectangle is removed, the ratio of the sides of the remaining rectangle is the same as the ratio of the sides of the original rectangle. If the shorter side of the original rectangle is 1, find the shorter side of the remaining rectangle (see Fig. 5). This number is called the golden ratio, and it turns up frequently in the history of mathematics.

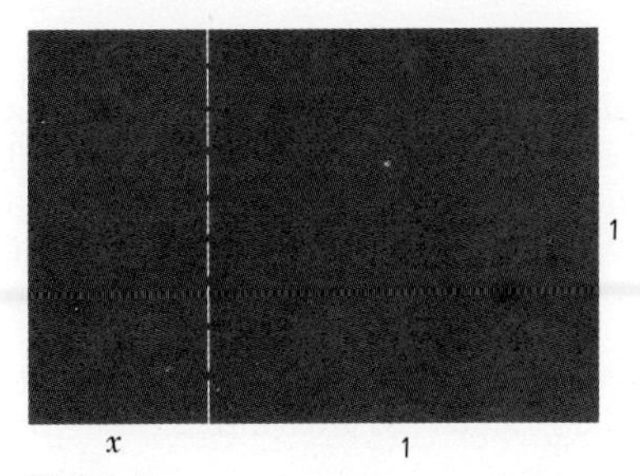

Figure 5

NUMBER PROBLEMS

10. Separate 21 into two parts whose product is 104.

11. Find two consecutive positive even integers whose product is 168.

12. Find all numbers with the property that when the number is added to itself, the sum is the same as when the number is multiplied by itself.

13. Find a number whose square exceeds itself by 56.

14. The sum of a number and its reciprocal is $\frac{10}{3}$. Find the number.

POLICE SCIENCE

15. Skid marks are often used to estimate the speed of a car in an accident. It is common practice for an officer to drive the car in question (if it is still running) at a speed of 20 to 30 mph and skid it to a stop near the original skid marks. It is known (from physics) that the speed of the car and the length of the skid marks are related by the formula

$$\frac{d_a}{v_a^2} = \frac{d_t}{v_t^2}$$

where d_a = length of accident car's skid marks
d_t = length of test car's skid marks
v_a = speed of accident car (to be found)
v_t = speed of test car

Estimate the speed of an accident vehicle if its skid marks are 120 ft and the test car driven at 30 mph produces skid marks of 36 ft.

16. At 20 mph a car collides with a stationary object with the same force it would have if it had been dropped $13\frac{1}{2}$ ft, that is, if it had been pushed off the roof of an average one-story house. In general, a car moving at r mph hits a stationary object with a force of impact that is equivalent to that force with which it would hit the ground when falling from a certain height h given by the formula $h = 0.0336r^2$. Approximately how fast would a car have to be moving if it crashed as hard as if it had been pushed off the top of a 12-story building 121 ft high? (Fig. 6)

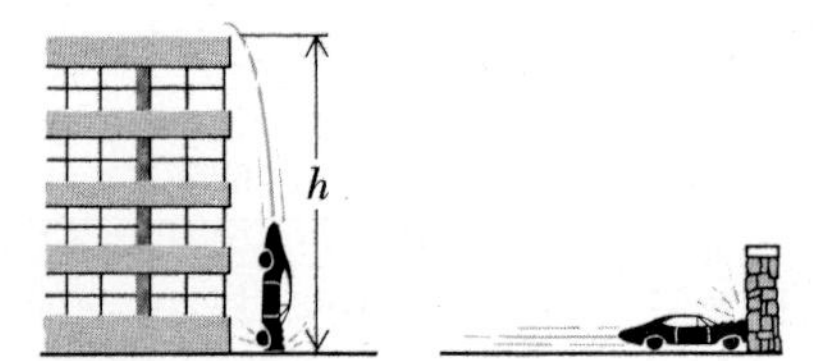

Figure 6 Energy and cars.

PHYSICS-ENGINEERING

17. In physics it is found that the illumination I in foot-candles on a surface d ft from a light source of c candlepower is given by the formula $I = c/d^2$. How far should a light of 20 candlepower be placed from a surface to produce the same illumination as a light of 10 candlepower at 10 ft? Write the answer in simplest radical form, and approximate it to two decimal places.

18. If an arrow is shot vertically in the air (from the ground) with an initial velocity of 176 fps, its distance y above the ground t sec after it is released (neglecting air resistance) is given by $y = 176t - 16t^2$.
(A) Find the times when y is 0, and interpret physically.
(B) Find the times when the arrow is 16 ft off the ground. Compute answers to two decimal places.

19. A *Pitot tube* is a device (based on Bernoulli's principles) that is used to measure the flow speed of a gas (Fig. 7). If mounted on an airplane, it indicates the plane's velocity

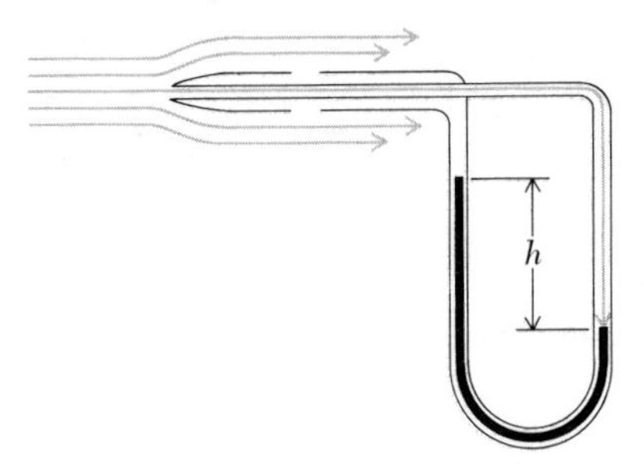

Figure 7

relative to the surrounding air and is known as an airspeed indicator. A simple equation relates the height differential of the fluid in the tube with the velocity of the gas. Thus,

$$v^2 = 2gkh$$

where v = velocity of gas (feet per second)
g = gravitational constant (approximately 32 ft per sec per sec)
k = ratio of the density of the fluid in the tube to the density of the gas
h = height differential (in feet)

Approximate an aircraft's speed in feet per second and miles per hour if $k = 625$ and $h = 0.81$.

RATE-TIME PROBLEMS

20. Two boats travel at right angles to each other after leaving the same dock at the same time. One hour later they are 13 miles apart. If one travels 7 mph faster than the other, what is the rate of each?

21. A new printing press can do a job in 1 hr less than an older press. Together they can do the same job in 1.2 hr. How long would it take each alone to do the job?

22. One pipe can fill a tank in 5 hr less than another; together they fill the tank in 5 hr. How long would it take each alone to fill the tank? Compute answer to two decimal places.

23. A speedboat takes 1 hr longer to go 24 miles up a river than to return. If the boat cruises at 10 mph in still water, what is the rate of the current?

Exercise 64 Chapter Review

A *Find all real solutions by factoring or square root methods.*

1. $x^2 = 25$

2. $x^2 - 3x = 0$

3. $(2x - 1)(x + 3) = 0$

4. $x^2 - 5x + 6 = 0$

5. $x^2 - 2x - 15 = 0$

6. Write $4x = 2 - 3x^2$ in standard form $ax^2 + bx + c = 0$ and identify a, b, and c.

7. Write down the quadratic formula associated with $ax^2 + bx + c = 0$.

8. Use the quadratic formula to solve $x^2 + 3x + 1 = 0$.

9. Solve $x^2 + 3x - 10 = 0$ by any method.

10. Find two positive numbers whose product is 27 if one is 6 more than the other.

B *Find all real solutions by factoring or square root methods.*

11. $3x^2 = 36$

12. $10x^2 = 20x$

13. $(x - 2)^2 = 16$

14. $3t^2 - 8t - 3 = 0$

15. $2x = \dfrac{3}{x} - 5$

16. Solve $x^2 - 6x - 3 = 0$ by completing-the-square method.

17. Solve $3x^2 = 2(x + 1)$ using the quadratic formula.

18. Solve $2x^2 - 2x = 40$ by any method.

19. Divide 18 into two parts so that their product is 72.

20. The perimeter of a rectangle is 22 in. If its area is 30 sq in. find the length of each side.

C *Find all real solutions by factoring or square root methods.*

21. $3x^2 + 27 = 0$

22 $(t - \frac{3}{2})^2 = \frac{3}{2}$

23. $\dfrac{8m^2 + 15}{2m} = 13$

24. Solve $2x^2 - 2x - 3 = 0$ by completing-the-square method.

25. Solve $3x - 1 = \dfrac{2(x + 1)}{x + 2}$ using the quadratic formula.

26. If $b^2 - 4ac > 0$, then the quadratic equation has two real solutions. [True (T) or false (F)?]

Appendix

SQUARES AND SQUARE ROOTS (0 TO 199)

n	n^2	$\sqrt{n}$	n	n^2	$\sqrt{n}$	n	n^2	$\sqrt{n}$	n	n^2	$\sqrt{n}$
0	0	0.000	50	2,500	7.071	100	10,000	10.000	150	22,500	12.247
1	1	1.000	51	2,601	7.141	101	10,201	10.050	151	22,801	12.288
2	4	1.414	52	2,704	7.211	102	10,404	10.100	152	23,104	12.329
3	9	1.732	53	2,809	7.280	103	10,609	10.149	153	23,409	12.369
4	16	2.000	54	2,916	7.348	104	10,816	10.198	154	23,716	12.410
5	25	2.236	55	3,025	7.416	105	11,025	10.247	155	24,025	12.450
6	36	2.449	56	3,136	7.483	106	11,236	10.296	156	24,336	12.490
7	49	2.646	57	3,249	7.550	107	11,449	10.344	157	24,649	12.530
8	64	2.828	58	3,364	7.616	108	11,664	10.392	158	24,964	12.570
9	81	3.000	59	3,481	7.681	109	11,881	10.440	159	25,281	12.610
10	100	3.162	60	3,600	7.746	110	12,100	10.488	160	25,600	12.649
11	121	3.317	61	3,721	7.810	111	12,321	10.536	161	25,921	12.689
12	144	3.464	62	3,844	7.874	112	12,544	10.583	162	26,244	12.728
13	169	3.606	63	3,969	7.937	113	12,769	10.630	163	26,569	12.767
14	196	3.742	64	4,096	8.000	114	12,996	10.677	164	26,896	12.806
15	225	3.873	65	4,225	8.062	115	13,225	10.724	165	27,225	12.845
16	256	4.000	66	4,356	8.124	116	13,456	10.770	166	27,556	12.884
17	289	4.123	67	4,489	8.185	117	13,689	10.817	167	27,889	12.923
18	324	4.243	68	4,624	8.246	118	13,924	10.863	168	28,224	12.961
19	361	4.359	69	4,761	8.307	119	14,161	10.909	169	28,561	13.000
20	400	4.472	70	4,900	8.367	120	14,400	10.954	170	28,900	13.038
21	441	4.583	71	5,041	8.426	121	14,641	11.000	171	29,241	13.077
22	484	4.690	72	5,184	8.485	122	14,884	11.045	172	29,584	13.115
23	529	4.796	73	5,329	8.544	123	15,129	11.091	173	29,929	13.153
24	576	4.899	74	5,476	8.602	124	15,376	11.136	174	30,276	13.191
25	625	5.000	75	5,625	8.660	125	15,625	11.180	175	30,625	13.229
26	676	5.099	76	5,776	8.718	126	15,876	11.225	176	30,976	13.266
27	729	5.196	77	5,929	8.775	127	16,129	11.269	177	31,329	13.304
28	784	5.292	78	6,084	8.832	128	16,384	11.314	178	31,684	13.342
29	841	5.385	79	6,241	8.888	129	16,641	11.358	179	32,041	13.379
30	900	5.477	80	6,400	8.944	130	16,900	11.402	180	32,400	13.416
31	961	5.568	81	6,561	9.000	131	17,161	11.446	181	32,761	13.454
32	1,024	5.657	82	6,724	9.055	132	17,424	11.489	182	33,124	13.491
33	1,089	5.745	83	6,889	9.110	133	17,689	11.533	183	33,489	13.528
34	1,156	5.831	84	7,056	9.165	134	17,956	11.576	184	33,856	13.565
35	1,225	5.916	85	7,225	9.220	135	18,225	11.619	185	34,225	13.601
36	1,296	6.000	86	7,396	9.274	136	18,496	11.662	186	34,596	13.638
37	1,369	6.083	87	7,569	9.327	137	18,769	11.705	187	34,969	13.675
38	1,444	6.164	88	7,744	9.381	138	19,044	11.747	188	35,344	13.711
39	1,521	6.245	89	7,921	9.434	139	19,321	11.790	189	35,721	13.748
40	1,600	6.325	90	8,100	9.487	140	19,600	11.832	190	36,100	13.784
41	1,681	6.403	91	8,281	9.539	141	19,881	11.874	191	36,481	13.820
42	1,764	6.481	92	8,464	9.592	142	20,164	11.916	192	36,864	13.856
43	1,849	6.557	93	8,649	9.644	143	20,449	11.958	193	37,249	13.892
44	1,936	6.633	94	8,836	9.659	144	20,736	12.000	194	37,636	13.928
45	2,025	6.708	95	9,025	9.747	145	21,025	12.042	195	38,025	13.964
46	2,116	6.782	96	9,216	9.798	146	21,316	12.083	196	38,416	14.000
47	2,209	6.856	97	9,409	9.849	147	21,609	12.124	197	38,809	14.036
48	2,304	6.928	98	9,604	9.899	148	21,904	12.166	198	39,204	14.071
49	2,401	7.000	99	9,801	9.950	149	22,201	12.207	199	39,601	14.107

n n^2 $\sqrt{n}$ n n^2 $\sqrt{n}$ n n^2 $\sqrt{n}$ n n^2 $\sqrt{n}$

Answers

EXERCISE 1

1. 6, 13 **3.** 67, 402 **5.** Even: 14, 28; odd: 9, 33 **7.** Even: 426; odd: 23, 105, 77 **9.** Composite: 6, 9; prime: 2, 11 **11.** Composite: 12, 27; prime: 17, 23 **13.** 20, 22, 24, 26, 28, 30 **15.** 21, 23, 25, 27, 29 **17.** 20, 21, 22, 24, 25, 26, 27, 28, 30 **19.** 23, 29 **21.** $2 \cdot 5$ **23.** $2 \cdot 3 \cdot 5$ **25.** $2 \cdot 3 \cdot 5 \cdot 7$ **27.** No, No, No, Yes **29.** No **31.** Finite **33.** Finite **35.** Infinite **37.** Infinite **39.** Finite

EXERCISE 2

1. 21 sq in. **3.** 20 in. **5.** 96 cu in. **7.** 387 miles **9.** 13 **11.** 2 **13.** 17 **15.** 2 **17.** 1 **19.** 10 **21.** 5 **23.** 10 **25.** 18 **27.** 4 **29.** 15 sq in. **31.** 16 in. **33.** 14 **35.** 7 **37.** 20 **39.** 8 **41.** 5 **43.** 3 **45.** 22 **47.** 33 **49.** Constants: 2 and 3; variables: x and y **51.** Constants: 2 and 3; variables: u and v **53.** $x + 5$ **55.** $x - 5$ **57.** $2x + 3$ **59.** $3(x - 3)$ **61.** 320 sq ft **63.** 16 **65.** 16 **67.** $3(t + 2)$ **69.** $t + (t + 2) + (t + 4)$ **71.** (*A*) $d = 2t$; (*B*) Constant: 2; variables: d and t; (*C*) 20 miles

EXERCISE 3

1. T **3.** F **5.** T **7.** T **9.** T **11.** $5 = x + 3$ **13.** $8 = x - 3$ **15.** $18 = 3x$ **17.** $49 = 2x + 7$ **19.** $4x = 3x + 3$ **21.** $x + (x + 1) + (x + 2) + (x + 3) = 54$ **23.** $x + 2x = 27$ **25.** $50 = x(x + 10)$ **27.** Incorrect use of equal sign. We cannot write 5 = prime number (why?). **29.** 1. Reflexive law, 2. Given, 3. Substitution principle

31.

$a + c = a + c$	reflexive law
$a = b$	given
$a + c = b + c$	substitution principle

EXERCISE 4

1. Closure axiom for addition **3.** Commutative axiom for addition **5.** Associative axiom for multiplication **7.** Commutative axiom for addition **9.** Closure axiom for multiplication **11.** Commutative axiom multiplication **13.** Closure axiom for addition **15.** Associative axiom for multiplication **17.** $x + 5$ **19.** $12x$ **21.** $x + 10$ **23.** $15x$ **25.** Commutative axiom for addition **27.** Commutative axiom for addition **29.** Commutative axiom for addition **31.** $x + y + 15$ **33.** $12xy$ **35.** $a + b + c + 10$ **37.** $48xyz$ **39.** (*A*) A; (*B*) A, *B* **41.** (*C*) is false since $9 - 7 \neq 7 - 9$; (*D*) is false since $14 \div 7 \neq 7 \div 14$ **43.** 1. Associative axiom for addition; 2. Substitution principle for equality **45.** Closure and associative axioms for addition

EXERCISE 5

1. xxx **3.** $aaaaa$ **5.** $2xxxyy$ **7.** $3wwxyyy$ **9.** x^3 **11.** $3a^6$ **13.** $2x^3y^2$ **15.** $3xy^2z^3$ **17.** x^5 **19.** u^{14} **21.** a^6 **23.** w^{19} **25.** 7^8 **27.** 4×10 **29.** 4×10^3 **31.** 47×10^6 **33.** 50 **35.** 500,000 **37.** 230,000 **39.** $6x^5$ **41.** $35u^{16}$ **43.** 6×10^9 **45.** 66×10^{10} **47.** x^7 **49.** $24x^9$ **51.** 24×10^8 **53.** a^3b^3 **55.** $12x^2y^2$ **57.** $6x^4y^4$ **59.** 93,000,000 **61.** 5,000,000,000 **63.** Diameter = 588×10^{15} miles; thickness = $1{,}176 \times 10^{14}$ miles **65.** 132×10^{23} lb

EXERCISE 6

1. Both 12 **3.** Both 45 **5.** $4x + 4y$ **7.** $7m + 7n$ **9.** $6x + 12$
11. $10 + 5m$ **13.** $3(x + y)$ **15.** $5(m + n)$ **17.** $a(x + y)$
19. $2(x + 2)$ **21.** $2x + 2y + 2z$ **23.** $3x + 3y + 3z$ **25.** $7(x + y + z)$
27. $2(m + n + 3)$ **29.** $x + x^2$ **31.** $y + y^3$ **33.** $6x^2 + 15x$
35. $2m^4 + 6m^3$ **37.** $6x^3 + 9x^2 + 3x$ **39.** $10x^3 + 15x^2 + 5x + 10$
41. $6x^5 + 9x^4 + 3x^3 + 6x^2$ **43.** $10x$ **45.** $11u$ **47.** $5xy$ **49.** $10x^2y$
51. $(7 + 2 + 5)x = 14x$ **53.** $x(x + 2)$ **55.** $u(u + 1)$ **57.** $2x(x^2 + 2)$
59. $x(x + y + z)$ **61.** $3m(m^2 + 2m + 3)$ **63.** $uv(u + v)$
65. $8m^5n^4 + 4m^3n^5$ **67.** $6x^3y^4 + 12x^3y + 3x^2y^3$ **69.** $uc + ud + vc + vd$
71. $x^2 + 5x + 6$ **73.** $abc(a + b + c)$ **75.** $4xyz(4x^2z + xy + 3yz^2)$

EXERCISE 7

1. 4 **3.** 8 **5.** 1 **7.** 1 **9.** 3 **11.** 2 **13.** 1 **15.** $3x, 4x$; $2y, 5y$ **17.** $6x^2, 3x^2, x^2$; $x^3, 4x^3$ **19.** $2u^2v, u^2v$; $3uv^2, 5uv^2$ **21.** $9x$
23. $4u$ **25.** $9x^2$ **27.** $10x$ **29.** $7x + 4y$ **31.** $3x + 5y + 6$
33. $m^2n, 5m^2n$; $4mn^2, mn^2$; $2mn, 3mn$ **35.** $6t^2$ **37.** $4x + 7y + 7z$
39. $11x^3 + 4x^2 + 4x$ **41.** $4x^2 + 3xy + 2y^2$ **43.** $6x + 9$
45. $4t^2 + 8t + 10$ **47.** $3x^3 + 3x^2y + 4xy^2 + 2y^3$ **49.** $8x + 31$
51. $3x^2 + 4x$ **53.** $11t^2 + 13t + 17$ **55.** $3y^3 + 3y^2 + 4y$
57. $6x^2 + 5xy + 6y^2$ **59.** $x + (x + 1) + (x + 2) + (x + 3)$; $4x + 6$
61. $9x^2y^2 + 5x^3y^3$ **63.** $6x^2 + 13x + 6$ **65.** $2x^2 + 5xy + 2y^2$
67. $x^3 + 5x^2 + 11x + 15$ **69.** $y(y + 2)$; $y^2 + 2y$ **71.** $x + x(x + 2) = 180$; $x^2 + 3x = 180$

EXERCISE 8

1. $7 > 3$ **3.** $11 < 12$ **5.** $x \geq 5$ **7.** $x \leq 12$ **9.** T **11.** T
13. F **15.** F **17.** T **19.** T **21.** {6} **23.** {2, 3, 4, 5}
25. {2, 3, 4, 5, 6} **27.** {2} **29.** {3, 4, 5, 6} **31.** {5, 6} **33.** {4, 5, 6}
35. {4, 5, 6} **37.** {4, 5, 6} **39.** 5 is greater than 2 **41.** 2 is less than 5
43. x is greater than or equal to 3 **45.** T **47.** T **49.** T
51. {6, 8, 10} **53.** {2, 4} **55.** {8, 10} **57.** {2, 3} **59.** {2, 3, 4}
61. {1, 2, 3, 4} **63.** (A) $v = (18 - 2x)(12 - 2x)x$; (B) {1, 2, 3, 4, 5}; (C) $1 \leq x \leq 5$, $x \in N$

EXERCISE 9

1. 3, 5, 8 **3.** 19, 27, 37

5.  1 2 3 4 5 6 7 8

7.  1 2 3 4 5 6 7 8 9

9. 1 2 3 4 5 6 7 8 9

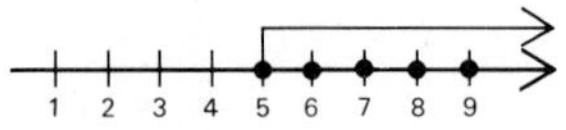

11.  1 2 3 4 5 6 7 8

13.  1 2 3 4 5 6 7 8

15.  1 2 3 4 5 6 7 8 9

17. 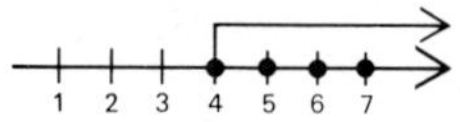 1 2 3 4 5 6 7

19.  30 35 40 45 50

21. 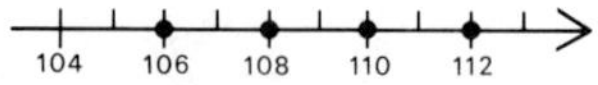 104 106 108 110 112

23. 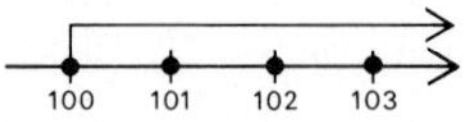 100 101 102 103

25.  40 42 44 46 48

27. 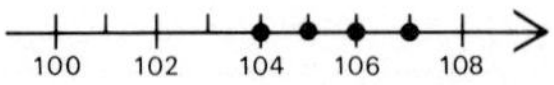 100 102 104 106 108

29. 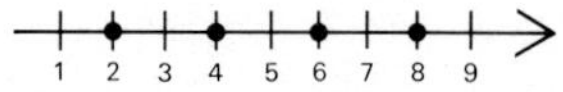 1 2 3 4 5 6 7 8 9

31. 1 2 3 4 5 6 7 8 9

33. $\{4, 5, 6, 7, 8, 9\}$

35. 1 2 3 4 5 6 7 8

37. 1 2 3 4 5 6 7 8 9

39. 1 2 3 4 5 6 7

EXERCISE 10

1. $\{11, 13, 15\}$ **2.** $\{10, 12, 14, 15\}$ **3.** (A) F, (B) T, (C) T **4.** 3 **5.** $8y$ **6.** (A) Commutative axiom for multiplication; (B) Associative axiom for addition **7.** (A) $x + 10$, (B) $10x$ **8.** (A) $7mmmnn$, (B) y^{32} **9.** (A) $3m + 3n$, (B) $5w + 5x + 5y$ **10.** (A) $3(m + n)$, (B) $8(u + v + w)$ **11.** (A) 3, (B) 1, (C) 1 **12.** (A) $9y$, (B) $5m + 5n$ **13.** (A) T, (B) T, (C) F **14.** (A) $\{4, 5\}$, (B) $\{1, 2, 3\}$, (C) $\{2, 3, 4\}$

15. 5 10 15

16. $P = \{29, 31\}$ **17.** $2 \cdot 2 \cdot 2 \cdot 3 \cdot 5$ **18.** 6 **19.** 6 **20.** $3x = x + 8$ **21.** (A) Commutative axiom for multiplication; (B) Closure axioms for addition and multiplication **22.** (A) $m + n + p + 14$, (B) $28xy$ **23.** $(12 \times 10^4)(3 \times 10^6) = 36 \times 10^{10}$ **24.** $6y^5 + 3y^4 + 15y^3$ **25.** $3m^2(m^3 + 2m^2 + 5)$ **26.** $8u^2 + 3u + 6$ **27.** $8x^2 + 22x$ **28.** (A) F, (B) T **29.** (A) $\{1, 2, 3, 4\}$, (B) $\{1, 2\}$

30. 1 2 3 4 5 6 7

31. $\{27, 51, 61\}$ **32.** $\{61\}$ **33.** 10 **34.** 49 **35.** $4x = (x + 2) + (x + 4)$ **36.** (A) Closure and commutative axioms for addition; (B) Closure and commutative axioms for multiplication **37.** (A) Q; (B) Neither **38.** 1. Commutative axiom for addition; 2. Associative axiom for addition; 3. Associative axiom for addition; 4. Substitution principle for equality; 5. Commutative axiom for addition; 6. Associative axiom for addition; 7. Right-hand distributive axiom; 8. Substitution principle for equality **39.** $10u^5v^4 + 5u^4v^3 + 10u^3v^2$ **40.** $5x^2y^2(4x + y + 3)$ **41.** $7x^5 + 11x^3 + 6x^2$ **42.** $8x^2 + 10x + 3$ **43.** (A) $P = 2x + 2(2x + 5)$; (B) $P = 6x + 10$ **44.** $\{2, 3, 4\}$ **45.** $C = \{x \in N \mid 3 \leq x < 7\}$

EXERCISE 11

1. $-8, -2, +3, +9$ **3.** -6 -4 -2 0 $+2$ $+4$ $+6$

5. -30 -25 -20 -15 -10 -5 0 $+5$ $+10$ $+15$ $+20$

7. $+4$ **9.** -10 **11.** -3 **13.** T **15.** F **17.** T **19.** F **21.** T **23.** T **25.** $+20{,}270$ **27.** -280 **29.** -5 **31.** $+27$ **33.** -3 **35.** $+25$ **37.** -10 **39.** -9 **41.** $+1$ **43.** $+17$ **45.** -3

EXERCISE 12

1. -9 **3.** $+2$ **5.** $+4$ **7.** $+6$ **9.** 0 **11.** Sometimes **13.** Never **15.** -11 **17.** -5 **19.** $+13$ **21.** $+2$ or -2 **23.** No solution **25.** $+6$ **27.** $+5$ **29.** -5 **31.** $+5$ **33.** -8 **35.** -7 **37.** $+5$ **39.** -7 **41.** -5 **43.** $+5$ **45.** $+2$

47. $\{+5\}$ **49.** $\{+3\}$ **51.** $\{-6, +6\}$ **53.** $\varnothing$ **55.** $\{0\}$ **57.** Set of all negative integers **59.** Set of all integers I **61.** Set of all nonnegative integers **63.** $\{0\}$

EXERCISE 13

1. $+11$ **3.** -3 **5.** -2 **7.** -8 **9.** $+3$ **11.** $+9$ **13.** -6 **15.** -9 **17.** -9 **19.** -2 **21.** -4 **23.** -5 **25.** -4 **27.** -12 **29.** -622 **31.** -38 **33.** -668 **35.** -36 **37.** -4 **39.** -4 **41.** -5 **43.** $+5$ **45.** -77 **47.** $+14$ **49.** -6 **51.** -2 **53.** $+3$ **55.** \$23 **57.** -1493 ft **59.** 0 **61.** $-m$ **63.** Commutative property, Associative property, Theorem 2, Definition of addition

EXERCISE 14

1. $+5$ **3.** $+13$ **5.** -5 **7.** -5 **9.** $+5$ **11.** -6 **13.** -8 **15.** -3 **17.** $+14$ **19.** -4 **21.** -6 **23.** -5 **25.** $+15$ **27.** $+87$ **29.** -315 **31.** -245 **33.** $+1$ **35.** -3 **37.** 0 **39.** $+7$ **41.** $+4$ **43.** $+2$ **45.** $+3$ **47.** $(+29{,}141) - (-35{,}800) = 64{,}941$ ft **49.** $(-245) - (-280) = +35$ ft **51.** True **53.** False; $(+7) - (-3) = +10$, $(-3) - (+7) = -10$ **55.** True **57.** False; $|(+9) + (-3)| = +6$, $|+9| + |-3| = +12$ **59.** Commutative property for addition; Associative property for addition, Theorem 2, Definition of addition

EXERCISE 15

1. $+14$ **3.** $+14$ **5.** -14 **7.** -14 **9.** 0 **11.** 0 **13.** -5 **15.** $+2$ **17.** -4 **19.** -6 **21.** -12 **23.** $+7$ **25.** Both $+7$ **27.** $+8$ **29.** -5 **31.** $8x^3 + 4x^2 + 12x$ **33.** $6x - 10$ **35.** -700 **37.** $+450$ **39.** $+72$ **41.** 0 **43.** $+34$ **45.** -10 **47.** -22 **49.** -10 **51.** $+70$ **53.** $+25$ **55.** -45 **57.** $+20$ **59.** Both $+5$ **61.** Both $+10$ **63.** $+25, -25$ **65.** $+10$ **67.** Never **69.** Sometimes **71.** $8t^4 + 12t^3 + 20t^2$ **73.** $6x^3y - 6xy^3$ **75.** $6a^2b^2 + 18ab^3 + 3a^2b^3 + 3ab^4$ **77.** $8u^5 - 12u^3v$ **79.** $+21$ **81.** 0 **83.** By Theorem 5 $(-a)(-b) = [(-1)a][(-1)b]$; by appropriate use of commutative and associative properties $[(-1)a][(-1)b] = [(-1)(-1)](ab)$; which is $(+1)(ab)$ by definition of multiplication; and by Theorem 5, $(+1)(ab) = ab$.

EXERCISE 16

1. $+5$ **3.** -5 **5.** 0 **7.** $+3$ **9.** -3 **11.** Not defined **13.** $+2$ **15.** -3 **17.** Not defined **19.** $+4$ **21.** 0 **23.** $+3$ **25.** 0 **27.** -9 **29.** -5 **31.** -8 **33.** -6 **35.** $+8$ **37.** 0 **39.** $+6$ **41.** 0 **43.** Sometimes; 0 cannot be divided by 0. **45.** $+8$ **47.** 0 **49.** No solution in the integers **51.** No solution **53.** $+1$ **55.** -3 **57.** -9 **59.** When both x and y are of the same sign, or $x = 0$ and $y \neq 0$.

EXERCISE 17

1. 12 **3.** -2 **5.** 5 **7.** -13 **9.** -3 **11.** -2 **13.** $10x$ **15.** $4x$ **17.** $8x$ **19.** $-2x$ **21.** $-2y$ **23.** $7x + 3y$ **25.** $-3x - 3y$ **27.** $-5x + 3y$ **29.** $6m - 2n$ **31.** $-7x + 4y$ **33.** $3x - 2y$ **35.** $-x + y$ **37.** $-x + 8y$ **39.** 1 **41.** -2 **43.** $6xy$ **45.** $-3x^2y$ **47.** $2x^2 + 2x - 3$ **49.** $-2x^2y - 6xy + 5xy^2$ **51.** $-a + 3b$ **53.** $-2x - y$ **55.** $2t - 20$ **57.** $-3y + 4$ **59.** $-10x$ **61.** $x - 14$ **63.** $3x - y$ **65.** $-3x + y$ **67.** $y + 2z$

69. $x - y + z$ **71.** $P = 2x + 2(x - 5) = 4x - 10$ **73.** 1 **75.** $-8x^2 - 16x$ **77.** $13x^2 - 26x + 10$ **79.** Value in cents $= 10x + 25(x - 4) = 35x - 100$

EXERCISE 18

1. 3 **3.** -3 **5.** -12 **7.** 5 **9.** -3 **11.** -13 **13.** 8 **15.** -4 **17.** -4 **19.** 3 **21.** 0 **23.** 3 **25.** 2 **27.** -3 **29.** 7 **31.** 4 **33.** 2 **35.** -4 **37.** 8 **39.** 7 **41.** 5 **43.** No solution **45.** 16 **47.** 16 **49.** -15 **51.** 4 **53.** 6 **55.** No solution **57.** $3x = 12$, $x = 4$ **59.** T **61.** F **63.** T **65.** T

67.

$a - c = a - c$	identity property of equality
$a = b$	given
$a - c = b - c$	substitution principle

EXERCISE 19

1. 25, 26, 27 **3.** 16, 18, 20 **5.** 8 hr **7.** 13 ft above and 104 ft below **9.** 239,000 miles **11.** 7, 9, 11 **13.** 10 ft by 23 ft **15.** 7 quarters and 10 dimes **17.** 5 sec **19.** 22,000 ft **21.** 2,331, 1,526, 1,918, 2,196, 2,279 **23.** 6 miles **25.** 70 hour (or 2 days and 22 hr); 1750 miles **27.** 130 min (or 2 hr, 10 min)

EXERCISE 20

1. $>$ **3.** $<$ **5.** $<$ **7.** $>$ **9.** $<$ **11.** $>$ **13.** $<$ **15.** $<$ **17.** $<$ **19.** $>$ **21.** $<$ **23.** $>$

25. $\{-1, 0, 1, 2\}$

27. $\{-5, -4, -3, -2, -1\}$

29. Right **31.** $>$ **33.** $<$ **35.** $<$ **37.** $\{-6, -4, -2, 0, 2\}$ **39.** $\varnothing$ **41.** $\{2, 4, 6\}$ **43.** $\{-6, -4, -2, 0, 2\}$ **45.** $\{-3, -2, -1, 0\}$

47.

49.

51. If $a < b$, then there exists a positive number p such that $a + p = b$. Subtracting a from each member, we obtain $b - a = p$, a positive integer. **53.** If $a < b$, then $a + p = b$ for p positive. If $b < c$, then $b + q = c$ for q positive. Therefore, $(a + p) + q = c$ or $a + (p + q) = c$ for $p + q$ positive. And it follows from the definition of $<$ that $a < c$.

EXERCISE 21

1.

2. (A) -4, (B) $+8$, (C) $+2$, (D) $+9$ **3.** (A) -245, (B) $+14,495$ **4.** (A) -5, (B) -13 **5.** (A) -3, (B) -3 **6.** (A) $+6$, (B) -3 **7.** (A) $+28$; (B) -18 **8.** (A) -4; (B) $+6$ **9.** (A) Not defined, (B) 0 **10.** $3m^2 - 12mn$ **11.** (A) $+4$; (B) $x - 4y$ **12.** $2m + 9n$ **13.** $x = -2$ **14.** 52, 53, 54 **15.** (A) $>$, (B) $<$, (C) $>$, (D) $<$ **16.** (A) T, (B) F, (C) T, (D) F, (E) T **17.** (A) $+8$, (B) $+5$ **18.** (A) -10, (B) -60 **19.** $+1$ **20.** (A) -5, $+5$; (B) -7 **21.** $+4$ **22.** -6 **23.** -1 **24.** $+15$ **25.** $6y^7 + 2y^5 + 10y^4$ **26.** $-2x^2y^2 - 5xy$ **27.** $10x - 24y$ **28.** $x = 2$ **29.** 10 nickels and 5 quarters

30. 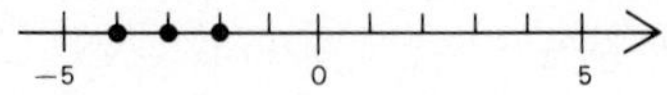

31. -45 **32.** All positive integers and 0 **33.** (*A*) a, (*B*) 0 **34.** (*A*) $+15$, (*B*) $+5$ **35.** (*A*) $+1$, (*B*) $+4$ **36.** -48 **37.** -7 **38.** $2y-3$ **39.** $-11x-8$ **40.** $\{-9\}$ **41.** 17 hr **42.** Less than

EXERCISE 22

1. $-\frac{9}{4}, -\frac{3}{4}, \frac{7}{4}$

3. 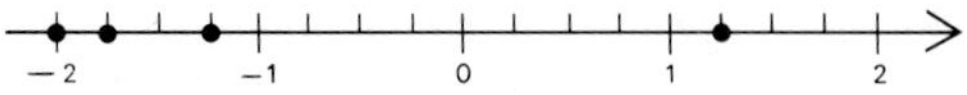

5. $\frac{6}{35}$ **7.** $\frac{-8}{15}$ or $-\frac{8}{15}$ **9.** $\frac{6}{35}$ **11.** $\frac{10x}{21y}$ **13.** $\frac{3x^2}{2y^3}$ **15.** 2

17. 15 **19.** 3 **21.** $9x^2$ **23.** $3x^2$ **25.** $\frac{3}{2}$ **27.** $-\frac{1}{4}$ **29.** $\frac{1}{4y}$

31. $\frac{4a}{b}$ **33.** $-\frac{y^2}{4x}$ **35.** $\frac{2}{3}$ **37.** $-\frac{3}{4}$ **39.** $\frac{1}{z}$ **41.** $\frac{3x}{2y}$

43. $\frac{3ad}{2c}$ **45.** 1 **47.** 2 **49.** 1. Theorem 2A, 2. Definition of multiplication, 3. Definition of multiplication in the integers

51.

$\frac{-a}{-b} = \frac{(-1)a}{(-1)b}$	Property of integers
$= \frac{(-1)}{(-1)} \frac{a}{b}$	Definition of multiplication
$= 1 \cdot \frac{a}{b}$	Theorem 2A
$= \frac{a}{b}$	Theorem 2B

EXERCISE 23

1. $\frac{3}{2}$ **3.** $\frac{7}{2}$ **5.** 5 **7.** $-\frac{1}{30}$ **9.** 4 **11.** y **13.** $2y^2$ **15.** $\frac{8}{7}$
17. $\frac{1}{3}$ **19.** $-\frac{3}{2}$ **21.** $\frac{2}{3}$ **23.** 10 **25.** 6 **27.** $\frac{4}{3}$ **29.** $-\frac{5}{9}$

31. $\frac{3v}{2u}$ **33.** $-\frac{2x^2}{3y}$ **35.** $\frac{81}{100}$ **37.** $\frac{ade}{bcf}$ **39.** $-\frac{3}{2}$ **41.** $-\frac{3}{2}$

43. $\frac{2}{3}$ **45.** $\frac{3}{4}$ **47.** $\frac{3}{5}$ **49.** $\frac{3}{5}$ hr **51.** $\frac{35}{4}$ min **53.** True **55.** False; use $a = 4$ and $b = 2$ **57.** True **59.** Commutative property for multiplication; Associative property for multiplication; Definition of multiplication, Theorem 3, Theorem 2B

EXERCISE 24

1. 2 **3.** $\frac{4}{5}$ **5.** $\frac{7}{8}$ **7.** $\frac{19}{15}$ **9.** $\frac{4}{11}$ **11.** $\frac{10}{11}$ **13.** $\frac{1}{8}$ **15.** $-\frac{1}{15}$

17. $\frac{-3}{5xy}$ **19.** $\frac{5y}{x}$ **21.** $\frac{6}{7y}$ **23.** $\frac{7}{6x}$ **25.** $\frac{13x}{6}$ **27.** $\frac{9-10x}{15x}$

29. $\dfrac{x^2 - y^2}{xy}$ **31.** $\dfrac{x - 2y}{y}$ **33.** $\dfrac{5x + 3}{x}$ **35.** $\dfrac{1 - 3x}{xy}$ **37.** $\dfrac{9 + 8x}{6x^2}$
39. $\dfrac{15 - 2m^2}{24m^3}$ **41.** $\dfrac{5}{3}$ **43.** $\dfrac{3x^2 - 4x - 6}{12}$ **45.** $\dfrac{18y - 16x + 3}{24xy}$
47. $\dfrac{18 + 4y + 3y^2 - 18y^3}{6y^3}$ **49.** $\dfrac{22y + 9}{252}$ **51.** $\dfrac{15x^2 + 10x - 6}{180}$
53. Property $a - b = a + (-b)$; Theorem 4, Sec. 3.2; Definition of addition; Property $a - b = a + (-b)$

EXERCISE 25

1. 8 **3.** 12 **5.** 6 **7.** -6 **9.** 36 **11.** $-\frac{4}{3}$ **13.** 20
15. 30 **17.** 15 **19.** $-\frac{5}{6}$ **21.** $\frac{27}{5}$ **23.** 9 **25.** 150 **27.** \$240
29. 75 ft **31.** \$2 **33.** $\frac{11}{5}$ **35.** 14,080 ft **37.** 84 yr

EXERCISE 26

1. 1:4 **3.** 5:1 **5.** 1:3 **7.** 8 **9.** 18 **11.** 4 **13.** 600
15. 60 ft **17.** $3\frac{3}{4}$ in. **19.** 1.4 in. **21.** \$95 **23.** 7.5 lb
25. 24,000 miles

EXERCISE 27

1. T **3.** T **5.** T **7.** T **9.** T **11.** T

13. -2 3 x

15. -2 3 x

17. T **19.** T **21.** F **23.** T **25.** F **27.** F **29.** T
31. F **33.** T **35.** T **37.** T **39.** F **41.** T

43. $-\frac{9}{4}$ $\frac{3}{2}$ x

45. $-\frac{15}{8}$ $-\frac{1}{8}$ x

47. $-\frac{7}{4}$ x

49. -4 6 x

51. -5 0 5 x

53. $-\sqrt{2}$ π

55. (A) $0.37\overline{500}$, (B) $2.\overline{55}$, (C) 0.538461538461 **57.** (A) $\frac{3}{11}$, (B) $\frac{106}{33}$

EXERCISE 28

1. -2 -1 0 1 2

2. $\dfrac{15x}{8y}$ **3.** $\dfrac{6}{5xy}$ **4.** $\dfrac{5}{6}$ **5.** $\dfrac{5y}{6}$ **6.** 60 **7.** 9 **8.** 660
9. (A) F, (B) T, (C) T

10. -3 4 x

11. $\dfrac{2}{5x^2}$ **12.** $\dfrac{9y^2}{10z^2}$ **13.** $-\dfrac{10}{9}$ **14.** $\dfrac{17}{18}$
15. $\dfrac{3xz + 18xy - 4yz - 24xyz}{12xyz}$ **16.** -12 **17.** 41 **18.** \$80

19. (*A*) T, (*B*) F, (*C*) T, (*D*) F

20. 

21. $-\frac{4}{9}$ **22.** $\frac{1}{8}x = -\frac{3}{2}$, $x = -12$ **23.** $\frac{20xyz + 135xy^3 + 6z^2}{18y^2z}$

24. $-\frac{13}{5}$ **25.** $\frac{6}{5}$ lb **26.** 5, 36 **27.** 

EXERCISE 29

1. $5x - 2$ **3.** $5x - 3$ **5.** $6x^2 - x + 3$ **7.** $3x + 4$ **9.** $-2x + 12$
11. $x^2 - 3x$ **13.** $2x^2 + x - 6$ **15.** $2x^3 - 7x^2 + 13x - 5$
17. $x^3 - 6x^2y + 10xy^2 - 3y^3$ **19.** 1 **21.** 3 **23.** 2 **25.** 6
27. $3x^3 + x^2 + x + 6$ **29.** $-x^3 + 2x^2 - 3x + 7$ **31.** $2x^2 - 2xy + 3y^2$
33. $a^3 + b^3$ **35.** $x^3 + 6x^2y + 12xy^2 + 8y^3$
37. $2x^4 - 5x^3 + 5x^2 + 11x - 10$
39. $2x^4 + x^3y - 7x^2y^2 + 5xy^3 - y^4$ **41.** $-x^2 + 17x - 11$
43. $2x^3 - 13x^2 + 25x - 18$

EXERCISE 30

1. $x^2 + 3x + 2$ **3.** $y^2 + 7y + 12$ **5.** $x^2 - 9x + 20$ **7.** $n^2 - 7n + 12$
9. $s^2 + 5s - 14$ **11.** $m^2 - 7m - 60$ **13.** $u^2 - 9$ **15.** $x^2 - 64$
17. $y^2 + 16y + 63$ **19.** $c^2 - 15c + 54$ **21.** $x^2 - 8x - 48$
23. $a^2 - b^2$ **25.** $x^2 + 4xy + 3y^2$ **27.** $3x^2 + 7x + 2$
29. $4t^2 - 11t + 6$ **31.** $3y^2 - 2y - 21$ **33.** $2x^2 + xy - 6y^2$
35. $6x^2 + x - 2$ **37.** $9y^2 - 4$ **39.** $5s^2 + 34s - 7$
41. $6m^2 - mn - 35n^2$ **43.** $12n^2 - 13n - 14$ **45.** $6x^2 - 13xy + 6y^2$
47. $x^2 + 6x + 9$ **49.** $4x^2 - 12x + 9$ **51.** $4x^2 - 20xy + 25y^2$
53. $16a^2 + 24ab + 9b^2$ **55.** $x^3 + x^2 - 3x + 1$
57. $6x^3 - 13x^2 + 14x - 12$ **59.** $x^4 - 2x^3 - 4x^2 + 11x - 6$

EXERCISE 31

1. $(x + 1)(x + 4)$ **3.** $(x + 2)(x + 3)$ **5.** $(x - 1)(x - 3)$ **7.** $(x - 2)(x - 5)$
9. Not factorable **11.** Not factorable **13.** $(x + 3y)(x + 5y)$
15. $(x - 3y)(x - 7y)$ **17.** Not factorable **19.** $(3x + 1)(x + 2)$
21. $(3x - 4)(x - 1)$ **23.** $(3x - 2y)(x - 3y)$ **25.** $(n - 4)(n + 2)$
27. $(x - 2y)(x + 6y)$ **29.** Not factorable **31.** $(3s + 1)(s - 2)$
33. $(6x - 1)(2x + 3)$ **35.** $(3u - 2v)(u + 3v)$ **37.** Not factorable
39. $(6x + y)(2x - 7y)$ **41.** $(12x - 5y)(x + 2y)$ **43.** 13, −13, 8, −8, 7, −7

EXERCISE 32

1. $3x^2(2x + 3)$ **3.** $u^2(u + 2)(u + 4)$ **5.** $x(x - 2)(x - 3)$
7. $(x - 2)(x + 2)$ **9.** $(2x - 1)(2x + 1)$ **11.** Not factorable
13. $2(x - 2)(x + 2)$ **15.** $(3x - 4y)(3x + 4y)$ **17.** $3uv^2(2u - v)$
19. $xy(2x - y)(2x + y)$ **21.** $3x^2(x^2 + 9)$ **23.** $6(x + 2)(x + 4)$
25. $3x(x^2 - 2x + 5)$ **27.** Not factorable **29.** $4x(3x - 2y)(x + 2y)$
31. $(x + y)(x + 3)$ **33.** $(x - 3)(x - y)$ **35.** $2xy(2x + y)(x + 3y)$
37. $5y^2(6x + y)(2x - 7y)$ **39.** $(x - 2)(x^2 + 2x + 4)$ **41.** $(x + 3)(x^2 - 3x + 9)$

EXERCISE 33

1. $\frac{x}{3}$ **3.** $\frac{1}{A}$ **5.** $\frac{1}{x + 3}$ **7.** $4(y - 5)$ **9.** $\frac{x}{3(x + 7)^2}$ **11.** $\frac{x}{2}$

13. $3 - y$ **15.** $\frac{1}{n}$ **17.** $\frac{2x - 1}{3x}$ **19.** $\frac{2x + 1}{3x - 7}$ **21.** $\frac{x + 2}{2x}$

23. $\dfrac{x-3}{x+3}$ **25.** $\dfrac{x-2}{x-3}$ **27.** $\dfrac{x+3}{2x+1}$ **29.** $3x^2-x+2$
31. $\dfrac{2+m-3m^2}{m}$ **33.** $2m^2-mn+3n^2$ **35.** $x+2$ **37.** $\dfrac{2x-3y}{2xy}$
39. $\dfrac{x+2}{x+y}$ **41.** $x-5$ **43.** $\dfrac{x+5}{2x}$ **45.** $\dfrac{x^2+2x+4}{x+2}$

EXERCISE 34

1. $x+2$ **3.** $2x-3$ **5.** $2x+3$, $R=5$ **7.** $m-2$ **9.** $2x+3$ **11.** $2x+5$, $R=-2$ **13.** $x+5$, $R=-2$ **15.** $x+2$ **17.** $m+3$, $R=2$ **19.** $5c-2$, $R=8$ **21.** $3x+2$, $R=-4$ **23.** $2y^2+y-3$ **25.** x^2+x+1 **27.** x^3-2x^2+4x-8 **29.** $2y^2-5y+13$, $R=-25$ **31.** $2x^3-3x^2-5$, $R=5$ **33.** $2x^2-3x+2$, $R=4$

EXERCISE 35

1. $5x^2+3x-4$ **2.** $2x^2-x+7$ **3.** $6x^2+11x-10$ **4.** $(x-2)(x-7)$ **5.** $(3x-4)(x-2)$ **6.** ~~$2xy(2x-y)$~~ 2xy (2x-3y) **7.** $x(x-2)(x-3)$ **8.** $3(u-2)(u+2)$ **9.** $\dfrac{x^2}{3}$ **10.** $3x+4$, $R=2$ **11.** $27x^4+63x^3-66x^2-28x+24$ **12.** $2x^2+5x+5$ **13.** (*A*) 5, (*B*) 3 **14.** Not factorable **15.** $(2x-3y)(x+y)$ **16.** $3y(2y-5)(y+3)$ **17.** $3xy(4x^2+9y^2)$ **18.** $(x-y)(x+4)$ **19.** $\dfrac{3x-4}{3x+4}$ **20.** $3x^2+2x-2$, $R=-2$ **21.** $-2x+20$ **22.** 0 **23.** $(5x-4y)(3x+8y)$ **24.** 9, −9, 6, −6 **25.** $3xy(6x-5y)(2x+3y)$ **26.** $4u^2(3u^2-3uv-5uv^2)$ **27.** $(2x+1)(4x^2-2x+1)$ **28.** $\dfrac{x-3}{2x(x+2y)}$ **29.** $4x^2-2x+3$, $R=-2$

EXERCISE 36

1. Chemist, 40 hr; assistant, 30 hr **3.** 75 percent **5.** \$80,000 **7.** 12 lb **9.** (*A*) 3.09 grams; (*B*) 324.31 grams **11.** 350 cc **13.** $T=80-5.5\left(\dfrac{h}{1{,}000}\right)$ or $T=80-0.0055h$; 10,000 ft **15.** 0.1 in., 107.4 ft **17.** (*A*) GNP = NNP + (0.1)NNP or GNP = (1.1)NNP, (*B*) 105.6; 52.8 **19.** 78 by 27 ft **21.** 30°, 60°, 90° **23.** $2\frac{1}{4}$ cups of milk and 3 cups of flour **25.** 16 **27.** 400 miles **29.** (*A*) 216 miles, (*B*) 225 miles **31.** 250 **33.** 315; 105; 105; 35 **35.** 264; 330 **37.** $\frac{1}{3}$ in. **39.** 200,000 mps **41.** 4 ft from the 6-lb end **43.** 400 lb **45.** (*A*) 15 ohms, (*B*) 0.06 ampere **47.** 150 cm **49.** 18 four-cent stamps and 12 five-cent stamps **51.** $5\frac{5}{11}$ minutes after 1 P.M.

EXERCISE 37

1. T **3.** T **5.** T **7.** T **9.** T **11.** T **13.** $x>7$ **15.** $x<-7$ **17.** $x>4$ **19.** $x<-4$ **21.** $x<-21$ **23.** $x>21$

25. $x<2$ 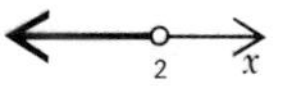

27. $x>2$ 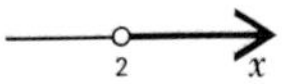

29. $x\le 5$ 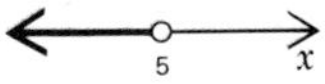

31. $y\le -2$ 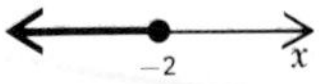

33. $y<2$ 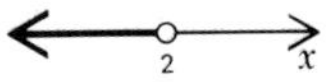

35. $-1<x<2$ 

37. $-2\le x\le 3$ 

39. $-2<x<3$ 

41. $x\ge 3$ 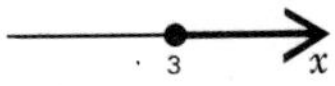

43. $x<10$ 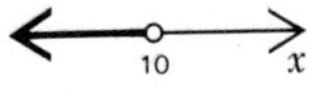

45. $m > \frac{19}{6}$ **47.** $x > -4$

49. $-20 \le C \le 20$ **51.** $14 \le F \le 77$

53. 1. Given, 2. definition of $>$, 3. definition of $<$, 4. property of $=$, 5. definition of $<$, 6. definition of $>$. **55.** 1. Given, 2. definition of $>$, 3. definition of $<$, 4. given, 5. property of $=$, 6. distributive property, 7. property of $=$, 8. definition of $<$ (since $-cp$ is positive). **57.** $(b - a)$ is negative

EXERCISE 38

1. $2x + 5 \le 7$, $x \le 1$ **3.** $3x - 5 \le 4x$, $x \ge -5$ **5.** $w - 30 \ge 125$, $w \ge 155$ **7.** $2(10) + 2w < 30$, $w < 5$ in. **9.** $100 + 6t > 133$, $t > 5.5$ yr **11.** $15{,}000 \le 15x \le 22{,}500$; $1{,}000 \le x \le 1{,}500$ **13.** $68 \le \frac{9}{5}c + 32 \le 77$, $20° \le c \le 25°$ **15.** $\frac{55 + 73 + x}{3} \ge 70$, $x \ge 82$ **17.** $1.7 \le \frac{v}{740} \le 2.4$, $1{,}258 \le v \le 1{,}776$ **19.** $\frac{W}{110} \le 30$, $W \le 3{,}300$ watts **21.** $5 \le \frac{d}{1{,}088} \le 10$; $5{,}440 \le d \le 10{,}880$ ft; 1.03 miles $\le d \le 2.06$ miles

EXERCISE 39

1. $A(5,2)$, $B(-2,3)$, $C(-4,-3)$, $D(3,-2)$, $E(2,0)$, $F(0,-4)$ **3.** $A(5,5)$, $B(8,2)$, $C(-5,5)$, $D(-3,8)$, $E(-5,-6)$, $F(-7,-8)$, $G(5,-5)$, $H(2,-2)$, $I(7,0)$, $J(-2,0)$, $K(0,-9)$, $L(0,4)$

5.

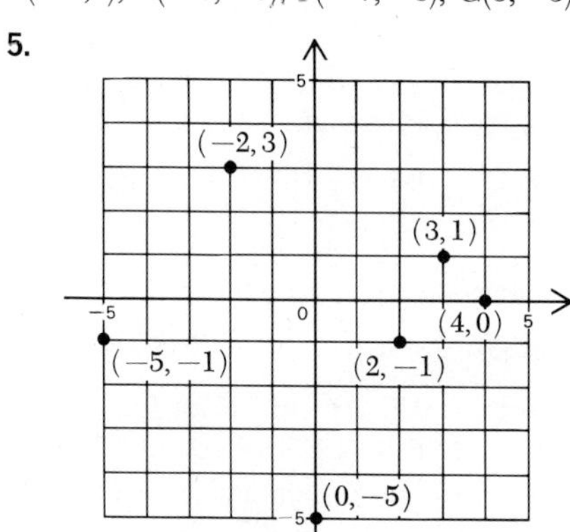

7.

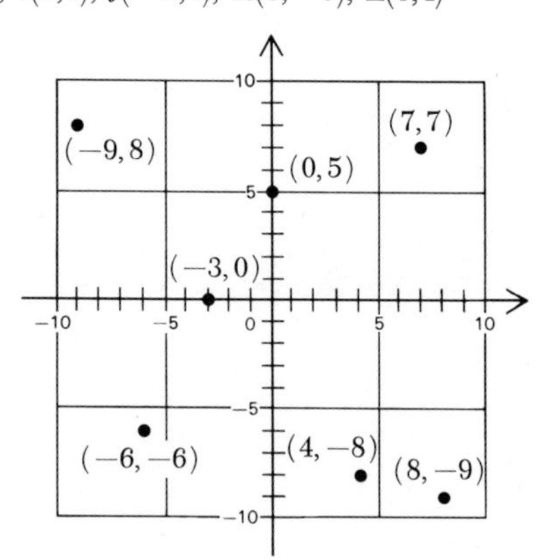

9. $A(2\frac{1}{2},1)$, $B(-2\frac{1}{2},3\frac{1}{2})$, $C(-2,-4\frac{1}{2})$, $D(3\frac{1}{4},-3)$, $E(1\frac{1}{4},2\frac{1}{4})$, $F(-3\frac{1}{4},0)$, $G(1\frac{1}{2},-4\frac{1}{4})$

11.

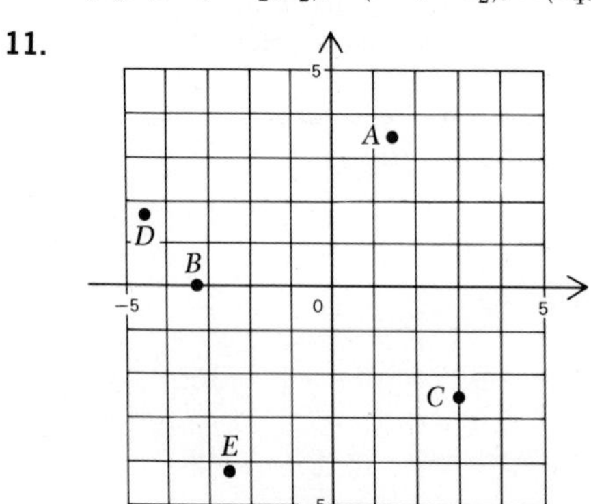

13. (A) III, (B) II, (C) I, (D) IV **15.** 5 **17.** 5 **19.** 3 **21.** 2

EXERCISE 40

1.

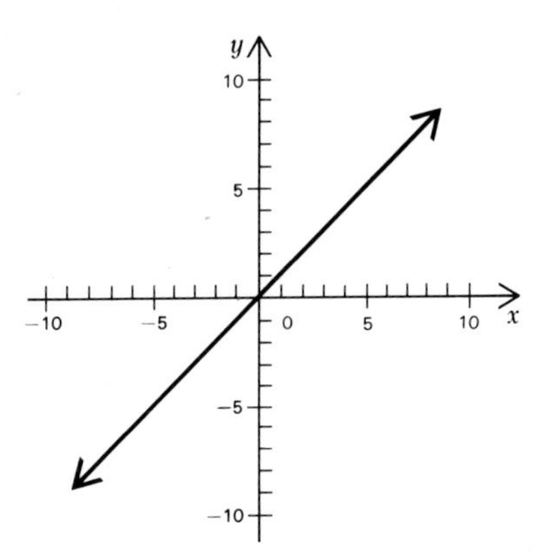

3.

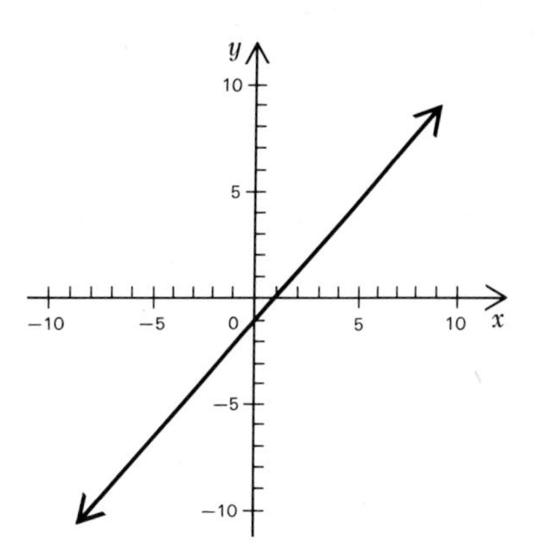

5.

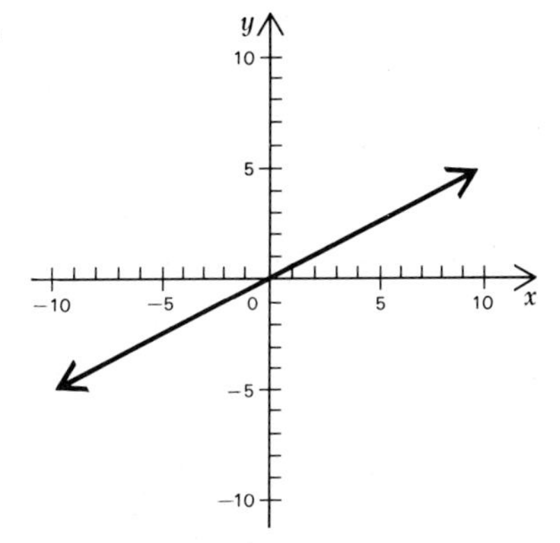

7.

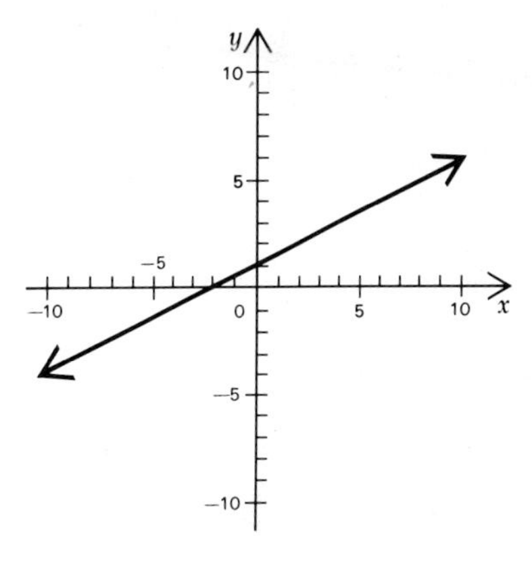

9.

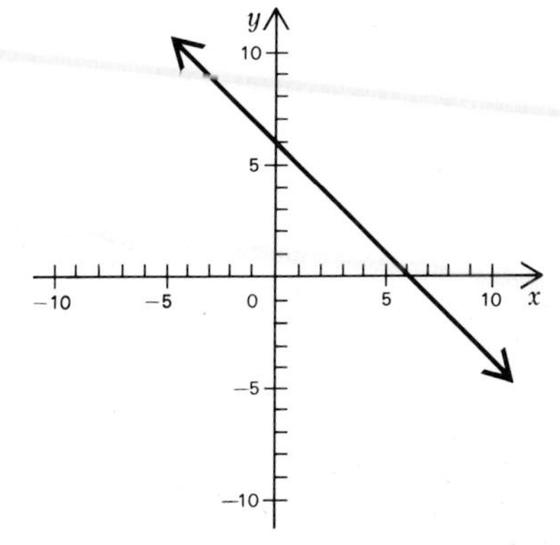

11.

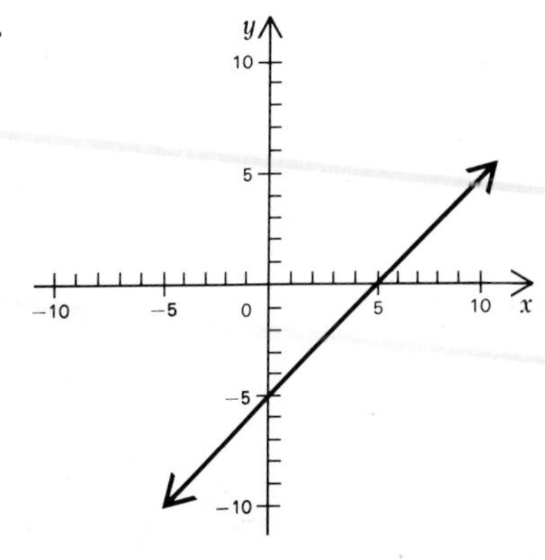

13.

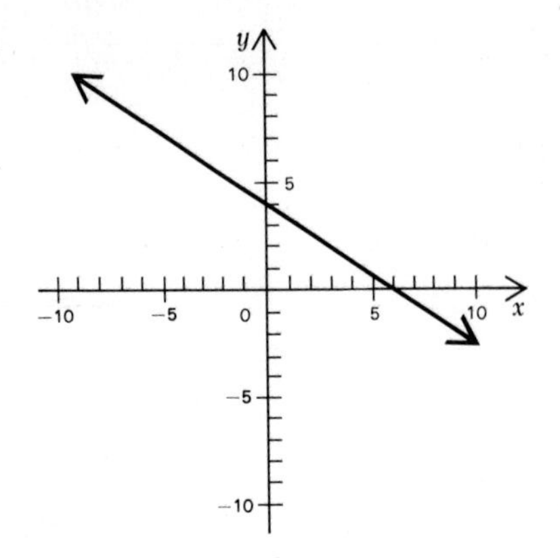

15.

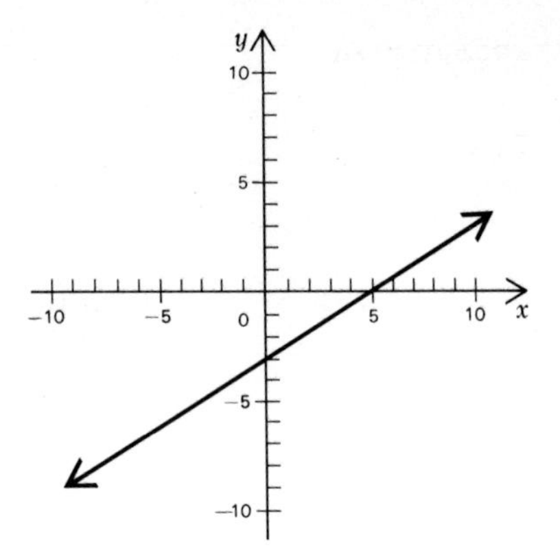

17.

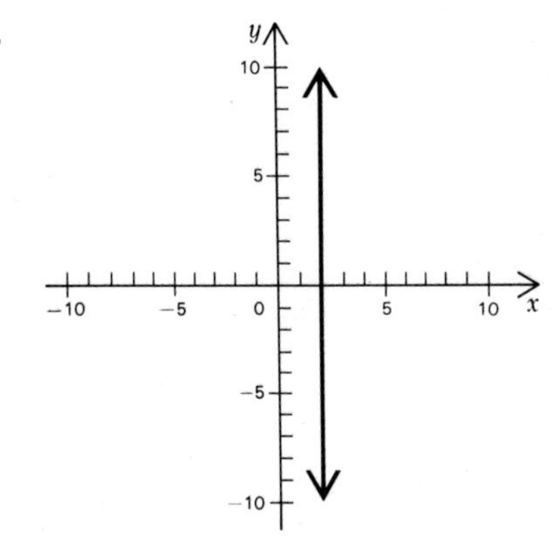

19.

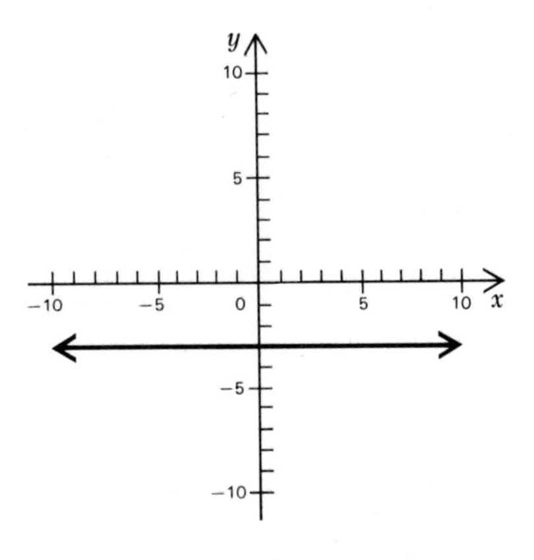

21.

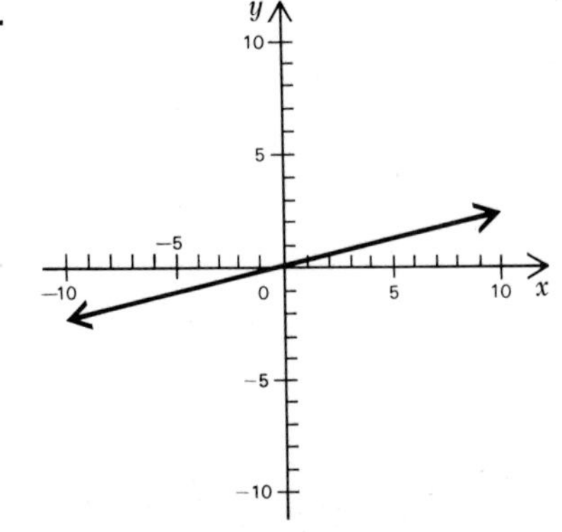

23.

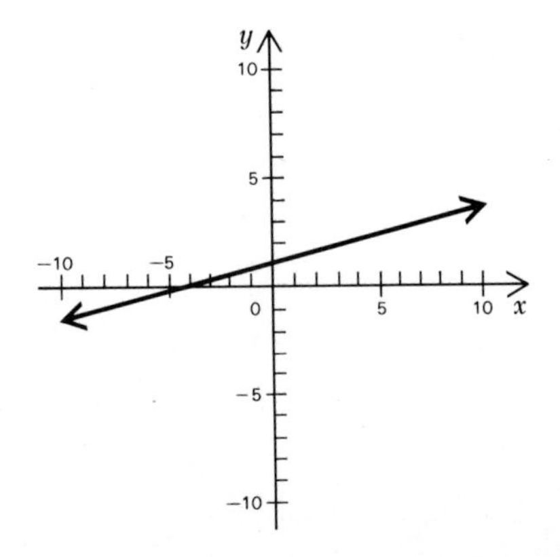

25.

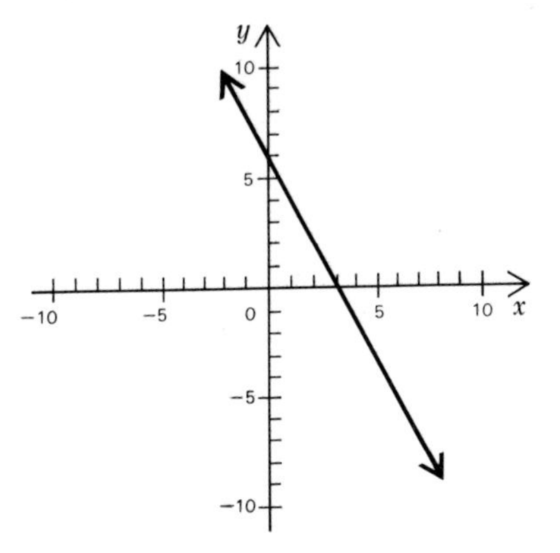

27.

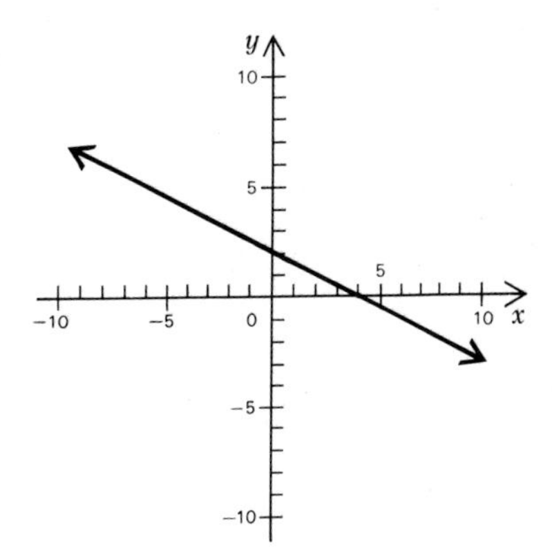

29.

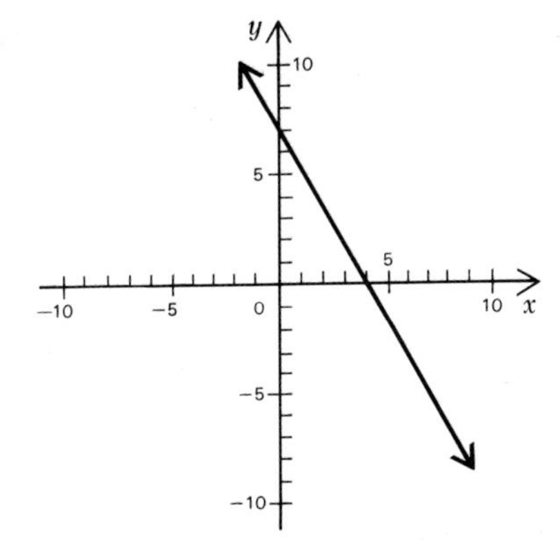

31.

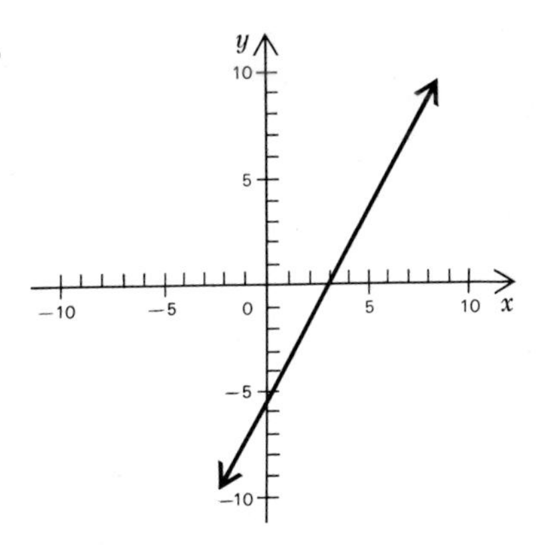

33.

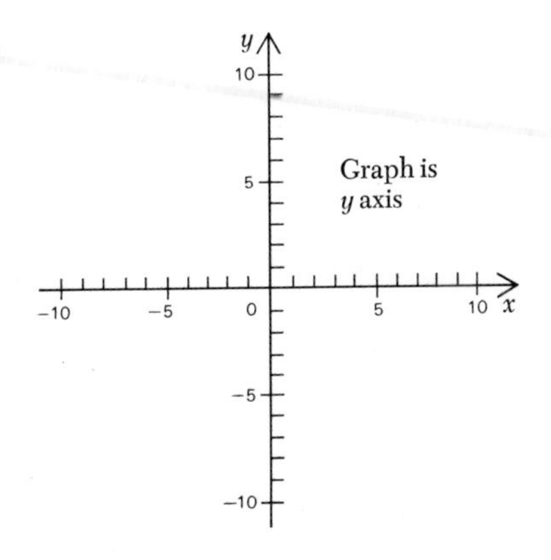

35.

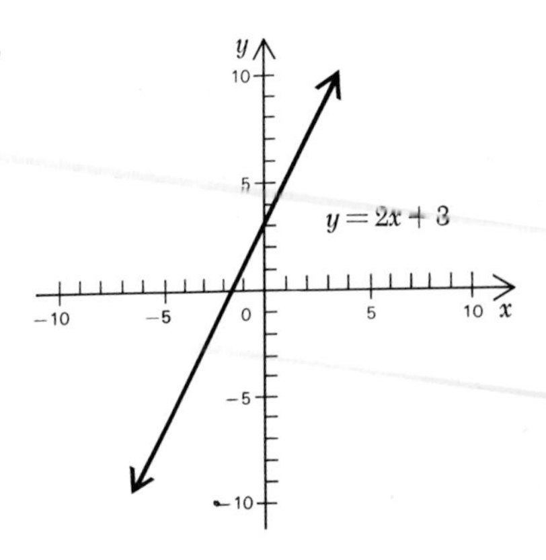

37.

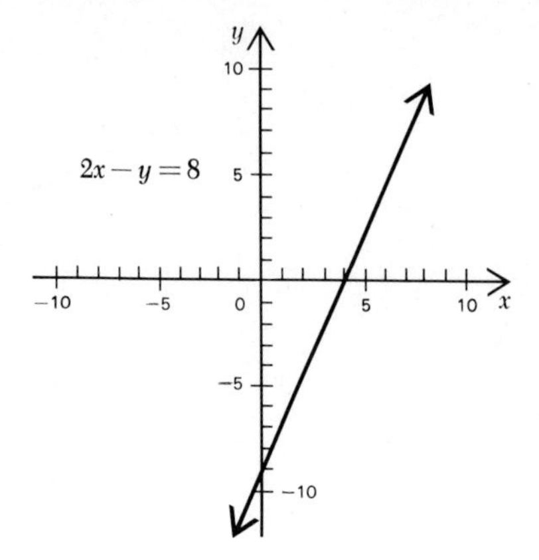

39.

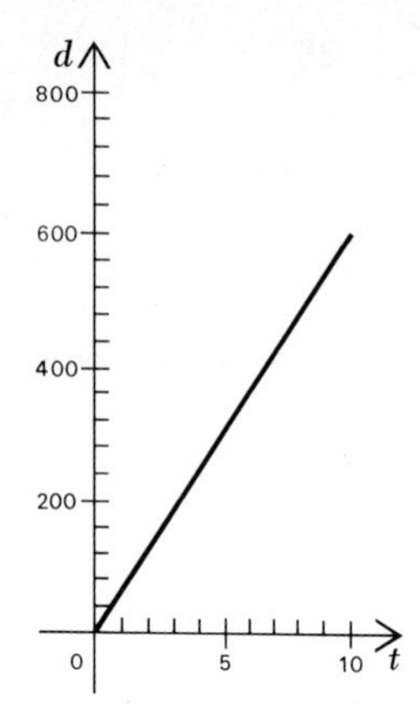

41.

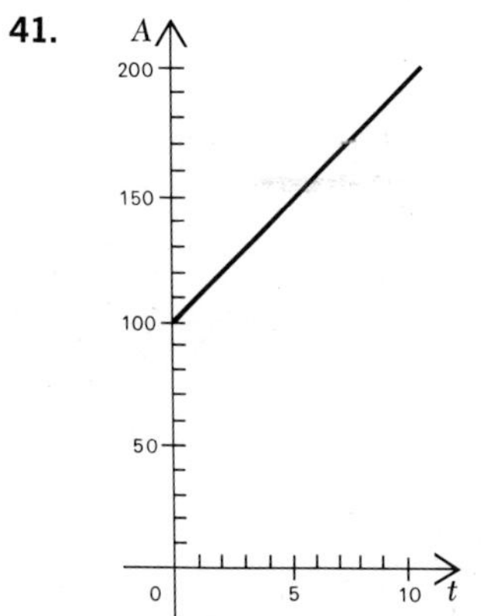

43.

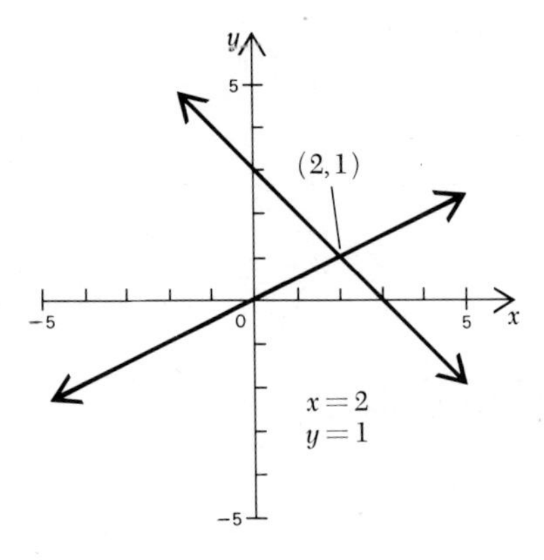

45.

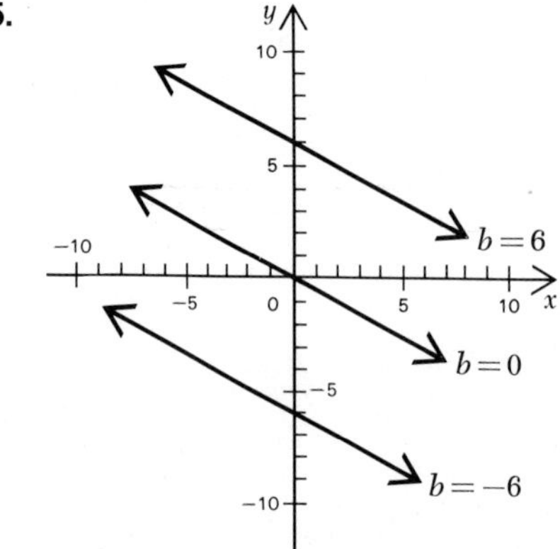

47.

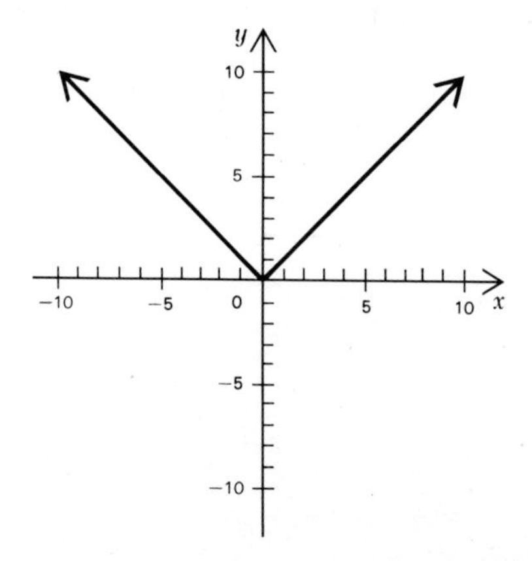

49.

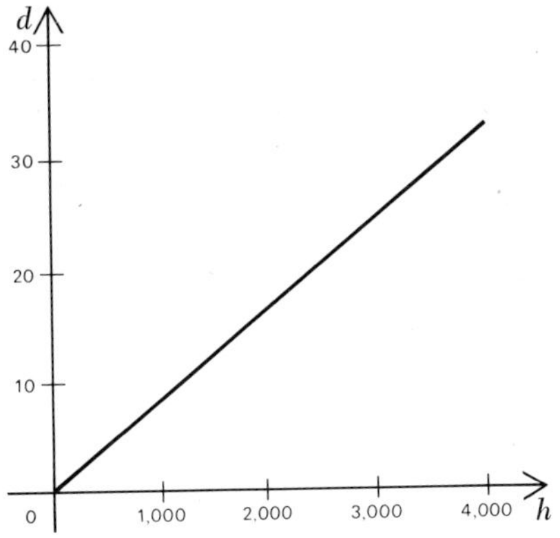

51.

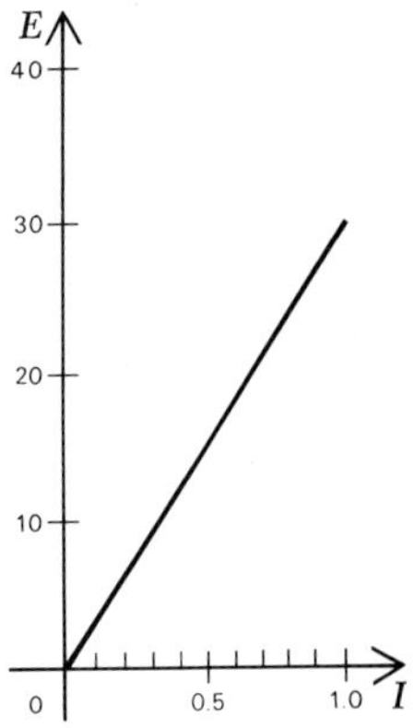

EXERCISE 41

1. $x = 3, y = 2$ **3.** $x = 3, y = 2$ **5.** $x = 9, y = 2$ **7.** $x = 2, y = 4$ **9.** $x = 6, y = 8$ **11.** $x = -4, y = -3$ **13.** No solution **15.** An infinite number of solutions. Any solution of one is a solution of the other. **17.** $x + 2y = 4$ becomes $y = -\frac{1}{2}x + 2$ and $2x + 4y = -8$ becomes $y = -\frac{1}{2}x - 2$. Since the coefficients of x are the same, both are parallel. **19.** If two lines are parallel and have a point in common, then they coincide.

EXERCISE 42

1. $x = 3, y = 2$ **3.** $x = 1, y = 4$ **5.** $x = 2, y = -4$ **7.** $x = 2, y = -1$ **9.** $x = 3, y = 2$ **11.** $x = -1, y = 2$ **13.** $x = 1, y = -5$ **15.** $x = -\frac{4}{3}, y = 1$ **17.** $x = \frac{3}{2}, y = -\frac{2}{3}$ **19.** No solution **21.** Infinitely many solutions **23.** $x = \frac{1}{3}, y = -2$ **25.** $m = -2, n = 2$ **27.** Limes: 11 cents each; lemons: 4 cents each **29.** $x = 1, y = 0.2$ **31.** $x = 6, y = 4$ **33.** $x = 6, y = -6$ **35.** $m(Aa + Bb + C) + n(Da + Eb + F) = m \cdot 0 + n \cdot 0 = 0$

EXERCISE 43

1. 84¼-lb packages; 60½-lb packages **3.** Typist: 8 hr, \$20; Stenographer: 6 hr, \$24 **5.** 60 cc of 80 percent solution and 40 cc of 50 percent solution **7.** 40 sec, 24 sec; 120 miles **9.** 11 ft, 7 ft **11.** 52°, 38° **13.** 700 student, 300 adult **15.** 927 brown, 309 blue **17.** 16 in., 20 in. **19.** 3 ft from 42-lb weight **21.** $d = 141$ cm (approximately); move away from goal box; move toward goal box **23.** 14 nickels, 8 dimes

EXERCISE 44

1. (A) **2.** 8 ft by 13 ft

3. $x \leq -3$

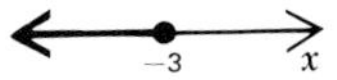

4. $-4 \leq x < 3$

5. $x \leq 3$

6.

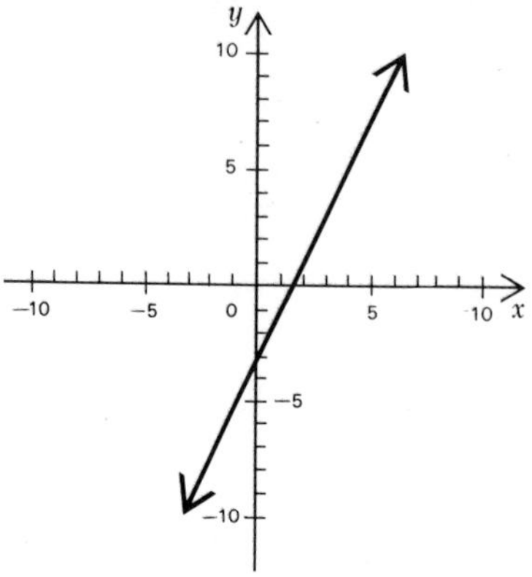

7.

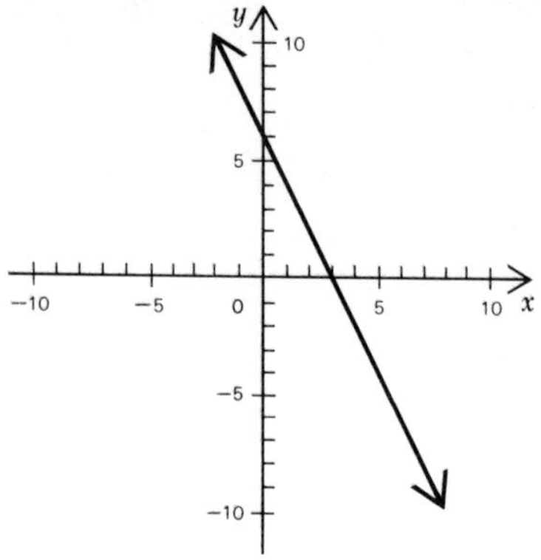

8. $x = 6, y = 1$ **9.** $x = 2, y = 1$ **10.** 16 dimes, 14 nickels **11.** One or none **12.** Eight 11-cent, eleven 8-cent

13. $x \geq -1$ 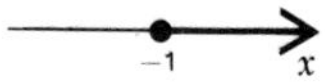 **14.** $x \geq -8$ 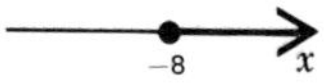 **15.** $15° \leq C \leq 30°$

16.

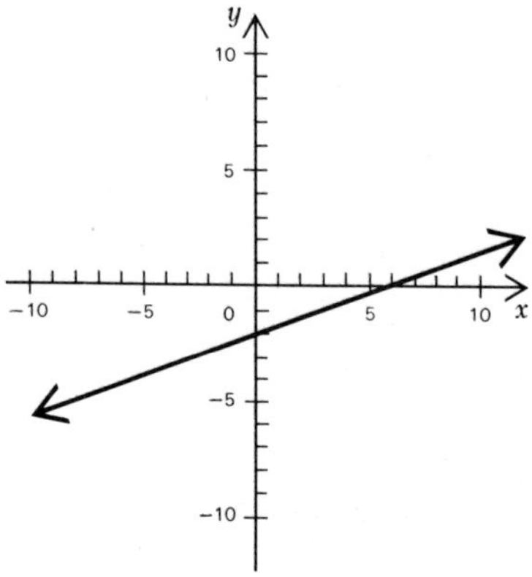

17.

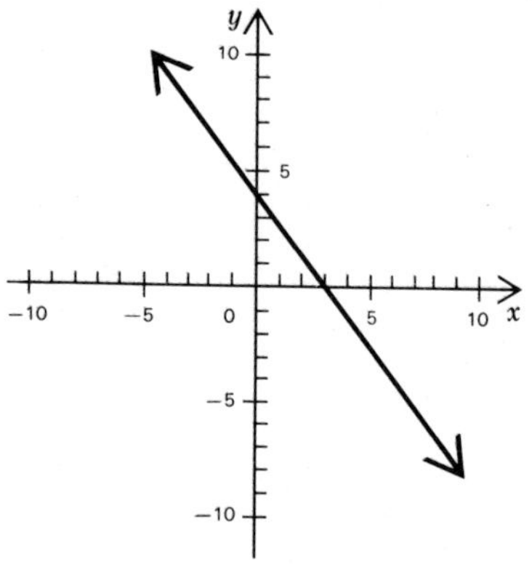

18. $x = 3, y = 3$ **19.** $m = -1, n = -3$ **20.** \$4,000 at 6% and \$2,000 at 10% **21.** $a(-b/a) + b = -b + b = 0$ **22.** 0.45 gal. **23.** (*A*) T, (*B*) T
24. $68 \leq F \leq 77$

25.

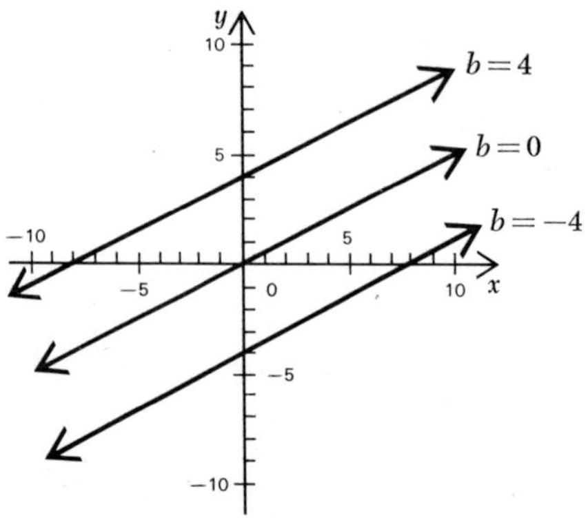

26. Lines coincide—infinite number of solutions. **27.** No solution **28.** 30 g of 70%, 70 g of 40%

EXERCISE 45

1. $\frac{5}{6}$ **3.** 2 **5.** $\frac{2y^3}{9u^2}$ **7.** $\frac{u^2w^2}{25y^2}$ **9.** $\frac{2}{x}$ **11.** $a+1$ **13.** $\frac{x-2}{2x}$
15. $8d^6$ **17.** $\frac{1}{y(x+4)}$ **19.** $\frac{1}{2y-1}$ **21.** $\frac{x}{x+5}$ **23.** $-3(x-2)$ or $6-3x$ **25.** $\frac{m+2}{m(m-2)}$ **27.** $\frac{x^2(x+y)}{(x-y)^2}$ **29.** All but one (namely, $x = 1$)
31. Obtain $\frac{x^2-(x+2)^2}{(x+1)(x+5)-(x+3)^2}$, which simplifies to $x+1$. Thus we get 108,642 for the answer.

EXERCISE 46

1. $\frac{5y+3}{4x}$ **3.** $\frac{3m-1}{2m^2}$ **5.** $\frac{6-x}{3x}$ **7.** $\frac{x^2+1}{x}$ **9.** $\frac{x^2-xy+y^2}{x^3}$
11. $\frac{x-1}{x}$ **13.** 5 **15.** $\frac{1}{y+3}$ **17.** $\frac{2x+1}{(x-2)(x+3)}$ **19.** $\frac{6-x}{2x(x+2)}$
21. $\frac{y^2+8}{8y^3}$ **23.** $\frac{1-b}{a^2}$ **25.** $\frac{2x+7}{(2x-3)(x+2)}$ **27.** $\frac{a+2}{a+1}$
29. $\frac{-x-7}{(2x+3)(x-1)}$ **31.** $\frac{2}{t-1}$ **33.** $\frac{2x^2-3x-6}{x+2}$
35. $\frac{3y^2-y-18}{(y+2)(y-2)}$ **37.** $\frac{3}{x+3}$ **39.** $\frac{5t-12}{3(t-4)(t+4)}$ **41.** $\frac{1}{x-1}$
43. $\frac{-17}{15(x-1)}$ **45.** $\frac{1}{x-3}$ **47.** $\frac{1}{y-x}$

EXERCISE 47

1. 9 **3.** 2 **5.** -4 **7.** 4 **9.** -4 **11.** No solution **13.** $-\frac{6}{5}$
15. 8 **17.** $\frac{53}{11}$ **19.** 1 **21.** -4

EXERCISE 48

1. $b = 60$, $m = 2$ **3.** 12 cm **5.** 1 sq in. **7.** 10 in. **9.** 400 mph
11. 3 mph **13.** $3\frac{3}{5}$ hr **15.** 1.5 hr **17.** $3\frac{1}{2}$ percent **19.** 36

EXERCISE 49

1. $I = A - P$ **3.** $r = \frac{d}{t}$ **5.** $t = \frac{I}{Pr}$ **7.** $\pi = \frac{C}{D}$ **9.** $x = -\frac{b}{a}$, $a \neq 0$
11. $t = \frac{s+5}{2}$ **13.** $y = \frac{3x-12}{4}$ or $y = \frac{3}{4}x - 3$ **15.** $E = IR$
17. $L = \frac{100B}{C}$ **19.** $m_1 = \frac{Fd^2}{Gm_2}$ **21.** $h = \frac{2A}{b_1 + b_2}$ **23.** $F = \frac{9}{5}C + 32$
25. $f = \frac{ab}{a+b}$ or $f = \frac{1}{\frac{1}{a}+\frac{1}{b}}$

EXERCISE 50

1. $\frac{2xy}{ab}$ **2.** $\frac{(d-2)^2}{d+2}$ **3.** $\frac{2x+11}{6x}$ **4.** $\frac{-1}{(x+2)(x+3)}$ **5.** $\frac{3x+2}{3x}$
6. $m = 5$ **7.** No solution **8.** $\frac{2}{3}$ **9.** $R = \frac{W}{I^2}$ **10.** $b = \frac{2A}{h}$

11. $\frac{9y^4}{9a^4}$ **12.** $\frac{x}{x+1}$ **13.** $\frac{2q^2 - 15pq + 8p^2}{20p^2q^2}$ **14.** $\frac{5x - 12}{3(x-4)(x+4)}$
15. $\frac{2}{m+1}$ **16.** $x = -2$ **17.** $x = -5$ **18.** 175 mph **19.** $W = \frac{E^2}{R}$
20. $L = \frac{2S}{n} - a$ or $L = \frac{2S - an}{n}$ **21.** -1 **22.** $\frac{y+4}{2x-y}$ **23.** $\frac{x-y}{x}$
24. No solution **25.** 1, 2 **26.** $f_1 = \frac{ff_2}{f_2 - f}$

EXERCISE 51

1. 12 **3.** 4 **5.** v^6 **7.** 2 **9.** 5 **11.** 3 **13.** 4 **15.** 7
17. 4 **19.** 7 **21.** 4 **23.** 7 **25.** $6x^{10}$ **27.** $2x^2$ **29.** $\frac{2}{u^4}$
31. $c^{12}d^{12}$ **33.** $\frac{x^6}{y^6}$ **35.** $6x^9$ **37.** 6×10^{15} **39.** 10^{20} **41.** y^{20}
43. x^8y^{12} **45.** $\frac{a^{12}}{b^8}$ **47.** $\frac{y^6}{3x^4}$ **49.** $3^3a^9b^6$ **51.** $2x^8y^4$ **53.** $\frac{x^6y^3}{8w^6}$
55. $\frac{y^3}{16x^4}$ **57.** $\frac{-x^2}{32}$ **59.** -1 **61.** $\frac{-1}{a^8}$

EXERCISE 52

1. 1 **3.** 1 **5.** $1/2^2$ **7.** $1/x^4$ **9.** 3^2 **11.** x^3 **13.** 10^2
15. x^4 **17.** 1 **19.** 10^{11} **21.** a^{12} **23.** $1/b^8$ **25.** $1/10^6$
27. 2^6 **29.** x^{10} **31.** x^3y^2 **33.** y^6/x^4 **35.** y^3/x^2 **37.** 1
39. 10^2 **41.** $1/x$ **43.** 10^{17} **45.** 4×10^2 **47.** x^6 **49.** $1/(2^3c^3d^6)$
51. $(3^2x^6)/y^4$ **53.** x^6/y^4 **55.** $2^6/3^4$ **57.** $1/10^4$ **59.** $(3n^4)/(4m^3)$
61. $(2x)/y^2$ **63.** n^2 **65.** x^{12}/y^8 **67.** $(4x^8)/y^6$ **69.** $1/(x+y)^2$
71. $\frac{1}{30}$ **73.** $144/7$ **75.** $\frac{36}{13}$ **77.** $1{,}000/11$

EXERCISE 53

1. 6×10 **3.** 6×10^2 **5.** 6×10^5 **7.** 6×10^{-2} **9.** 6×10^{-5}
11. 3.5×10 **13.** 7.2×10^{-1} **15.** 2.7×10^2 **17.** 3.2×10^{-2}
19. 5.2×10^3 **21.** 7.2×10^{-4} **23.** 500 **25.** 0.08 **27.** 6,000,000
29. 0.00002 **31.** 7,100 **33.** 0.00086 **35.** 8,800,000 **37.** 0.0000061
39. 4.27×10^7 **41.** 7.23×10^{-5} **43.** 5.87×10^{12} **45.** 3×10^{-23}
47. 3,460,000,000 **49.** 0.000000623 **51.** 93,000,000 **53.** 0.000075
55. 8×10^2 **57.** 8×10^{-3} **59.** 3×10^3 **61.** 3×10^7 **63.** 2×10^4 or 20,000 **65.** 2×10^{-4} or 0.0002 **67.** 3.3×10^{18} **69.** 10^7; 6×10^8

EXERCISE 54

1. 4 **3.** -9 **5.** x **7.** $3m$ **9.** $2\sqrt{2}$ **11.** $x\sqrt{x}$ **13.** $3y\sqrt{2y}$
15. $\frac{1}{2}$ **17.** $-\frac{2}{3}$ **19.** $\frac{1}{x}$ **21.** $\frac{\sqrt{3}}{3}$ **23.** $\frac{\sqrt{3}}{3}$ **25.** $\frac{\sqrt{x}}{x}$
27. $\frac{\sqrt{x}}{x}$ **29.** $5xy^2$ **31.** $2x^2y\sqrt{xy}$ **33.** $2x^3y^3\sqrt{2x}$ **35.** $\frac{\sqrt{3y}}{3y}$
37. $2x\sqrt{2y}$ **39.** $\frac{2}{3}x\sqrt{3xy}$ **41.** $\frac{\sqrt{6}}{3}$ **43.** $\frac{\sqrt{6mn}}{2n}$ **45.** $\frac{2a\sqrt{3ab}}{3b}$
47. 4.242 **49.** 0.447 **51.** 4.062 **53.** $\frac{\sqrt{2}}{2}$ **55.** In simplest radical

form **57.** $x\sqrt{x^2 - 2}$ **59.** Because the square of any real number cannot be negative **61.** Yes, No **63.** $5^{\frac{1}{2}} = \sqrt{5}$ **65.** If $a^2 = b^2$, then a does not necessarily equal b. For example, let $a = 2$ and $b = -2$.

EXERCISE 55

1. $8\sqrt{2}$ **3.** $3\sqrt{x}$ **5.** $4\sqrt{7} - 3\sqrt{5}$ **7.** $-3\sqrt{y}$ **9.** $4\sqrt{5}$ **11.** $4\sqrt{x}$ **13.** $2\sqrt{2} - 2\sqrt{3}$ **15.** $2\sqrt{x} + 2\sqrt{y}$ **17.** $\sqrt{2}$ **19.** $-3\sqrt{3}$ **21.** $2\sqrt{2} + 6\sqrt{3}$ **23.** $-\sqrt{x}$ **25.** $2\sqrt{6} + \sqrt{3}$ **27.** $-\sqrt{6}/6$ **29.** $5\sqrt{2xy}/2$ **31.** $2\sqrt{3} - \frac{\sqrt{2}}{2}$ **33.** $3\sqrt{2}$

EXERCISE 56

1. $4\sqrt{5} + 8$ **3.** $10 - 2\sqrt{2}$ **5.** $2 + 3\sqrt{2}$ **7.** $5 - 4\sqrt{5}$ **9.** $2\sqrt{3} - 3$ **11.** $x - 3\sqrt{x}$ **13.** $3\sqrt{m} - m$ **15.** $2\sqrt{3} - \sqrt{6}$ **17.** $5\sqrt{2} + 5$ **19.** $2\sqrt{2} - 1$ **21.** $x - \sqrt{x} - 6$ **23.** $9 + 4\sqrt{5}$ **25.** $2 - 11\sqrt{2}$ **27.** $6x - 13\sqrt{x} + 6$ **31.** $\frac{2 + \sqrt{2}}{3}$ **33.** $\frac{-1 - 2\sqrt{5}}{3}$ **35.** $2 - \sqrt{2}$ **37.** $\frac{\sqrt{11} + 3}{2}$ **39.** $\frac{\sqrt{5} - 1}{2}$ **41.** $\frac{y - 3\sqrt{y}}{y - 9}$ **43.** $\frac{7 + 4\sqrt{3}}{-1}$ or $-7 - 4\sqrt{3}$ **45.** $\frac{x + 5\sqrt{x} + 6}{x - 9}$

EXERCISE 57

1. (*A*) 16, (*B*) $\frac{1}{9}$ **2.** (*A*) 1, (*B*) 9 **3.** $\frac{4x^4}{9y^6}$ **4.** $\frac{y^3}{x^2}$ **5.** (*A*) 6×10^{-2}, (*B*) 0.06 **6.** -5 **7.** $2xy^2$ **8.** $\frac{5}{y}$ **9.** $-3\sqrt{x}$ **10.** $5 + 2\sqrt{5}$ **11.** (*A*) 9, (*B*) $\frac{1}{100}$ **12.** (*A*) $\frac{1}{27}$, (*B*) 1 **13.** $\frac{4x^4}{y^6}$ **14.** $\frac{m^2}{2n^5}$ **15.** (*A*) 2×10^{-3}, (*B*) 0.002 **16.** $6x^2y^3\sqrt{y}$ **17.** $\frac{\sqrt{2y}}{2y}$ **18.** $\frac{\sqrt{6xy}}{2y}$ **19.** $\frac{5\sqrt{6}}{6}$ **20.** $1 + \sqrt{3}$ **21.** $\frac{n^{10}}{9m^{10}}$ **22.** $\frac{xy}{x + y}$ **23.** 1.036 cc **24.** $\frac{n^2\sqrt{6m}}{3}$ **25.** $2x\sqrt{x^2 + 4}$ **26.** $\frac{x - 4\sqrt{x} + 4}{x - 4}$ **27.** $a^2 = b$ **28.** All real numbers

EXERCISE 58

1. 3, 4 **3.** -6, 5 **5.** -4, $\frac{2}{3}$ **7.** $-\frac{3}{4}$, $\frac{2}{5}$ **9.** 0, $\frac{1}{4}$ **11.** 1, 5 **13.** 1, 3 **15.** -2, 6 **17.** 0, 3 **19.** 0, 2 **21.** -5, 5 **23.** $\frac{1}{2}$, -3 **25.** $\frac{2}{3}$, 2 **27.** -4, 8 **29.** $-\frac{2}{3}$, 4 **31.** Not factorable in the integers **33.** 3, -4 **35.** -3, 3 **37.** 11 by 3 in. **39.** 5, -3 **41.** $\frac{1}{2}$, 2 **43.** $\frac{2}{3}$ or $\frac{3}{2}$ **45.** 1 ft

EXERCISE 59

1. ± 4 **3.** ± 8 **5.** $\pm\sqrt{3}$ **7.** $\pm\sqrt{5}$ **9.** $\pm\sqrt{18}$ or $\pm 3\sqrt{2}$ **11.** $\pm\sqrt{12}$ or $\pm 2\sqrt{3}$ **13.** $\pm\frac{2}{3}$ **15.** $\pm\frac{4}{3}$ **17.** $\pm\frac{2}{3}$ **19.** $\pm\frac{2}{5}$ **21.** $\pm\sqrt{\frac{3}{4}}$ or $\pm\frac{\sqrt{3}}{2}$ **23.** $\pm\sqrt{\frac{5}{2}}$ or $\pm\frac{\sqrt{10}}{2}$ **25.** $\pm\sqrt{\frac{1}{3}}$ or $\pm\frac{\sqrt{3}}{3}$ **27.** -1, 5 **29.** 3, -7 **31.** $2 \pm \sqrt{3}$ **33.** -1, 2 **35.** No real solution **37.** $\frac{3 \pm \sqrt{6}}{2}$ **39.** $b = \pm\sqrt{c^2 - a^2}$ **41.** 70 mph

EXERCISE 60

1. $x^2 + 4x + 4 = (x + 2)^2$ **3.** $x^2 - 6x + 9 = (x - 3)^2$
5. $x^2 + 12x + 36 = (x + 6)^2$ **7.** $-2 \pm \sqrt{2}$ **9.** $3 \pm 2\sqrt{3}$
11. $x^2 + 3x + \frac{9}{4} = \left(x + \frac{3}{2}\right)^2$ **13.** $u^2 - 5u + \frac{25}{4} = \left(u - \frac{5}{2}\right)^2$
15. $\frac{-1 \pm \sqrt{5}}{2}$ **17.** $\frac{5 \pm \sqrt{17}}{2}$ **19.** No real solution **21.** $\frac{2 \pm \sqrt{2}}{2}$
23. $\frac{-3 \pm \sqrt{17}}{4}$ **25.** $\frac{-m \pm \sqrt{m^2 - 4n}}{2}$

EXERCISE 61

1. $a = 1, b = 4, c = 2$ **3.** $a = 1, b = -3, c = -2$ **5.** $a = 3, b = -2, c = 1$ **7.** $a = 2, b = 3, c = -1$ **9.** $a = 2, b = -5, c = 0$
11. $-2 \pm \sqrt{2}$ **13.** $3 \pm 2\sqrt{3}$ **15.** $\frac{-3 \pm \sqrt{13}}{2}$ **17.** $\frac{3 \pm \sqrt{3}}{2}$
19. $\frac{-1 \pm \sqrt{13}}{6}$ **21.** No real solution **23.** $\frac{3 \pm \sqrt{33}}{4}$ **25.** $\frac{4 \pm \sqrt{11}}{5}$
27. $\frac{2\sqrt{3}}{3}, -\frac{\sqrt{3}}{3}$ **29.** T **31.** T

EXERCISE 62

1. $-2, 3$ **3.** $0, -7$ **5.** 4 **7.** $-1 \pm \sqrt{3}$ **9.** 0, 2 **11.** $1 \pm \sqrt{2}$
13. $\frac{3 \pm \sqrt{3}}{2}$ **15.** 0, 1 **17.** $\frac{1}{3}, -\frac{1}{2}$ **19.** $\pm 5\sqrt{2}$ **21.** $2 \pm \sqrt{3}$
23. No real solution **25.** $-\frac{2}{3}, 3$ **27.** $-50, 2$ **29.** $t = \sqrt{\frac{2d}{g}}$
31. $r = -1 + \sqrt{\frac{A}{P}}$

EXERCISE 63

1. 20 **3.** (*A*) 2,000 and 8,000; (*B*) 5,000 **5.** $p = \$2$ **7.** 3 in., 4 in., 5 in., **9.** $\frac{-1 \pm \sqrt{5}}{2}$ **11.** 12, 14 **13.** -7 or 8 **15.** 55 mph
17. $10\sqrt{2}$ ft, 14.14 ft **19.** 180 fps, 122.76 mph **21.** 2 hr, 3 hr **23.** 2 mph

EXERCISE 64

1. 5 **2.** 0, 3 **3.** $-3, \frac{1}{2}$ **4.** 2, 3 **5.** $-3, 5$ **6.** $a = 3, b = 4, c = -2$ **7.** $x = \frac{-b \pm \sqrt{b^2 - 4ac}}{2a}$ **8.** $\frac{-3 \pm \sqrt{5}}{2}$ **9.** $-5, 2$ **10.** 3, 9
11. $\pm 2\sqrt{3}$ **12.** 0, 2 **13.** $-2, 6$ **14.** $-\frac{1}{3}, 3$ **15.** $\frac{1}{2}, -3$
16. $3 \pm 2\sqrt{3}$ **17.** $\frac{1 \pm \sqrt{7}}{3}$ **18.** $5, -4$ **19.** 6, 12 **20.** 6 by 5 in.
21. No real solutions **22.** $\frac{3 \pm \sqrt{6}}{2}$ **23.** $\frac{3}{4}, \frac{5}{2}$ **24.** $\frac{1 \pm \sqrt{7}}{2}$
25. $\frac{-3 \pm \sqrt{57}}{6}$ **26.** T

Index

Index